Preventing the Adverse Effects of SARS-CoV-2 Infection and COVID-19 through Diet, Supplements and Lifestyle

Preventing the Adverse Effects of SARS-CoV-2 Infection and COVID-19 through Diet, Supplements and Lifestyle

Editors

William B. Grant
Ronan Lordan

MDPI • Basel • Beijing • Wuhan • Barcelona • Belgrade • Manchester • Tokyo • Cluj • Tianjin

Editors
William B. Grant
Sunlight, Nutrition, and
Health Research Center
USA

Ronan Lordan
University of Pennsylvania
USA

Editorial Office
MDPI
St. Alban-Anlage 66
4052 Basel, Switzerland

This is a reprint of articles from the Special Issue published online in the open access journal *Nutrients* (ISSN 2072-6643) (available at: https://www.mdpi.com/journal/nutrients/special_issues/Preventing_SARS_CoV_2_and_COVID_19).

For citation purposes, cite each article independently as indicated on the article page online and as indicated below:

LastName, A.A.; LastName, B.B.; LastName, C.C. Article Title. *Journal Name* **Year**, *Volume Number*, Page Range.

ISBN 978-3-0365-3889-1 (Hbk)
ISBN 978-3-0365-3890-7 (PDF)

Contents

About the Editors

William B. Grant has a Ph.D. in physics from the University of California, Berkeley (1971). He is the director of Sunlight, Nutrition and Health Research Center, a non-profit organization founded in San Francisco in 2004., He has made 300 publications, including articles, reviews, editorials, and letters to the editor, regarding vitamin D listed at pubmed.gov. His main interest now is the study of solar UV exposure to produce vitamin D and nitric oxide for optimal health. His minor interest is proper nutrition to reduce the risk of chronic diseases. Twitter handle: @wbgrant2

Ronan Lordan is a researcher of circadian biology and nutrition at the Institute for Translational Medicine and Therapeutics, Perelman School of Medicine, at the University of Pennsylvania. He received his BSc (Hons) in Biological Sciences with concurrent Education at the University of Limerick, Ireland. Here he continued his studies and obtained a PhD (Nutrition and Biochemistry). Ronan has also lectured in nutrition, genetics, and physiology at the University of Limerick. He has published several peer-reviewed articles, book chapters, and is an active reviewer and editor for several journals. His research interests include: (a) circadian biology and metabolism, (b) functional foods and nutraceuticals, (c) the role of nutritional status in COVID-19; and (d) discerning the mechanisms of platelet-activating factor-induced systemic inflammation in noncommunicable diseases. Twitter Handle: @el_ronan

Editorial

Preventing the Adverse Effects of SARS-CoV-2 Infection and COVID-19 through Diet, Supplements, and Lifestyle

Ronan Lordan [1,*] and William B. Grant [2]

[1] Institute for Translational Medicine and Therapeutics (ITMAT), Perelman School of Medicine, University of Pennsylvania, Philadelphia, PA 19104-5158, USA

[2] Sunlight, Nutrition, and Health Research Center, P.O. Box 641603, San Francisco, CA 94164-1603, USA; williamgrant08@comcast.net

* Correspondence: ronan.lordan@ul.ie

Severe acute respiratory syndrome coronavirus 2 (SARS-CoV-2), the pathogen responsible for the coronavirus disease 2019 (COVID-19) and the ongoing worldwide pandemic, has cost the lives of almost 5.4 million people and infected over 276 million worldwide as of December 2021 [1]. While great strides have been made to produce and repurpose therapeutics, develop novel vaccines, and promote non-pharmacological interventions to reduce disease burden, parts of the world are now entering their fifth pandemic wave. An underappreciated mitigation strategy has been the role of preventing the adverse effects of COVID-19 by promoting healthy lifestyle patterns in conjunction with non-pharmacological interventions. This is even more important in parts of the world that are disadvantaged by their lack of access to vaccines. Moreover, additional protections via dietary and lifestyle changes may improve overall health. Indeed, it is well established that poor host nutritional status is a potential risk factor for severe respiratory diseases and comorbidities such as diabetes, hypertension, and obesity, which all increase the risk for severe disease, hospitalization, and death in COVID-19 patients [2,3].

In this Special Issue, we provided a platform for scientists to submit their research investigating nutritional status, the potential of dietary and lifestyle alterations, and the use of supplements in relation to COVID-19 outcomes. In this editorial, we present the advances this Special Issue has brought to fruition in the battle against the coronavirus pandemic.

At the beginning of the pandemic, multiple lines of evidence suggested a potential link between vitamin D and SARS-CoV-2 infection and COVID-19. AlSafar and colleagues [4] examined 25(OH)D levels in serum samples taken upon admission in 464 hospitalized COVID-19 patients in relation to patient outcomes in the United Arab Emirates (UAE). They determined that 25(OH)D < 12 ng/mL was significantly associated with 2.58-fold (95% CI, 1.01, 6.62) increased risk of COVID-19 mortality following adjustment for age, comorbidities or sex ($p = 0.048$). Indeed, Karanova et al. [5] presents further data in accordance with AlSafar et al. [4], which supports the notion that 25(OH)D deficiency is common among hospitalized COVID-19 patients. In their study, 90 out of 133 Russian COVID-19 patients were either 25(OH)D deficient or insufficient. Karanova et al. [5] also determined that 25(OH)D levels between approximately 11–12 ng/mL was the threshold value for increased risk of severe COVID-19 disease and mortality. Both studies add to a growing literature demonstrating that having sufficient 25(OH)D levels may be of critical importance and a predictor of coronavirus patient outcomes.

Considering these findings, one of the most important issues regarding COVID-19 is how to treat patients. In the article from Saudi Arabia, a small-scale randomized controlled trial was conducted to assess the effects of vitamin D supplementation on COVID-19 patients shortly after symptoms arose [6]. The trial involved 69 SARS-CoV-2-positive patients hospitalized with mild to moderate COVID-19 between 29 July and early September 2020. They were randomized to receive either 1000 or 5000 IU of vitamin D3 daily for 14 days.

Citation: Lordan, R.; Grant, W.B. Preventing the Adverse Effects of SARS-CoV-2 Infection and COVID-19 through Diet, Supplements, and Lifestyle. *Nutrients* **2022**, *14*, 115. https://doi.org/10.3390/nu14010115

Received: 15 December 2021
Accepted: 20 December 2021
Published: 28 December 2021

Most of the descriptive characteristics and symptoms were not significantly different between the two arms. Mean baseline 25(OH)D concentration was 63 ± 3 nmol/L in the 1000 IU arm and 53 ± 3 nmol/L in the 5000 IU arm, while mean 25(OH)D concentrations of 60 nmol/L were achieved in the 1000 IU arm and 63 nmol/L in the 5000 IU arm. However, those in the 5000 IU arm were younger (46 ± 15 vs. 54 ± 12 years, $p = 0.03$) and had lower BMI (28 ± 7 vs. 32 ± 7 kg/m^2, $p = 0.02$). Two COVID-19 symptoms were resolved faster with 5000 IU/d: cough (6 ± 1 vs. 9 ± 1 days, $p = 0.007$) and ageusia (loss of taste) (11 ± 1 vs. 17 ± 2 days, $p = 0.04$). There were no significant differences in pre- and post-clinical parameters between the two arms other than D-dimer concentrations, which decreased in the 1000 IU arm but increased in the 5000 IU arm. One reason for limited beneficial effects of vitamin D supplementation in this trial may be that the low doses did not raise 25(OH)D concentrations rapidly enough to significantly affect the course of the disease. It is thought that providing much higher doses [7] or using calcifediol [25(OH)D] may have achieved more favorable clinical benefits [8].

Golabi and colleagues investigated the association between vitamin D and zinc status and progression of clinical symptoms among 53 outpatients in Iran infected with SARS-CoV-2 as well as 53 potentially non-infected participants [9]. Infected patients had non-significantly lower 25(OH)D concentrations than non-infected ones (26 ± 17 vs. 29 ± 16 ng/mL). There was a trend for lower 25(OH)D among patients with moderate illness than with asymptomatic or mild illness (19 ± 12 vs. 29 ± 18 ng/mL, $p = 0.054$). In terms of progression, patients with 25(OH)D > 20 ng/mL had a reduced progression compared to those with <12 ng/mL (OR = 0.19, $p < 0.001$) as did those with 25(OH)D between 12 and 20 ng/mL compared to <12 ng/mL (OR = 0.3, $p = 0.007$). Infected patients had significantly lower serum zinc concentrations than non-infected patients (101 ± 18 vs. 114 ± 13 μg/dL). However, the difference between zinc concentrations for those with mild or no sign vs. moderate severity was not significant ($p = 0.41$). A study involving four SARS-CoV-2 or COVID-19 patients treated with high-dose zinc salts (23 to 150 mg/d) found significant improvements starting one day after treatment [10]. On the other hand, a larger study involving 58 patients given 50 mg/day zinc, 58 patients given 50 mg/d zinc plus 8 g/d ascorbic acid, and 48 patients given 8 g/d ascorbic acid, did not find any significant difference in secondary outcomes compared to 50 patients given the standard of care [6].

Many during the pandemic turned to seeking additional protections against SARS-CoV-2 and COVID-19 via the use of dietary supplement and nutraceuticals [11]. In this Special Issue, Puscion-Jakubik et al. examined the consumption of food supplements during the first three COVID-19 waves in Poland (spring and autumn 2020 and winter 2021) [12]. Approximately 300 participants responded to each questionnaire, with 80% of the respondents being women, the mean age was approximately 29 ± 10 years, and approximately 50% having medical or related education, and nearly all either working in an office or studying as a student. Thus, the survey does not relate to the general public. The authors reported that vitamin D was the most popular supplement during the second wave that started in September 2020, taken by 23% of the respondents during the first wave, 38% in the second wave, and 33% in the third wave. The mixture of office workers vs. students changed from 50% vs. 34% in the first wave to 39% vs. 51% in the second wave and 36% vs. 54% in the third wave, which seems to have affected the vitamin D supplementation findings.

Scientists have also had a keen interest to pursue the development of novel supplements or nutraceuticals to quell the spread of the pandemic and provide effective treatments for patients. In this Special Issue, two natural products, *Glycyrrhiza glabra* extract and hesperidin, have been assessed with a focus on inhibiting viral entry via angiotensin-converting enzyme 2 (ACE2) and transmembrane serine protease 2 (TMPRSS2), the key cellular proteins required for SARS-CoV-2 entry into mammalian cells.

Jezova et al. [13] aimed to harness the antiviral potential of glycyrrhizin, a saponin type molecule responsible for the sweet taste in *Glycyrrhiza glabra* (licorice) root. The authors show that a *Glycyrrhiza glabra* extract reduces ACE2 expression via inhibiting the activity of 11-β-hydroxysteroid dehydrogenase type-2 (11-β-HSD2) leading to the activation of the mineralocorticoid receptor (MR). The authors showed that supplementation of *Glycyrrhiza glabra* extract in a stress model in Sprague Dawley rats reduced the expression of ACE2 in target tissues where ACE2 is co-expressed with 11-β-HSD2 and MR, such as the ileum, versus tissues where co-expression of these proteins does not occur, such as the brain cortex. Although the authors were not able to conduct in vivo challenge studies with SARS-CoV-2 to assess if reduced expression of ACE2 in these target tissues reduces viral RNA copy number or disease severity, this study does show promise for further development and research.

Cheng et al. [14] investigated the binding capacity of hesperidin, a flavanone glycoside that naturally occurs in citrus fruits, and hesperitin, an aglycone metabolite of hesperidin, to ACE2 and TMPRSS2. Notably, the authors were able to show that both molecules could suppress the infection of VeroE6 cells by lentiviral pseudo-particles of wild type SARS-CoV-2 and variants with the D614G and 501Y.v2 (beta variant) mutations. Indeed, there was also a suppression of ACE2 and TMPRSS2 expression. In contrast, neither molecule affected SARS-CoV-2 viral proteases papain-like protease (PLpro) and 3-chymotrypsin-like protease (Mpro) despite molecular docking predictions. Despite these promising findings, it is important to interpret with caution as these experiments were conducted in silico and in vitro, which is not always translatable to efficacy in humans. Indeed, while veroE6 cells are commonly used for SARS-CoV-2 studies, they are not representative of the respiratory tract, the primary site of SARS-CoV-2 infection, and so findings should be interpreted with caution. Despite these caveats, it has been shown that 1 g daily of hesperidin given to symptomatic non-vaccinated COVID-19 patients in a randomized, double-blind, placebo-controlled trial appeared to modestly reduce some symptoms of COVID-19, including fever, cough, shortness of breath and anosmia, but much further study is required [15].

Other authors that contributed to this Special Issue took a much broader view of the potential nutritional requirements and supply chain issues that occurred during the pandemic. Currently more than 4.2 million children and 138,000 adults receive nutritious meals and snacks in the U.S. through the Child and Adult Care Food Program (CACFP) [16]. In the latest month for which data are available, August 2021, the total program cost was nearly $250 million [17]. In a review by Stephens and colleagues in this Special Issue [18], comparisons were made on how the COVID-19 pandemic affected operations for CACFP and non-CACFP in Arizona and Pennsylvania. Not surprisingly, CACFP sites were much more likely to offer "grab and go" meals, meal delivery, and distributed food boxes.

Indeed, maintaining a functional, hygienic, and sustainable food supply and distribution network during the pandemic has been vastly underappreciated. Filip et al. provide an in-depth review of the literature concerning the clinical outcomes of COVID-19 patients and how nutritional status and changes to the food chain, food hygiene, food security, and people's dietary patterns during the coronavirus pandemic are interrelated and affect health [19]. Although fomites are no longer thought to significantly contribute to SARS-CoV-2 transmission, the authors also provide cautionary guidelines to the food industry for the processing, packaging, and distribution of food with the intention to limit the spread of SARS-CoV-2.

Another topic that was highlighted in this Special Issues was that responses to the COVID-19 pandemic can have unintended consequences. An example of this is found in the study of weight gain among school teachers in Long Island, NY, who switched from in person teaching to online teaching [20]. Teachers for grades kindergarten to 5 gained a mean weight of 4 ± 8 pounds, those who taught in middle school did not change mean weight (0 ± 11 pounds), while those who taught in high school lost weight (-1 ± 9 pounds). Some of the weight gain appears to be associated with emotional eating due to nothing to do, being bored, depressed, or discouraged, irritated, feeling anxious, feeling lonely, etc. Junk food including chips and ice cream had strong associations with weight gain. Exercise was associated with weight loss. This study should lead to additional studies that examine changes in weight due to changing environmental conditions and how to modify the effects. Another article examined emotional eating during the COVID-19 pandemic in Norway, finding that 54% of the respondents to an electronic survey reported emotional eating, with higher rates among women [21]. That led to higher intake of high-sugar foods and beverages. A paper published in 1996 reported that Type A women in Northern Ireland had a weak positive association with sugar and alcohol intake, as opposed to men, who had a significant association with fat and protein intake as well as beef, cheese, yoghurt, and chips [22]. An earlier article in Nutrients studied the effects of weight gain during pregnancy associated with emotional eating [23]. The authors suggested the need for psychosocial and nutritional education and interventions during pregnancy checkups.

As 2022 draws closer, the COVID-19 pandemic is not over and will likely affect everyday life for the foreseeable future. Despite the successful development and distribution of vaccines in the Western world, many regions have yet to receive adequate supply of vaccines. Therefore, the implementation of efficacious non-pharmacological interventions coupled with the promotion of healthy dietary and lifestyle patterns may promote overall health and reduce one's risk of infection, disease, and death as a result of the SARS-CoV-2 pandemic. Evidentially, further extensive, and broad-ranging research is required to understand how the majority of the adverse effects of SARS-CoV-2 infection and COVID-19 could be prevented through diet, supplements, and lifestyle.

Acknowledgments: W.B.G. and R.L. would like to thank all of the contributing authors to our Special Issue "Preventing the Adverse Effects of SARS-CoV-2 Infection and COVID-19 through Diet, Supplements, and Lifestyle".

Conflicts of Interest: W.B.G.'s non-profit organization, Sunlight, Nutrition and Health Research Center, receives funding from Bio-Tech Pharmacal, Inc. R.L has no conflicts of interest to declare. R.L. provided editorial commentary on the articles authored by W.B.G in this Special Issue.

References

1. John Hopkins University. John Hopkins University & Medicine: Coronavirus Resource Center. Available online: https://coronavirus.jhu.edu/map.html (accessed on 8 December 2020).
2. Zabetakis, I.; Lordan, R.; Norton, C.; Tsoupras, A. COVID-19: The inflammation link and the role of nutrition in potential mitigation. *Nutrients* **2020**, *12*, 1466. [CrossRef] [PubMed]
3. Im, J.H.; Je, Y.S.; Baek, J.; Chung, M.-H.; Kwon, H.Y.; Lee, J.-S. Nutritional status of patients with COVID-19. *Int. J. Infect. Dis.* **2020**, *100*, 390–393. [CrossRef] [PubMed]
4. AlSafar, H.; Grant, W.B.; Hijazi, R.; Uddin, M.; Alkaabi, N.; Tay, G.; Mahboub, B.; Al Anouti, F. COVID-19 Disease Severity and Death in Relation to Vitamin D Status among SARS-CoV-2-Positive UAE Residents. *Nutrients* **2021**, *13*, 1714. [CrossRef] [PubMed]
5. Karonova, T.L.; Andreeva, A.T.; Golovatuk, K.A.; Bykova, E.S.; Simanenkova, A.V.; Vashukova, M.A.; Grant, W.B.; Shlyakhto, E.V. Low 25(OH)D Level Is Associated with Severe Course and Poor Prognosis in COVID-19. *Nutrients* **2021**, *13*, 3021. [CrossRef] [PubMed]
6. Sabico, S.; Enani, M.A.; Sheshah, E.; Aljohani, N.J.; Aldisi, D.A.; Alotaibi, N.H.; Alshingetti, N.; Alomar, S.Y.; Alnaami, A.M.; Amer, O.E.; et al. Effects of a 2-Week 5000 IU versus 1000 IU Vitamin D3 Supplementation on Recovery of Symptoms in Patients with Mild to Moderate COVID-19: A Randomized Clinical Trial. *Nutrients* **2021**, *13*, 2170. [CrossRef] [PubMed]
7. Gönen, M.S.; Alaylıoğlu, M.; Durcan, E.; Özdemir, Y.; Şahin, S.; Konukoğlu, D.; Nohut, O.K.; Ürkmez, S.; Küçükece, B.; Balkan, İ.İ.; et al. Rapid and Effective Vitamin D Supplementation May Present Better Clinical Outcomes in COVID-19 (SARS-CoV-2) Patients by Altering Serum INOS1, IL1B, IFNg, Cathelicidin-LL37, and ICAM1. *Nutrients* **2021**, *13*, 4047. [CrossRef] [PubMed]
8. Entrenas Castillo, M.; Entrenas Costa, L.M.; Vaquero Barrios, J.M.; Alcalá Díaz, J.F.; López Miranda, J.; Bouillon, R.; Quesada Gomez, J.M. Effect of calcifediol treatment and best available therapy versus best available therapy on intensive care unit admission and mortality among patients hospitalized for COVID-19: A pilot randomized clinical study. *J. Steroid Biochem. Mol. Biol.* **2020**, *203*, 105751. [CrossRef] [PubMed]
9. Golabi, S.; Adelipour, M.; Mobarak, S.; Piri, M.; Seyedtabib, M.; Bagheri, R.; Suzuki, K.; Ashtary-Larky, D.; Maghsoudi, F.; Naghashpour, M. The Association between Vitamin D and Zinc Status and the Progression of Clinical Symptoms among Outpatients Infected with SARS-CoV-2 and Potentially Non-Infected Participants: A Cross-Sectional Study. *Nutrients* **2021**, *13*, 3368. [CrossRef] [PubMed]
10. Finzi, E. Treatment of SARS-CoV-2 with high dose oral zinc salts: A report on four patients. *Int. J. Infect. Dis.* **2020**, *99*, 307–309. [CrossRef]
11. Lordan, R. Dietary supplements and nutraceuticals market growth during the coronavirus pandemic—Implications for consumers and regulatory oversight. *Pharma Nutr.* **2021**, *18*, 100282. [CrossRef]

12. Puścion-Jakubik, A.; Bielecka, J.; Grabia, M.; Mielech, A.; Markiewicz-Żukowska, R.; Mielcarek, K.; Moskwa, J.; Naliwajko, S.K.; Soroczyńska, J.; Gromkowska-Kępka, K.J.; et al. Consumption of Food Supplements during the Three COVID-19 Waves in Poland—Focus on Zinc and Vitamin D. *Nutrients* **2021**, *13*, 3361. [CrossRef] [PubMed]
13. Jezova, D.; Karailiev, P.; Karailievova, L.; Puhova, A.; Murck, H. Food Enrichment with Glycyrrhiza glabra Extract Suppresses ACE2 mRNA and Protein Expression in Rats—Possible Implications for COVID-19. *Nutrients* **2021**, *13*, 2321. [CrossRef] [PubMed]
14. Cheng, F.-J.; Huynh, T.-K.; Yang, C.-S.; Hu, D.-W.; Shen, Y.-C.; Tu, C.-Y.; Wu, Y.-C.; Tang, C.-H.; Huang, W.-C.; Chen, Y.; et al. Hesperidin Is a Potential Inhibitor against SARS-CoV-2 Infection. *Nutrients* **2021**, *13*, 2800. [CrossRef] [PubMed]
15. Dupuis, J.; Laurin, P.; Tardif, J.-C.; Hausermann, L.; Rosa, C.; Guertin, M.-C.; Thibaudeau, K.; Gagnon, L.; Cesari, F.; Robitaille, M.; et al. Fourteen-days Evolution of COVID-19 Symptoms During the Third Wave in Non-vaccinated Subjects and Effects of Hesperidin Therapy: A randomized, double-blinded, placebo-controlled study. *medRxiv* **2021**. [CrossRef]
16. United States Department of Agriculture: Food and Nutrition Service. Child and Adult Care Food Program: Ensuring Children and Adults Have Access to Nutritious Meals and Snacks. Available online: https://www.fns.usda.gov/cacfp (accessed on 8 December 2021).
17. United States Department of Agriculture. U.S. Summary, FY 2020–FY 2021—Generated from National Data Bank Version 8.2 PUBLIC on 11/12/2021 August 2021. Available online: https://fns-prod.azureedge.net/sites/default/files/data-files/keydata-august-2021.pdf (accessed on 8 December 2021).
18. Stephens, L.; Rains, C.; Benjamin-Neelon, S.E. Connecting Families to Food Resources amid the COVID-19 Pandemic: A Cross-Sectional Survey of Early Care and Education Providers in Two U.S. States. *Nutrients* **2021**, *13*, 3137. [CrossRef] [PubMed]
19. Filip, R.; Anchidin-Norocel, L.; Gheorghita, R.; Savage, W.K.; Dimian, M. Changes in Dietary Patterns and Clinical Health Outcomes in Different Countries during the SARS-CoV-2 Pandemic. *Nutrients* **2021**, *13*, 3612. [CrossRef] [PubMed]
20. Silverman, J.R.; Wang, B.Z. Impact of School Closures, Precipitated by COVID-19, on Weight and Weight-Related Risk Factors among Schoolteachers: A Cross-Sectional Study. *Nutrients* **2021**, *13*, 2723. [CrossRef] [PubMed]
21. Bemanian, M.; Mæland, S.; Blomhoff, R.; Rabben, Å.K.; Arnesen, E.K.; Skogen, J.C.; Fadnes, L.T. Emotional Eating in Relation to Worries and Psychological Distress Amid the COVID-19 Pandemic: A Population-Based Survey on Adults in Norway. *Int. J. Environ. Res. Public Health* **2021**, *18*, 130. [CrossRef] [PubMed]
22. Barker, M.E.; Thompson, K.A.; McClean, S.I. Do Type As Eat Differently? A Comparison of Men and Women. *Appetite* **1996**, *26*, 277–286. [CrossRef] [PubMed]
23. Zhang, J.; Zhang, Y.; Huo, S.; Ma, Y.; Ke, Y.; Wang, P.; Zhao, A. Emotional Eating in Pregnant Women during the COVID-19 Pandemic and Its Association with Dietary Intake and Gestational Weight Gain. *Nutrients* **2020**, *12*, 2250. [CrossRef] [PubMed]

Article

COVID-19 Disease Severity and Death in Relation to Vitamin D Status among SARS-CoV-2-Positive UAE Residents

Habiba AlSafar [1,2,3,†], William B. Grant [4], Rafiq Hijazi [5], Maimunah Uddin [6], Nawal Alkaabi [6], Guan Tay [1,7,8], Bassam Mahboub [9] and Fatme Al Anouti [10,*,†]

1 Center for Biotechnology, Khalifa University of Science and Technology, Abu Dhabi 127788, United Arab Emirates; habiba.alsafar@ku.ac.ae (H.A.); guan.tay@uwa.edu.au (G.T.)
2 Department of Biomedical Engineering, College of Engineering, Khalifa University of Science and Technology, Abu Dhabi 127788, United Arab Emirates
3 Department of Genetics and Molecular Biology, College of Medicine and Health Sciences, Khalifa University of Science and Technology, Abu Dhabi 127788, United Arab Emirates
4 Sunlight, Nutrition and Health Research Center, P.O. Box 641603, San Francisco, CA 94164-1603, USA; williamgrant08@comcast.net
5 Department of Mathematics and Statistics, College of Natural and Health Sciences, Zayed University, Abu Dhabi 144534, United Arab Emirates; Rafiq.Hijazi@zu.ac.ae
6 Department of Pediatric Infectious Disease, Sheikh Khalifa Medical City, Abu Dhabi 51900, United Arab Emirates; muddin@seha.ae (M.U.); nalkaabi@seha.ae (N.A.)
7 Division of Psychiatry, Faculty of Health and Medical Sciences, University of Western Australia, Crawley, Western Australia, Australia
8 School of Medical and Health Sciences, Edith Cowan University, Joondalup, Western Australia, Australia
9 Dubai Health Authority, Rashid Hospital, Dubai, United Arab Emirates; drbassam_mahboub@yahoo.com
10 Department of Health Sciences, College of Natural and Health Sciences, Zayed University, Abu Dhabi 144534, United Arab Emirates
* Correspondence: Fatme.AlAnouti@zu.ac.ae
† These authors contributed equally to this work.

Citation: AlSafar, H.; Grant, W.B.; Hijazi, R.; Uddin, M.; Alkaabi, N.; Tay, G.; Mahboub, B.; Al Anouti, F. COVID-19 Disease Severity and Death in Relation to Vitamin D Status among SARS-CoV-2-Positive UAE Residents. *Nutrients* **2021**, *13*, 1714. https://doi.org/10.3390/nu13051714

Academic Editor: Karl Michaëlsson

Received: 5 May 2021
Accepted: 17 May 2021
Published: 19 May 2021

Abstract: Insufficient blood levels of the neurohormone vitamin D are associated with increased risk of COVID-19 severity and mortality. Despite the global rollout of vaccinations and promising preliminary results, the focus remains on additional preventive measures to manage COVID-19. Results conflict on vitamin D's plausible role in preventing and treating COVID-19. We examined the relation between vitamin D status and COVID-19 severity and mortality among the multiethnic population of the United Arab Emirates. Our observational study used data for 522 participants who tested positive for SARS-CoV-2 at one of the main hospitals in Abu Dhabi and Dubai. Only 464 of those patients were included for data analysis. Demographic and clinical data were retrospectively analyzed. Serum samples immediately drawn at the first hospital visit were used to measure serum 25-hydroxyvitamin D [25(OH)D] concentrations through automated electrochemiluminescence. Levels < 12 ng/mL were significantly associated with higher risk of severe COVID-19 infection and of death. Age was the only other independent risk factor, whereas comorbidities and smoking did not contribute to the outcomes upon adjustment. Sex of patients was not an important predictor for severity or death. Our study is the first conducted in the UAE to measure 25(OH)D levels in SARS-CoV-2-positive patients and confirm the association of levels < 12 ng/mL with COVID-19 severity and mortality.

Keywords: vitamin D; COVID-19; SARS-CoV-2; severity; mortality; United Arab Emirates

1. Introduction

COVID-19 is a complex respiratory syndrome caused by the severe acute respiratory syndrome coronavirus 2 (SARS-CoV-2), an enveloped RNA virus extremely transmissible through respiratory aerosols [1]. This virus, which can lead to pulmonary failure and fatality, has a noticeable genetic similarity to the beta-coronaviruses that cause SARS and Middle East respiratory syndrome [2]. The clinical indications of COVID-19 disease range from asymptomatic to mild to severe. Although most affected patients develop mild symptoms, about 5% of cases might progress to acute respiratory distress syndrome, requiring hospitalization and intensive care [3]. Elevated oxidative stress levels, exaggerated immune response due to the cytokine storm, and uncontrollable liberation of proinflammatory cytokines, along with the activation of pre-coagulating factors, all contribute to severe inflammation, which is exaggerated in acute respiratory distress syndrome [4,5]. Vitamin D is a fat-soluble prohormone steroid that has endocrine, paracrine, and autocrine functions [6]. Recent studies demonstrated that vitamin D could mediate antiviral activity by many actions, including enhancing apoptosis and autophagy as well as by inducing antimicrobial peptides [7,8].

Accumulating evidence has shown that severe disease is high among vulnerable populations: the elderly and patients with chronic diseases such as asthma, cancer, chronic obstructive pulmonary disease, diabetes, and hypertension. People who are obese or belong to ethnic groups with darker skin also experience more severe disease [9,10].

To fully decipher the mechanism underlying COVID-19 disease susceptibility, researchers are considering several possible contributing factors [11]. Vitamin D deficiency has emerged as a leading candidate [7,12,13]. Although concrete evidence about vitamin D's therapeutic role in COVID-19 has yet to be confirmed through randomized controlled trials (RCTs), vitamin D is associated with protective effects [7]. Such effects arise because vitamin D, as an essential prohormone that maintains bone homeostasis, also mediates many important non-skeletal functions, including modulating the immune system [14].

Several studies have documented the correlation between vitamin D deficiency and severity of viral infections such as influenza [15]. A study among children and adolescents indicated a higher risk of viral respiratory tract infections with deficient and insufficient serum 25-hydroxyvitamin D [25(OH)D] levels [16]. Moreover, a meta-analysis by Martineau and colleagues of RCTs across the globe including 11,321 participants showed that vitamin D-deficient patients had better protection against respiratory tract infections after supplementation with vitamin D (odds ratio (OR) = 0.30; 95% confidence interval (CI), 0.17, 0.53) [17]. Recently, vitamin D was identified by genomics-guided tracing research to be involved in regulating gene expression with potential to alleviate SARS-CoV-2 infection upon binding to the vitamin D response element [18]. The well-established role of vitamin D as an anti-inflammatory agent explains the beneficial effect of vitamin D in both the innate and adaptive immune responses and in producing antimicrobial agents cathelicidin (LL-37) and human β-defensin 2 [19,20].

Moreover, vitamin D regulates the renin–angiotensin system and expression of angiotensin-converting enzyme 2 (ACE2), and the corresponding cell receptor, which mediates coronavirus infection (ACE2 and the ACE2 receptor are distinct, and ACE2 seems able to bind SARS-CoV-2, preventing it from attaching to the ACE2 receptor). Elevated expression of ACE2 had been linked to a protective effect in the lungs during acute injury. Higher expression also reduces infectivity of SARS-CoV-2 by attenuating attachment to ACE2 receptors in target cells [21,22].

A previous study that examined the expression pattern for ACE2 in a mouse model in the context of aging and sex showed a significantly downregulated expression for ACE2 in older female and male rats, by 67% and 78%, respectively [23]. That decrease of ACE2 protein accords with the reported higher risk of COVID-19 infection and severity of disease among males [22]. Vitamin D also strengthens the epithelial physical barrier through its effect on E-cadherin, which tightens the cellular junctions to be tight and effective in impeding viral particles from penetrating the lungs [20]. Evidence from studies in 20 European countries showed that 25(OH)D concentrations and COVID-19 mortality were inversely associated, as well as that vitamin D deficiency was a poor prognostic factor for COVID-19. Severe vitamin D deficiency was remarkably evident among the elderly [24].

A systematic review and meta-analysis of 14 studies from an observational prospective and retrospective investigation with 999,179 participants indicated that low serum 25(OH)D was associated with higher susceptibility for COVID-19 infection and more severe disease and mortality [25]. Ongoing clinical trials for assessing the role of vitamin D supplementation in treating COVID-19 infections are under way, and the results so far have shown potential for using vitamin D supplementation, particularly for intensive care patients [26].

Mounting evidence from retrospective studies conducted in the United States and Europe indicates that lower vitamin D levels are commonly associated with risk of acquiring, and dying from, COVID-19 among hospitalized patients. Low levels may have some role in determining severity and outcome of COVID-19 [27]. Moreover, vitamin D deficiency is highly prevalent among critically ill patients and could aggravate the clinical outcome of those vulnerable people by increasing infection rates and mortality [28,29]. Supplementation with vitamin D for those susceptible people plays a pivotal role in helping them recover through supporting the immune system [30,31]. Despite the global rollout of vaccinations, the focus is still on additional promising preventive measures, such as using vitamin D to manage COVID-19 [32]. Vitamin D deficiency is a major public health burden in the Middle East, including the United Arab Emirates (UAE), despite abundant year-round sunlight [33–35].

Our objective was to assess the association of vitamin D status with COVID-19 disease severity and mortality in a sample of SARS-CoV-2-positive people from the UAE population. The multiethnic differences among the UAE population together with the unique pattern of COVID-19 mortality and severity in the country merit further investigation.

We used *Our World in Data* (Stats. WHO 2021), an online interactive dashboard hosted by Johns Hopkins University, to track reported COVID-19 cases in real time (https://coronavirus.jhu.edu/map.html accessed on 26 April 2021).

2. Materials and Methods

2.1. Participants and Collecting Samples

This study was a multicenter observational study with data collected between August 2020 and February 2021. We recruited 522 participants in the UAE who tested positive for SARS-CoV-2 during a COVID-19 screening at either Sheikh Khalifa Medical Centre in Abu Dhabi, or Rashed hospital in Dubai. We obtained written informed consent from all participants. Inclusion criteria were UAE residency and an age of 18 years or older. Blood samples and nasal swabs were collected for examination. Of 522 participants, 58 had missing data for body mass index (BMI; kilograms per square meter of body surface area), so only 464 were included for data analysis. A flow chart for the selection of participants is shown in Figure 1. This study was approved by the Abu Dhabi Health COVID-19 Research Ethics Committee (DOH/DQD/2020/538), the Dubai Scientific Research Ethics Committee (DSREC-04/2020_09), and the SEHA Research Ethics Committee (SEHA-IRB-005).

2.2. Collecting Demographic Data

Demographic, clinical, and outcome data of COVID-19 patients were gathered from questionnaires administered by medical staff at the hospital. Health care providers assessed patients for discharge, including determining disease severity (mild, moderate, or severe). Smoking status was coded as current smoker or nonsmoker. Chest X-ray and/or computed tomography scans were performed in all COVID-19 patients. Concerning immunomodulatory therapy, patients were selected for tocilizumab according to our institutional protocol. The endpoint variable for COVID-19 severity was defined as admission to the intensive care unit, requirement of mechanical ventilation, or death.

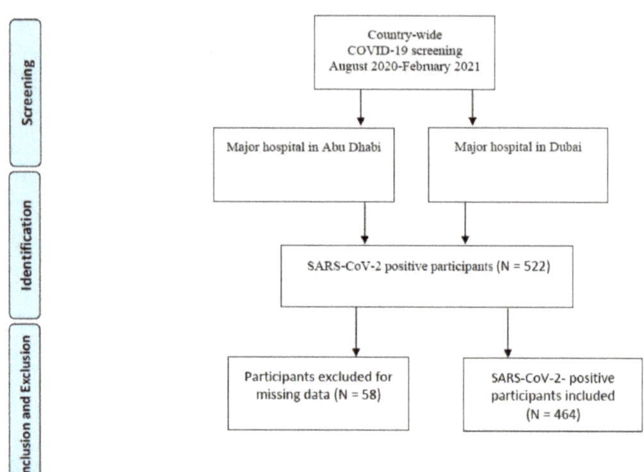

Figure 1. Flowchart diagram for selection of participants.

2.3. *Extracting and Quantifying SARS-CoV-2 Viral RNA*

An experienced phlebotomy nurse collected blood. A total of 2 mL of blood was collected from the cubital vein by using a gold-top (serum separator) tube. Samples were stored in a sealed biohazard bag and transported at 4 °C in a cool transport container to the Khalifa University Center for Biotechnology's laboratory for a second confirmatory testing for SARS-CoV-2. Viral RNA was extracted from swab by using the Miracle-AutoXT Automated Nucleic Acid Extraction System (iNtRON Biotechnology, Seoul, South Korea). Genesig from the Primerdesign reverse transcription-PCR COVID-19 detection kit (Watchmoor Point, UK) was used to quantify the viral RNA. PCRs were performed according to the manufacturer's instructions. Quantitative reverse transcription-PCR was performed using the Magnetic Induction Cycler PCR Machine (MiC) (Bio Molecular Systems, Queensland, Australia).

2.4. *Measuring Serum 25(OH)D Levels*

To assess vitamin D status, we measured the levels of total 25(OH)D. At recruitment, serum samples (that were collected from participants immediately upon arrival to testing centers) were cryopreserved at –80 °C in gel tubes and were used to measure 25(OH)D concentrations with automated electrochemiluminescence (Elecsys 2010; Roche Diagnostics, GmbH, Mannheim, Germany). The detection limit of serum 25(OH)D was 4 ng/mL. The intra-assay coefficient of variation was 5%, and the interassay coefficient was 7.5%.

2.5. *Defining Severity of Infection*

Clinical assessments of participants included determining the severity (mild, moderate, or severe) and diagnosis of pneumonia, confirmed using a chest X-ray. Although participants who presented with mild or no symptoms did not require hospitalization, they were included in the study because they tested positive for SARS-CoV-2. The moderate group had symptoms such as fever, cough, and pneumonia, requiring hospitalization. The severe group presented with critical clinical features, such as high temperature, cough, pneumonia, and shortness of breath, requiring intensive care (World Health Organization, 2020).

2.6. *Statistical Analysis*

Data were analyzed using SPSS version 27.0 (IBM, Armonk, NY, USA). Categorical variables were presented as frequencies and percentages. Continuous variables were presented as mean ± standard deviation. Patients were grouped into three categories of serum 25(OH)D levels: <12 ng/mL, 12–20 ng/mL, and ≥20 ng/mL. Differences according to SARS-CoV-2 severity of infection (asymptomatic, mild, moderate, high) were compared using the chi-square test for categorical variables and one-way analysis of variance for continuous variables. Moreover, differences based on mortality (deceased, alive) were explored using the chi-square test or Fisher's exact test for categorical variables and independent-samples *t*-test for continuous variables. Simple and multivariate ordered logistic regression models were constructed to determine predictors of infection severity. Simple and multivariate binary logistic regression models were considered to identify variables associated with mortality. Associations between risk factors and outcomes were presented as ORs and 95% CIs, with $p < 0.05$ considered statistically significant.

3. Results

We included 464 participants for data analysis and excluded 58 because of missing BMI data. Table 1 summarizes the main demographic and clinical characteristics of participants according to SARS-CoV-2 severity of infection and mortality. The mean age was 47 ± 15 years, with more than 60% of patients being male. Significant differences were observed for age, nationality, chronic disease (type 2 diabetes mellitus (T2D), cardiac disease, and renal disease), smoking, and BMI. The main comorbidities, T2D (32.8%), cardiac disease (11.6%), and renal disease (8.8%), were more prevalent among patients in the severe category. Similarly, those patients were older and more obese than others who had either asymptomatic or mild or moderate COVID-19. About 59% of patients who had vitamin D deficiency and severe vitamin D deficiency had severe symptoms of COVID-19 infection.

A total of 155 (33.4%) patients were vitamin D sufficient, whereas others were either deficient or severely deficient. In total, 65 (14%) UAE nationals were included in the study, of whom 25 (38.5%) had severe infection. The number of Southeast Asian patients was 276 and accounted for 59.5% of all patients, with 72 (26.1%) severely affected by COVID-19. In addition, patients who had severe infection were older and more obese than others ($p < 0.001$). We also evaluated differences in demographics and clinical investigations for patients according to mortality (Table 1). The baseline features differed significantly only in terms of age and major comorbidities. A total of 26 (5.6%) of 464 subjects died.

In Table 2, predictors for severity of infection were determined using multivariate ordered logistic regression analysis with both the adjusted and unadjusted models. To adjust for confounding factors, we used two models: model 1 adjusted for age, sex, and smoking, whereas model 2 adjusted for age, sex, smoking, and comorbidities. BMI > 30 kg/m^2 (obesity) was significant in the unadjusted model (OR = 2.42 (95% CI, 1.68, 3.49); $p < 0.001$). That factor was not included in the adjusted model. Patients' sex was not a significant risk factor, whereas smoking and comorbidities lost significance of effect upon adjustment. By contrast, age stood out as a strong independent predictor in both models 1 (OR = 1.08 (95% CI, 1.07, 1.10); $p < 0.001$) and 2 (OR = 1.07 (95% CI, 1.06, 1.09); $p < 0.001$). Serum 25(OH)D levels of < 12 ng/mL in model 1 (OR = 1.79 (95% CI, 1.21, 2.64); $p = 0.003$) and model 2 (OR = 1.76 (95% CI, 1.19, 2.61); $p = 0.005$) were strongly associated with severity of COVID-19.

Table 1. Characteristics of COVID-19 patients according to disease severity and mortality.

		Severity					Mortality		
	N	Asymptomatic	Mild	Moderate	High	*p*	Alive	Deceased	*p*
Total	464	91 (19.6)	99 (21.3)	129 (27.8)	145 (31.3)		438 (94.4)	26 (5.6)	
Age (years)	46.6 ± 14.9	34.3 ± 7.2	41.6 ± 13.7	49.0 ± 13.0	55.7 ± 14.1	**<0.001**	45.7 ± 14.5	62.5 ± 13.1	**<0.001**
Sex									
Female	92 (19.8)	27 (29.3)	20 (21.7)	14 (15.2)	31 (33.7)	**0.006**	89 (96.7)	3 (3.3)	0.28
Male	372 (80.2)	64 (17.2)	79 (21.2)	115 (30.9)	114 (30.6)		349 (93.8)	23 (6.2)	
BMI (kg/m^2)	28.1 ± 5.9	25.6 ± 4.0	27.0 ± 5.4	28.3 ± 5.4	30.2 ± 6.8	**<0.001**	28.0 ± 5.7	30.2 ± 8.5	0.07
Obesity (BMI > 30 kg/m^2)									
Obese	136 (29.3)	11 (8.1)	23 (16.9)	44 (32.4)	58 (42.6)	**<0.001**	128 (94.1)	8 (5.9)	0.89
Not obese	328 (70.7)	80 (24.4)	76 (23.2)	85 (25.9)	87 (26.5)		310 (94.5)	18 (5.5)	
Nationality									
UAE	65 (14)	0 (0)	13 (20)	27 (41.5)	25 (38.5)	**<0.001**	59 (90.8)	6 (9.2)	0.25
Arab (Middle Eastern)	103 (22.2)	8 (7.8)	41 (39.8)	17 (16.5)	37 (35.9)		97 (94.2)	6 (5.8)	
Asian	276 (59.5)	81 (29.3)	44 (15.9)	79 (28.6)	72 (26.1)		264 (95.7)	12 (4.3)	
Others	20 (4.3)	2 (10)	1 (5)	6 (30)	11 (55)		18 (90)	2 (10)	
Current Smoker									
Yes	50 (10.8)	13 (26)	23 (46)	11 (22)	3 (6)	**<0.001**	50 (100)	0 (0)	0.10
No	414 (89.2)	78 (18.8)	76 (18.4)	118 (28.5)	142 (34.3)		388 (93.7)	26 (6.3)	
25(OH)D Level (ng/mL)									
<12	127 (27.4)	21 (16.5)	28 (22)	34 (26.8)	44 (34.6)	0.07	117 (92.1)	10 (7.9)	0.32
12–20	182 (39.2)	48 (26.4)	36 (19.8)	53 (29.1)	45 (24.7)		175 (96.2)	7 (3.8)	
≥20	155 (33.4)	22 (14.2)	35 (22.6)	42 (27.1)	56 (36.1)		146 (94.2)	9 (5.8)	
Cardiac Disease									
Yes	54 (11.6)	0 (0)	10 (18.5)	16 (29.6)	28 (51.9)	**<0.001**	44 (81.5)	10 (18.5)	**<0.001**
No	410 (88.4)	91 (22.2)	89 (21.7)	113 (27.6)	117 (28.5)		394 (96.1)	16 (3.9)	

Table 1. *Cont.*

	N	Severity					Mortality		
		Asymptomatic	**Mild**	**Moderate**	**High**	*p*	**Alive**	**Deceased**	*p*
Chronic Lung Disease									
Yes	24 (5.2)	0 (0)	2 (8.3)	8 (33.3)	14 (58.3)	**0.004**	20 (83.3)	4 (16.7)	**0.04**
No	439 (94.8)	91 (20.7)	97 (22.1)	120 (27.3)	131 (29.8)		417 (95)	22 (5)	
Diabetes									
Yes	152 (32.8)	4 (2.6)	23 (15.1)	56 (36.8)	69 (45.4)	**<0.001**	136 (89.5)	16 (10.5)	**0.001**
No	312 (67.2)	87 (27.9)	76 (24.4)	73 (23.4)	76 (24.4)		302 (96.8)	10 (3.2)	
Renal Disease									
Yes	41 (8.8)	0 (0)	3 (7.3)	12 (29.3)	26 (63.4)	**<0.001**	33 (80.5)	8 (19.5)	**<0.001**
No	423 (91.2)	91 (21.5)	96 (22.7)	117 (27.7)	119 (28.1)		405 (95.7)	18 (4.3)	
Metabolic Disease									
Yes	27 (5.8)	1 (3.7)	4 (14.8)	8 (29.6)	14 (51.9)	**0.04**	23 (85.2)	4 (14.8)	0.06
No	437 (94.2)	90 (20.6)	95 (21.7)	121 (27.7)	131 (30)		415 (95)	22 (5)	
Liver Disease									
Yes	7 (1.5)	0 (0)	1 (14.3)	2 (28.6)	4 (57.1)	0.41	6 (85.7)	1 (14.3)	0.33
No	457 (98.5)	91 (19.9)	98 (21.4)	127 (27.8)	141 (30.9)		432 (94.5)	25 (5.5)	

Data are presented as *N* (%); data presented as mean ± SD; BMI, body mass index; UAE, United Arab Emirates; 25(OH)D, serum 25-hydroxyvitamin D; $p < 0.05$ considered significant (shown in boldface).

Table 2. Predictors for COVID-19 severity using multivariate ordered logistic regression analysis.

			Model 1		Model 2	
	Unadjusted		Adjusted		Adjusted	
Predictor	OR (95% CI)	p	OR (95% CI)	p	OR (95% CI)	p
Age	1.08 (1.07, 1.10)	**<0.001**	1.08 (1.07, 1.10)	**<0.001**	1.07 (1.06, 1.09)	**<0.001**
Male	1.39 (0.91, 2.14)	0.13	1.22 (0.78, 1.91)	0.38	1.23 (0.78, 1.94)	0.38
Smoker	0.34 (0.21, 0.57)	**<0.001**	0.60 (0.35, 1.02)	0.06	0.60 (0.35, 1.02)	0.06
Obese (BMI > 30 kg/m^2)	2.42 (1.68, 3.49)	**<0.001**				
Cardiac Disease	3.11 (1.84, 5.26)	**<0.001**			0.72 (0.38, 1.37)	0.32
Chronic Lung Disease	3.96 (1.81, 8.67)	**0.001**			1.64 (0.68, 3.93)	0.27
Diabetes	3.68 (2.56, 5.29)	**<0.001**			1.27 (0.82, 1.97)	0.28
Renal Disease	5.13 (2.71, 9.73)	**<0.001**			1.66 (0.80, 3.48)	0.18
Metabolic Disease	2.79 (1.35, 5.75)	**0.005**			1.34 (0.60, 2.99)	0.45
Liver Disease	3.35 (0.81, 13.85)	0.10			2.99 (0.54, 16.52)	0.21
25(OH)D < 12 ng/mL	1.22 (0.84, 1.76)	0.29	1.79 (1.21, 2.64)	**0.003**	1.76 (1.19, 2.61)	**0.005**
25(OH)D < 20 ng/mL	0.71 (0.50, 1.00)	0.051	1.17 (0.80, 1.71)	0.41	1.14 (0.78, 1.66)	0.51

Model 1 is adjusted for age, sex, and smoking status. Model 2 is adjusted for age, sex, smoking status, and comorbidities. Data are presented as frequencies (%) and odds ratio (OR) (95% CI); BMI, body mass index; 25(OH)D, serum 25-hydroxyvitamin D; $p < 0.05$ considered significant (shown in boldface).

Predictors for mortality, obtained using binary regression analysis with the outcomes of deceased or alive, are shown in Table 3. Age was strongly associated with risk of mortality. The only other significant predictor in the adjusted model was serum 25(OH)D levels < 12 ng/mL, which were associated with 2.55 times higher risk for death upon adjustment for age and sex (OR = 2.55 (95% CI, 1.03, 6.33); $p = 0.04$) and 2.58 times higher risk for death upon adjustment for age, sex, and comorbidities (OR = 2.58 (95% CI, 1.01, 6.62); $p = 0.048$). Major comorbidities were risk factors in the unadjusted models only. No deceased case patients were smokers; hence, smoking as a risk factor for mortality was not applicable. BMI > 30 kg/m^2 was not correlated with risk of death in the unadjusted model. For model 2, obesity was utilized in the unadjusted model only and excluded from the adjusted because there is a strong inverse correlation between 25(OH)D and BMI, and confounding factors that affect the factor of interest should not be included for adjustment [36,37].

Table 3. Significant predictors of mortality, using binary logistic regression analysis.

			Model (1)		Model (2)	
	Unadjusted		Adjusted		Adjusted	
Predictor	OR (95% CI)	p	OR (95% CI)	p	OR (95% CI)	p
Age	1.08 (1.05, 1.11)	**<0.001**	1.08 (1.05, 1.12)	**0.001**	1.07 (1.03, 1.11)	**0.001**
Male	1.96 (0.57, 6.66)	0.28	1.67 (0.46, 6.02)	0.43	1.84 (0.47, 7.25)	0.38
Smoker			NA			
Obese (BMI > 30 kg/m^2)	1.08 (0.46, 2.54)	0.87				
Cardiac Disease	5.60 (2.39, 13.08)	**<0.001**			1.66 (0.57, 4.83)	0.35
Chronic Lung Disease	3.79 (1.19, 12.04)	0.02			1.12 (0.28, 4.41)	0.87
Diabetes	3.55 (1.57, 8.03)	**0.002**			0.99 (0.38, 2.58)	0.98
Renal Disease	5.45 (2.21, 13.49)	**<0.001**			1.33 (0.45, 3.95)	0.60
Metabolic Disease	3.28 (1.04, 10.31)	**0.04**			2.45 (0.64, 9.34)	0.19
Liver Disease	2.88 (0.33, 24.85)	0.34			1.66 (0.16, 17.41)	0.67
25(OH)D < 12 ng/mL	1.71 (0.76, 3.89)	0.20	2.55 (1.03, 6.33)	**0.04**	2.58 (1.01, 6.62)	**0.048**
25(OH)D < 20 ng/mL	0.94 (0.41, 2.17)	0.89	1.72 (0.68, 4.34)	0.25	1.71 (0.66, 4.43)	0.27

Model 1 is adjusted for age and sex. Model 2 is adjusted for age, sex, and comorbidities. Data are presented as frequencies (%) and odds ratio (OR) (95% CI); BMI, body mass index; 25(OH)D, serum 25-hydroxyvitamin D; $p < 0.05$ considered significant (shown in boldface).

4. Discussion

Several recent reviews highlighted the important role of micronutrients in supporting the immune system and, hence, potentially reducing the risk of COVID-19 infection. Among those, vitamin D is the most attractive for research [38–40].

In our study, serum 25(OH)D levels were associated with severity of COVID-19 infection after adjustment for the main confounding factors, namely, age and sex. The protective effect for vitamin D supplementation against viral respiratory infections has been well established, and similar results have started to emerge for COVID-19 [7]. Age was strongly associated with severity and mortality. Our results are in accordance with other observational studies indicating that vitamin D deficiency is significantly associated with COVID-19 severity and death [7]. Age, obesity, and vitamin D deficiency have been well-established risk factors for COVID-19 infection [41]. Aging, also known as senescence, is a complex phenomenon involving many changes in all physiological systems [42–44]. The immune system is one system that exhibits several changes during the life span [45–48]. Being an intricate system that protects the body from external and internal invaders makes it of particular interest to study in the context of aging. Immunosenescence is a term that encompasses the major changes that happen to the immune system during aging, which is characterized by a drop in various immune variables. Recent studies suggest that the most featured changes that happen during aging in the adaptive immune system define the state of immunosenescence [49,50]. One of the most important perceptions in recent years is that the innate immune system has a type of memory, called trained innate immune memory, which at least partially illustrates some of the immune-related features of aging [51,52]. The suggestion of trained innate memory may clarify why aging innate immune cells stay activated [53]. Another study suggests that this state of activation is maintained even in the absence of a specific challenge [47]. The chronic low-grade inflammation (inflamm-aging) is responsible for maintaining immune cells in the activation state. In addition, anti-inflammatory molecules are needed during aging to balance that state, the destruction of which may destroy the whole creature [54].

The association between vitamin D status at time of hospitalization and sequels of acute inflammatory illness could be bidirectional. Even though inflammation could lower the level of serum 25(OH), the immunomodulatory effects of vitamin D are probably the results of its long-term rather than short-term actions [55]. Serum 25(OH)D concentrations decrease near the onset of acute inflammatory illnesses. However, the effect appears short-lived, perhaps only for a few hours [56]. Obesity is another notable factor that has been profoundly associated with COVID-19 risk [57–62]. However, recent studies indicate that BMI should be used in the models with confounders to interpret COVID-19 outcomes [57].

The UAE has abundant sunlight throughout most of the year, yet the population is mostly deficient owing to several risk factors, including style of dress and avoidance of sun exposure [33,34]. A retrospective study of 60,979 people from the UAE reported the mean value for serum 25(OH)D to be 48.89 nmol/L. Overall, 82% of those examined presented with hypovitaminosis, of whom 26% of females and 18% of males had severe deficiency. That research showed the serious magnitude of this public health burden among the UAE population [34]. The use of vitamin D supplementation with different attitudes toward medical screening and sun exposure upon the incidental identification of vitamin D deficiency during treatment for other conditions among middle-aged and older adults prompted a robust recommendation for supplementation [61]. UAE health care professionals regularly prescribe supplementation for patients with chronic illnesses, including T2D, cardiovascular diseases, and hypertension. Recently, UAE health insurers have excluded vitamin D tests from coverage among the annual health screen for apparently healthy people, resulting in a resurgence of vitamin D deficiency among young adults in comparison with older adults [61]. Our results are in accordance with previous data reported by Haq and colleagues [34]. Their results showed lower mean serum 25(OH)D for the 33–44 age group than for participants 45 and older—most likely because that subpopulation uses supplementation prescribed by health professionals during medical consultation visits.

Our findings reveal strong implications for vitamin D supplementation not only as a preventive strategy against COVID-19 infection, but also to boost immunity during infection. Accumulating positive results about vitamin D supplementation from several RCTs and intervention-based studies prove that supplementation goes beyond simply addressing vitamin D deficiency to being a protective and maybe even therapeutic measure [26]. Many observational studies have reported the strong link between vitamin D status and risk of disease severity among COVID-19 patients. A meta-analysis of 27 studies reported that vitamin D deficiency in patients with COVID-19 was significantly associated with higher risks of severe infection (OR = 1.64; 95% CI, 1.30, 2.09), hospitalization (OR = 1.81; 95% CI, 1.42, 2.21), and mortality (OR = 1.92; 95% CI, 1.06, 2.58) [63]. Many studies worldwide have investigated the same research question but reached inconsistent and non-decisive results, possibly due to different patient characteristics and research designs. A retrospective observational study to determine the positivity rate for SARS-CoV-2 among more than 190,000 patients in the United States estimated seroprevalence to be 9.3% among the population and revealed a significant inverse association with serum 25(OH)D levels independent of latitude, ethnicity, age, and sex [64]. One plausible explanation for the putative protective role of vitamin D and adequate serum 25(OH) against COVID-19 was linked to the compound nitric oxide (NO), which is an important component of the body's antiviral defense mechanism [65]. NO inhibits replication of SARS-CoV-2 [66] and inactivates or modifies viral replicating proteins [67]. Calcitriol is a direct transcriptional regulator of endothelial NO synthase, the primary source of NO in the blood. NO reduces risk of arterial stiffness, an important risk factor for hypertension [68]. Hypertension is an important risk factor for COVID-19, and UV exposure can reduce blood pressure [69]. Similar research from the Middle East region and Gulf countries is limited. A comprehensive study in Israel among 14,000 participants showed that vitamin D deficiency was a strong risk factor for COVID-19 infection [70]. Another study among 73 seropositive Iranian patients showed that vitamin D deficiency correlated with mortality [71]. A recent investigation in Saudi Arabia showed a robust association for severe vitamin D deficiency with death but not severity of disease [72]. In a different case–control study in the same country, 138 mildly affected patients were matched with 82 negative controls, and serum 25(OH)D levels were significantly lower in affected people but were not a predictor of disease outcome. That finding called for additional large population-based RCTs to further confirm the results [73].

5. Strengths and Limitations

To our knowledge, this is the first study to evaluate 25(OH)D levels in patients who tested positive for SARS-CoV-2 and to examine their association with COVID-19 severity and mortality among a sample of affected people within the UAE. Our findings offer promising results that warrant further research to examine whether vitamin D supplementation could help reduce COVID-19 severity and risk of infection in this population. Vitamin D deficiency is often associated with several comorbidities such as cardiometabolic disorders, T2D, and obesity. The use of multivariate analysis to control for confounding variables and the fact that we recruited subjects from two main hospitals in the UAE's two main cities (Abu Dhabi and Dubai) strengthened our investigation. Moreover, all nationalities were included to reflect the multiethnic UAE population.

Some limitations are worth noting, however. The small number of deaths in our study most likely affected the analysis. Examining a larger sample to include more mortalities could offer more conclusive results about the relation between vitamin D status and the death outcome from COVID-19 infection in the UAE. The socioeconomic status for all participants was not assessed, but could have affected the dietary habits and availability of fortified foods, which in turn could have affected vitamin D status along with any use of supplements and sun exposure that were not recorded. In addition, the optimal concentration of serum 25(OH)D for overall health remains controversial and using different cutoffs might slightly change results. The bone-centric guidelines recommend a target 25(OH)D concentration of 20 ng/mL (50 nmol/L) and age-dependent daily vitamin D doses of 400–800 IU. The guidelines focused on the pleiotropic effects of vitamin D recommend a target 25(OH)D concentration of 30 ng/mL (75 nmol/L) and age-, body weight-, disease status-, and ethnicity-dependent vitamin D doses between 400 and 2000 IU/day [74]. However, mounting evidence indicates that optimal 25(OH)D levels are 40–60 ng/mL, as seen in the SARS-CoV-2 seropositivity study by Kaufman and colleagues [64], an open-label vitamin D supplementation-breast cancer incidence study [75], and an open-label vitamin D supplementation-blood pressure study [76].

6. Conclusions

Our data showed that serum 25(OH)D levels <12 ng/mL are strongly associated with COVID-19 severity and mortality among a sample of affected people in the UAE. Such findings suggest important implications that vitamin D supplementation could help reduce the severity of COVID-19 disease and risk of infection. Further larger observational studies and RCTs are needed to furnish a comprehensive picture about the link between vitamin D and COVID-19 severity and death among the UAE population.

Author Contributions: F.A.A.: Conceptualization, project administration, data analysis and interpretation, writing of original draft. H.A.: Conceptualization, project administration, writing and editing. R.H.: Conceptualization, data analysis and interpretation. W.B.G.: Data analysis, conceptualization, writing and editing. N.A.: Writing (review and editing). M.U.: Data collection and review. B.M.: Writing (review and editing). G.T.: Writing (review and editing). All authors have read and agreed to the published version of the manuscript.

Funding: The project was funded by internal funds provided by Khalifa University awarded to Habiba Al Safar and by the special grant R20104 to support the Country's effort in COVID-19 related research awarded by Zayed University, Research Office, United Arab Emirates to Fatme Al Anouti. The funding body was not involved in the design of the study and collection, analysis, and interpretation of data or in writing the manuscript.

Institutional Review Board Statement: The study was conducted according to the guidelines of the Declaration of Helsinki, and approved by Abu Dhabi Health COVID-19 Research Ethics Committee (DOH/DQD/2020/538), the Dubai Scientific Research Ethics Committee (DSREC-04/2020_09), and the SEHA Research Ethics Committee (SEHA-IRB-005).

Informed Consent Statement: Informed consent was obtained from all subjects involved in the study.

Data Availability Statement: Data can be available upon request from the first and corresponding authors.

Acknowledgments: We thank the participants of the study for their generosity in providing samples to advance our understanding of COVID-19 infection. We acknowledge the assistance of the health care workers at the front line of the COVID-19 pandemic, without whose assistance this study would not have been possible. We also are grateful to Hussein Kannout, who helped process samples in the laboratory.

Conflicts of Interest: W.B.G. receives funding from Bio-Tech Pharmacal Inc. (Fayetteville, AR, USA). All other authors declare that they have no competing interests.

References

1. Hussein, T.; Löndahl, J.; Thuresson, S.; Alsved, M.; Al-Hunaiti, A.; Saksela, K.; Aqel, H.; Junninen, H.; Mahura, A.; Kulmala, M. Indoor Model Simulation for COVID-19 Transport and Exposure. *Int. J. Environ. Res. Public Health* **2021**, *18*, 2927. [CrossRef]
2. Wimalawansa, S.J. Global epidemic of coronavirus—Covid-19: What can we do to minimize risks. *Eur. J. Biomed. Pharm. Sci.* **2020**, *7*, 432–438.
3. Thompson, R. Pandemic potential of 2019-nCoV. *Lancet Infect. Dis.* **2020**, *20*, 280. [CrossRef]
4. Butt, Y.; Kurdowska, A.; Allen, T.C. Acute Lung Injury: A Clinical and Molecular Review. *Arch. Pathol. Lab. Med.* **2016**, *140*, 345–350. [CrossRef] [PubMed]
5. Hughes, K.T.; Beasley, M.B. Pulmonary Manifestations of Acute Lung Injury: More Than Just Diffuse Alveolar Damage. *Arch. Pathol. Lab. Med.* **2016**, *141*, 916–922. [CrossRef]

6. Dattola, A.; Silvestri, M.; Bennardo, L.; Passante, M.; Scali, E.; Patruno, C.; Nisticò, S.P. Role of Vitamins in Skin Health: A Systematic Review. *Curr. Nutr. Rep.* **2020**, *9*, 226–235. [CrossRef] [PubMed]
7. Mercola, J.; Grant, W.B.; Wagner, C.L. Evidence Regarding Vitamin D and Risk of COVID-19 and Its Severity. *Nutrients* **2020**, *12*, 3361. [CrossRef]
8. Teymoori-Rad, M.; Shokri, F.; Salimi, V.; Marashi, S.M. The interplay between vitamin D and viral infections. *Rev. Med. Virol.* **2019**, *29*, e2032. [CrossRef]
9. Hooper, M.W.; Nápoles, A.M.; Pérez-Stable, E.J. COVID-19 and Racial/Ethnic Disparities. *JAMA* **2020**, *323*, 2466. [CrossRef]
10. Weiss, P.; Murdoch, D.R. Clinical course and mortality risk of severe COVID-19. *Lancet* **2020**, *395*, 1014–1015. [CrossRef]
11. Dastoli, S.; Bennardo, L.; Patruno, C.; Nisticò, S.P. Are erythema multiforme and urticaria related to a better outcome of COVID-19? *Dermatol. Ther.* **2020**, *33*, e13681. [CrossRef]
12. Grant, W.B.; Lahore, H.; McDonnell, S.L.; Baggerly, C.A.; French, C.B.; Aliano, J.L.; Bhattoa, H.P. Evidence that vitamin D supplementation could reduce risk of influenza and COVID-19 infections and deaths. *Nutrients* **2020**, *12*, 988. [CrossRef]
13. Rhodes, J.M.; Subramanian, S.; Laird, E.; Kenny, R.A. Editorial: Low population mortality from COVID-19 in countries south of latitude 35 degrees North supports vitamin D as a factor determining severity. *Aliment. Pharmacol. Ther.* **2020**, *51*, 1434–1437. [CrossRef] [PubMed]
14. Bordelon, P.; Ghetu, M.V.; Langan, R.C. Recognition and management of vitamin D deficiency. *Am. Fam. Physician* **2009**, *80*, 841–846.
15. Watkins, R.R.; Lemonovich, T.L.; Salata, R.A. An update on the association of vitamin D deficiency with common infectious diseases. *Can. J. Physiol. Pharmacol.* **2015**, *93*, 363–368. [CrossRef] [PubMed]
16. Science, M.; Maguire, J.L.; Russell, M.L.; Smieja, M.; Walter, S.D.; Loeb, M. Low serum 25-hydroxyvitamin D level and risk of upper respiratory tract infection in children and adolescents. *Clin. Infect. Dis.* **2013**, *57*, 392–397. [CrossRef]
17. Martineau, A.R.; Jolliffe, D.A.; Hooper, R.L.; Greenberg, L.; Aloia, J.F.; Bergman, P.; Dubnov-Raz, G.; Esposito, S.; Ganmaa, D.; Ginde, A.A.; et al. Vitamin D supplementation to prevent acute respiratory tract infections: Systematic review and meta-analysis of individual participant data. *BMJ* **2017**, *356*, i6583. [CrossRef] [PubMed]
18. Glinsky, G.V. Tripartite combination of candidate pandemic mitigation agents: Vitamin D, quercetin, and estradiol manifest properties of medicinal agents for targeted mitigation of the COVID-19 pandemic defined by genomics-guided tracing of SARS-CoV-2 targets in human cells. *Biomedicines* **2020**, *8*, 129.
19. Gois, P.H.F.; Ferreira, D.; Olenski, S.; Seguro, A.C. Vitamin D and Infectious Diseases: Simple Bystander or Contributing Factor? *Nutrients* **2017**, *9*, 651. [CrossRef]
20. Hossein-Nezhad, A.; Holick, M.F. Vitamin D for Health: A Global Perspective. In *Proceedings of the Mayo Clinic Proceedings*; Elsevier BV: Amsterdam, The Netherlands, 2013; Volume 88, pp. 720–755.
21. Kuba, K.; Imai, Y.; Penninger, J.M. Angiotensin-converting enzyme 2 in lung diseases. *Curr. Opin. Pharmacol.* **2006**, *6*, 271–276. [CrossRef]
22. Rodríguez-Puertas, R. ACE2 activators for the treatment of COVID 19 patients. *J. Med. Virol.* **2020**, *92*, 1701–1702. [CrossRef]
23. Xudong, X.; Junzhu, C.; Xingxiang, W.; Furong, Z.; Yanrong, L. Age- and gender-related difference of ACE2 expression in rat lung. *Life Sci.* **2006**, *78*, 2166–2171. [CrossRef] [PubMed]
24. Ilie, P.C.; Stefanescu, S.; Smith, L. The role of vitamin D in the prevention of coronavirus disease 2019 infection and mortality. *Aging Clin. Exp. Res.* **2020**, *32*, 1195–1198. [CrossRef]
25. Sasabe, J.; Suzuki, M.; Imanishi, N.; Aiso, S. Activity of D-amino acid oxidase is widespread in the human central nervous system. *Front. Synaptic Neurosci.* **2014**, *6*, 14. [CrossRef]
26. Castillo, M.E.; Costa, L.M.E.; Barrios, J.M.V.; Díaz, J.F.A.; Miranda, J.L.; Bouillon, R.; Gomez, J.M.Q. Effect of calcifediol treatment and best available therapy versus best available therapy on intensive care unit admission and mortality among patients hospitalized for COVID-19: A pilot randomized clinical study. *J. Steroid Biochem. Mol. Biol.* **2020**, *203*, 105751. [CrossRef]
27. Charoenngam, N.; Shirvani, A.; Reddy, N.; Vodopivec, D.M.; Apovian, C.M.; Holick, M.F. Association of Vitamin D Status With Hospital Morbidity and Mortality in Adult Hospitalized Patients With COVID-19. *Endocr. Pr.* **2021**, *27*, 271–278. [CrossRef]
28. Braun, A.; Chang, D.; Mahadevappa, K.; Gibbons, F.K.; Liu, Y.; Giovannucci, E.; Christopher, K.B. Association of low serum 25-hydroxyvitamin D levels and mortality in the critically ill. *Crit. Care Med.* **2011**, *39*, 671–677. [CrossRef]
29. De Haan, K.; Groeneveld, A.J.; de Geus, H.R.; Egal, M.; Struijs, A. Vitamin D deficiency as a risk factor for infection, sepsis and mortality in the critically ill: Systematic review and meta-analysis. *Crit. Care* **2014**, *18*, 1–8. [CrossRef]
30. Beard, J.A.; Bearden, A.; Striker, R. Vitamin D and the anti-viral state. *J. Clin. Virol.* **2011**, *50*, 194–200. [CrossRef]
31. Gunville, C.F.; Mourani, P.M.; Ginde, A.A. The role of vitamin D in prevention and treatment of infection. *Inflamm. Allergy-Drug Targets* **2013**, *12*, 239–245. [CrossRef]
32. Graham, B.S. Rapid COVID-19 vaccine development. *Science* **2020**, *368*, 945–946. [CrossRef]
33. Alanouti, F.; Thomas, J.; Abdel-Wareth, L.; Rajah, J.; Grant, W.B.; Haq, A. Vitamin D deficiency and sun avoidance among university students at Abu Dhabi, United Arab Emirates. *Dermato-Endocrinol.* **2011**, *3*, 235–239. [CrossRef] [PubMed]
34. Haq, A.; Svobodová, J.; Imran, S.; Stanford, C.; Razzaque, M.S. Vitamin D deficiency: A single centre analysis of patients from 136 countries. *J. Steroid Biochem. Mol. Biol.* **2016**, *164*, 209–213. [CrossRef] [PubMed]
35. Palacios, C.; Gonzalez, L. Is vitamin D deficiency a major global public health problem? *J. Steroid Biochem. Mol. Biol.* **2014**, *144*, 138–145. [CrossRef] [PubMed]

36. Drincic, A.T.; Armas, L.A.; van Diest, E.E.; Heaney, R.P. Volumetric Dilution, Rather Than Sequestration Best Explains the Low Vitamin D Status of Obesity. *Obesity* **2012**, *20*, 1444–1448. [CrossRef] [PubMed]
37. Van der Weele, T.J. Principles of confounder selection. *Eur. J. Epidemiol.* **2019**, *34*, 211–219. [CrossRef] [PubMed]
38. Gombart, A.F.; Pierre, A.; Maggini, S. A Review of Micronutrients and the Immune System—Working in Harmony to Reduce the Risk of Infection. *Nutrients* **2020**, *12*, 236. [CrossRef]
39. Gröber, U.; Holick, M.F. The coronavirus disease (COVID-19)—A supportive approach with selected micronutrients. *Int. J. Vitam. Nutr. Res.* **2021**, 1–22. [CrossRef]
40. Rossetti, M.; Martucci, G.; Starchl, C.; Amrein, K. Micronutrients in Sepsis and COVID-19: A Narrative Review on What We Have Learned and What We Want to Know in Future Trials. *Medicina* **2021**, *57*, 419. [CrossRef]
41. Biesalski, H.K. Vitamin D deficiency and co-morbidities in COVID-19 patients—A fatal relationship? *NFS J.* **2020**, *20*, 10–21. [CrossRef]
42. Zierer, J.; Menni, C.; Kastenmüller, G.; Spector, T.D. Integration of 'omics' data in aging research: From biomarkers to systems biology. *Aging Cell* **2015**, *14*, 933–944. [CrossRef] [PubMed]
43. Jazwinski, S.M.; Yashin, A.I. Aging and health—A systems biology perspective. Introduction. *Interdiscip. Top. Gerontol.* **2015**, *40*, VII–XII.
44. Cohen, A.A. Complex systems dynamics in aging: New evidence, continuing questions. *Biogerontology* **2016**, *17*, 205–220. [CrossRef]
45. Fulop, T.; McElhaney, J.; Pawelec, G.; Cohen, A.A.; Morais, J.A.; Dupuis, G.; Baehl, S.; Camous, X.; Witkowski, J.M.; Larbi, A. Frailty, Inflammation and Immunosenescence. *Frailty Aging* **2015**, *41*, 26–40.
46. Pawelec, G. Does the human immune system ever really become "senescent"? *F1000Research* **2017**, *6*. [CrossRef]
47. Franceschi, C.; Salvioli, S.; Garagnani, P.; de Eguileor, M.; Monti, D.; Capri, M. Immunobiography and the Heterogeneity of Immune Responses in the Elderly: A Focus on Inflammaging and Trained Immunity. *Front. Immunol.* **2017**, *8*, 982. [CrossRef]
48. Xu, W.; Larbi, A. Markers of T Cell Senescence in Humans. *Int. J. Mol. Sci.* **2017**, *18*, 1742. [CrossRef]
49. Goronzy, J.J.; Fang, F.; Cavanagh, M.M.; Fengqin, F.; Weyand, C.M. Naive T Cell Maintenance and Function in Human Aging. *J. Immunol.* **2015**, *194*, 4073–4080. [CrossRef]
50. Yanes, R.E.; Gustafson, C.E.; Weyand, C.M.; Goronzy, J.J. Lymphocyte generation and population homeostasis throughout life. *Semin. Hematol.* **2017**, *54*, 33–38. [CrossRef]
51. Kleinnijenhuis, J.; Quintin, J.; Preijers, F.; Joosten, L.A.B.; Ifrim, D.C.; Saeed, S.; Jacobs, C.; van Loenhout, J.; de Jong, D.; Stunnenberg, H.G.; et al. Bacille Calmette-Guerin induces NOD2-dependent nonspecific protection from reinfection via epigenetic reprogramming of monocytes. *Proc. Natl. Acad. Sci. USA* **2012**, *109*, 17537–17542. [CrossRef]
52. Netea, M.G.; van der Meer, J.W. Trained immunity: An ancient way of remembering. *Cell Host Microbe* **2017**, *21*, 297–300. [CrossRef] [PubMed]
53. Fülöp, T., Jr.; Foris, G.; Worum, I.; Leövey, A. Age-dependent alterations of Fc gamma receptor-mediated effector functions of human polymorphonuclear leucocytes. *Clin. Exp. Immunol.* **1985**, *61*, 425. [PubMed]
54. Fülöp, T.; Dupuis, G.; Witkowski, J.M.; Larbi, A. The Role of Immunosenescence in the Development of Age-Related Diseases. *Rev. Investig. Clin.* **2016**, *68*, 84–91.
55. Bayramoğlu, E.; Akkoç, G.; Ağbaş, A.; Akgün, Ö.; Yurdakul, K.; Duru, H.N.S.; Elevli, M. The association between vitamin D levels and the clinical severity and inflammation markers in pediatric COVID-19 patients: Single-center experience from a pandemic hospital. *Eur. J. Nucl. Med. Mol. Imaging* **2021**, 1–7. [CrossRef]
56. Smolders, J.; Ouweland, J.V.D.; Geven, C.; Pickkers, P.; Kox, M. Letter to the Editor: Vitamin D deficiency in COVID-19: Mixing up cause and consequence. *Metab. Clin. Exp.* **2021**, *115*, 154434. [CrossRef]
57. Dalamaga, M.; Christodoulatos, G.S.; Karampela, I.; Vallianou, N.; Apovian, C.M. Understanding the Co-Epidemic of Obesity and COVID-19: Current Evidence, Comparison with Previous Epidemics, Mechanisms, and Preventive and Therapeutic Perspectives. *Curr. Obes. Rep.* **2021**, 1–30. [CrossRef]
58. Ekiz, T.; Pazarlı, A.C. Relationship between COVID-19 and obesity. *Diabetes Metab. Syndr. Clin. Res. Rev.* **2020**, *14*, 761–763. [CrossRef]
59. Hamer, M.; Gale, C.R.; Kivimäki, M.; Batty, G.D. Overweight, obesity, and risk of hospitalization for COVID-19: A community-based cohort study of adults in the United Kingdom. *Proc. Natl. Acad. Sci. USA* **2020**, *117*, 21011–21013. [CrossRef]
60. Sattar, N.; McInnes, I.B.; McMurray, J.J. Obesity is a risk factor for severe COVID-19 infection: Multiple potential mechanisms. *Circulation* **2020**, *142*, 4–6. [CrossRef]
61. Al Saleh, Y.; Beshyah, S.A.; Hussein, W.; Almadani, A.; Hassoun, A.; Al Mamari, A.; Ba-Essa, E.; Al-Dhafiri, E.; Hassanein, M.; Fouda, M.A.; et al. Diagnosis and management of vitamin D deficiency in the Gulf Cooperative Council (GCC) countries: An expert consensus summary statement from the GCC vitamin D advisory board. *Arch. Osteoporos.* **2020**, *15*, 1–8. [CrossRef]
62. Simonnet, A.; Chetboun, M.; Poissy, J.; Raverdy, V.; Noulette, J.; Duhamel, A.; Labreuche, J.; Mathieu, D.; Pattou, F.; Jourdain, M.; et al. High prevalence of obesity in severe acute respiratory syndrome coronavirus-2 (SARS-CoV-2) requiring invasive mechanical ventilation. *Obesity* **2020**, *28*, 1195–1199. [CrossRef]
63. Kazemi, A.; Mohammadi, V.; Aghababaee, S.K.; Golzarand, M.; Clark, C.C.; Babajafari, S. Association of Vitamin D status with SARS-CoV-2 infection or COVID-19 severity: A systematic review and meta-analysis. *Adv. Nutr.* **2021**. [CrossRef]

64. Kaufman, H.W.; Niles, J.K.; Kroll, M.H.; Bi, C.; Holick, M.F. SARS-CoV-2 positivity rates associated with circulating 25-hydroxyvitamin D levels. *PLoS ONE* **2020**, *15*, e0239252. [CrossRef]
65. Guan, S.P.; Seet, R.C.S.; Kennedy, B.K. Does eNOS derived nitric oxide protect the young from severe COVID-19 complications? *Ageing Res. Rev.* **2020**, *64*, 101201. [CrossRef]
66. Åkerström, S.; Mousavi-Jazi, M.; Klingström, J.; Leijon, M.; Lundkvist, A.; Mirazimi, A. Nitric Oxide Inhibits the Replication Cycle of Severe Acute Respiratory Syndrome Coronavirus. *J. Virol.* **2005**, *79*, 1966–1969. [CrossRef]
67. Abdul-Cader, M.S.; Amarasinghe, A.; Abdul-Careem, M.F. Activation of toll-like receptor signaling pathways leading to nitric oxide-mediated antiviral responses. *Arch. Virol.* **2016**, *161*, 2075–2086. [CrossRef]
68. Andrukhova, O.; Slavic, S.; Zeitz, U.; Riesen, S.C.; Heppelmann, M.S.; Ambrisko, T.D.; Markovic, M.; Kuebler, W.M.; Erben, R.G. Vitamin D Is a Regulator of Endothelial Nitric Oxide Synthase and Arterial Stiffness in Mice. *Mol. Endocrinol.* **2014**, *28*, 53–64. [CrossRef] [PubMed]
69. Weller, R.B.; Wang, Y.; He, J.; Maddux, F.W.; Usvyat, L.; Zhang, H.; Feelisch, M.; Kotanko, P. Does Incident Solar Ultraviolet Radiation Lower Blood Pressure? *J. Am. Hear. Assoc.* **2020**, *9*, e013837. [CrossRef]
70. Merzon, E.; Tworowski, D.; Gorohovski, A.; Vinker, S.; Cohen, A.G.; Green, I.; Frenkel-Morgenstern, M. Low plasma 25(OH) vitamin D level is associated with increased risk of COVID-19 infection: An Israeli population-based study. *FEBS J.* **2020**, *287*, 3693–3702. [CrossRef]
71. Abrishami, A.; Dalili, N.; Torbati, P.M.; Asgari, R.; Arab-Ahmadi, M.; Behnam, B.; Sanei-Taheri, M. Possible association of vitamin D status with lung involvement and outcome in patients with COVID-19: A retrospective study. *Eur. J. Nutr.* **2020**, 1–9. [CrossRef]
72. Alguwaihes, A.M.; Sabico, S.; Hasanato, R.; Al-Sofiani, M.E.; Megdad, M.; Albader, S.S.; Alsari, M.H.; Alelayan, A.; Alyusuf, E.Y.; Alzahrani, S.H.; et al. Severe vitamin D deficiency is not related to SARS-CoV-2 infection but may increase mortality risk in hospitalized adults: A retrospective case–control study in an Arab Gulf country. *Aging Clin. Exp. Res.* **2021**, *33*, 1415–1422. [CrossRef] [PubMed]
73. Al-Daghri, N.M.; Amer, O.E.; Alotaibi, N.H.; Aldisi, D.A.; Enani, M.A.; Sheshah, E.; Aljohani, N.J.; Alshingetti, N.; Alomar, S.Y.; Alfawaz, H.; et al. Vitamin D status of Arab Gulf residents screened for SARS-CoV-2 and its association with COVID-19 infection: A multi-centre case-control study. *J. Transl. Med.* **2021**, *19*, 1–8. [CrossRef]
74. Pludowski, P.; Holick, M.F.; Grant, W.B.; Konstantynowicz, J.; Mascarenhas, M.R.; Haq, A.; Povoroznyuk, V.; Balatska, N.; Barbosa, A.P.; Karonova, T.; et al. Vitamin D supplementation guidelines. *J. Steroid Biochem. Mol. Biol.* **2018**, *175*, 125–135. [CrossRef]
75. McDonnell, S.L.; Baggerly, C.A.; French, C.B.; Baggerly, L.L.; Garland, C.F.; Gorham, E.D.; Hollis, B.W.; Trump, D.L.; Lappe, J.M. Breast cancer risk markedly lower with serum 25-hydroxyvitamin D concentrations ≥60 vs <20 ng/mL (150 vs 50 nmol/L): Pooled analysis of two randomized trials and a prospective cohort. *PLoS ONE* **2018**, *13*, e0199265. [CrossRef]
76. Mirhosseini, N.; Vatanparast, H.; Kimball, S.M. The Association between Serum 25(OH)D Status and Blood Pressure in Participants of a Community-Based Program Taking Vitamin D Supplements. *Nutrients* **2017**, *9*, 1244. [CrossRef]

MDPI

Article

Low 25(OH)D Level Is Associated with Severe Course and Poor Prognosis in COVID-19

Tatiana L. Karonova [1,*], Alena T. Andreeva [1], Ksenia A. Golovatuk [1], Ekaterina S. Bykova [1], Anna V. Simanenkova [1], Maria A. Vashukova [2], William B. Grant [3] and Evgeny V. Shlyakhto [1]

1 Clinical Endocrinology Laboratory, Department of Endocrinology, Almazov National Medical Research Centre, 194021 Saint-Petersburg, Russia; arabicaa@gmail.com (A.T.A.); ksgolovatiuk@gmail.com (K.A.G.); bykova160718@gmail.com (E.S.B.); annasimanenkova@mail.ru (A.V.S.); e.shlyakhto@almazovcentre.ru (E.V.S.)
2 Botkin Clinical Infectious Hospital, 195067 Saint-Petersburg, Russia; mavashukova@yahoo.com
3 Sunlight, Nutrition, and Health Research Center, P.O. Box 641603, San Francisco, CA 94164-1603, USA; williamgrant08@comcast.net
* Correspondence: karonova@mail.ru; Tel.: +7-921-310-60-41

Abstract: We evaluated associations between serum 25-hydroxyvitamin D [25(OH)D] level and severity of new coronavirus infection (COVID-19) in hospitalized patients. We assessed serum 25(OH)D level in 133 patients aged 21–93 years. Twenty-five (19%) patients had severe disease, 108 patients (81%) had moderate disease, and 18 (14%) patients died. 25(OH)D level ranged from 3.0 to 97.0 ng/mL (median, 13.5 [25%; 75%, 9.6; 23.3] ng/mL). Vitamin D deficiency was diagnosed in 90 patients, including 37 with severe deficiency. In patients with severe course of disease, 25(OH)D level was lower (median, 9.7 [25%; 75%, 6.0; 14.9] ng/mL), and vitamin D deficiency was more common than in patients with moderate course (median, 14.6 [25%; 75%, 10.6; 24.4] ng/mL, $p = 0.003$). In patients who died, 25(OH)D was 9.6 [25%; 75%, 6.0; 11.5] ng/mL, compared with 14.8 [25%; 75%, 10.1; 24.3] ng/mL in discharged patients ($p = 0.001$). Severe vitamin D deficiency was associated with increased risk of COVID-19 severity and fatal outcome. The threshold for 25(OH)D level associated with increased risk of severe course was 11.7 ng/mL. Approximately the same 25(OH)D level, 10.9 ng/mL, was associated with increased risk of mortality. Thus, most COVID-19 patients have vitamin D deficiency; severe vitamin D deficiency is associated with increased risk of COVID-19 severity and fatal outcome.

Keywords: vitamin D deficiency; 25(OH)D; obesity; COVID-19

Citation: Karonova, T.L.; Andreeva, A.T.; Golovatuk, K.A.; Bykova, E.S.; Simanenkova, A.V.; Vashukova, M.A.; Grant, W.B.; Shlyakhto, E.V. Low 25(OH)D Level Is Associated with Severe Course and Poor Prognosis in COVID-19. *Nutrients* **2021**, *13*, 3021. https://doi.org/10.3390/nu13093021

Academic Editor: Andrea Fabbri

Received: 23 July 2021
Accepted: 26 August 2021
Published: 29 August 2021

1. Introduction

Several studies conducted since the onset of the COVID-19 pandemic have shown that vitamin D deficiency can increase the incidence and worsen the course of acute respiratory viral infection caused by SARS-CoV-2 [1,2]. The immunomodulatory effects of vitamin D are well studied and are associated largely with the expression of the CYP27B1 enzyme and the presence of the vitamin D receptor in immune system cells [3–5]. Through several mechanisms, vitamin D may reduce the risk of bacterial and viral infection by creating a barrier, involving adaptive and humoral immunity [6]. Vitamin D is a potent stimulator of the monocyte/macrophage responses to bacterial infection, and thus it is an important participant in the innate immune response [7]. By contrast, inducing antimicrobial peptides, cathelicidin LL-37 [8,9] and defensins [10], vitamin D enhances cellular immunity, increases the synthesis of the NF-κB inhibitor IκBα, and decreases expression of proinflammatory genes [11]. Vitamin D also modulates humoral immunity [6,12,13] by suppressing interleukin 2 (IL-2) and interferon production and stimulating type 2 T-helper cytokine's production [12–14]. The optimal vitamin D status can promote immunoregulatory functions in conditions of viral respiratory infection and overall can influence the altered immune-inflammatory COVID-19 reactivity at least by down-regulating overly exuberant cytokine responses that comprise pathological cytokine storm [15].

At the same time, considering the role of vitamin D in the activity of the renin–angiotensin–aldosterone system, researchers believe that it controls the amount of mRNA and the expression of angiotensin-converting enzyme 2, which determines the protective function against various respiratory infections. Moreover, vitamin D can suppress DPP-4/CD26, the putative adhesion molecule for SARS-CoV-2 to enter the cell [16,17].

Analyzing data on risk factors for COVID-19, we note their similarity to factors contributing to vitamin D deficiency. Those factors include age, sex, and race [18–21]; seasonality of incidence of acute respiratory viral infection, which accounts for the lowest concentration of 25-hydroxyvitamin D [25(OH)D] [22,23]; presence of obesity and type 2 diabetes mellitus [24,25]; and smoking [26].

The vitamin D level in patients with COVID-19 can be judged based on the results of single studies. Positive PCR tests for COVID-19 were more common in individuals with lower 25(OH)D levels [27]. Previously, we published data concerning the high incidence of vitamin D deficiency in residents of the northwest region of the Russian Federation [28]. That analysis was a prerequisite for this study to assess serum 25(OH)D level in hospitalized COVID-19 patients.

Objective of study: To assess vitamin D status in patients with community-acquired viral pneumonia with a confirmed diagnosis of COVID-19 and to match 25(OH)D value with disease severity.

2. Materials and Methods

We analyzed records of 161 patients with a new COVID-19 infection, hospitalized between April and December 2020 at Botkin Clinical Infectious Hospital (St. Petersburg, Russia; latitude, 59° N). Demographic data, information on the clinical course and infection severity, presence of concomitant diseases and drug therapy, and results of computed tomography (CT) and laboratory examinations were collected. Alcohol abuse was an exclusion criterion.

Pneumonia was established by means of chest CT without intravenous contrast enhancement. Volume of lung tissue lesions was described as follows: CT-1, lesion volume <25%; CT-2, lesion volume 25–50%; CT-3, lesion volume 50–75%; CT-4, lesion volume >75%.

Serum 25(OH)D level was detected by chemiluminescence immunoassay on micro particles (Abbott Architect c8000, Chicago, IL, USA, intra-assay CV of 1.60–5.92%, inter-assay CV ranged from 2.15 to 2.63%). According to Russian and international guidelines [29,30], normal vitamin D status was considered to be 25(OH)D $\geq$30 ng/mL ($\geq$75 nmol/L); for insufficiency, $\geq$20 and <30 ng/mL ($\geq$50 and <75 nmol/L); for deficiency, <20 ng/mL (<50 nmol/L), and for severe vitamin D deficiency, less than 10 ng/mL (<25 nmol/L). The 25(OH)D level between 10 to 20 ng/mL was assessed as mild deficiency in this work. The reference interval for serum 25(OH)D level determination was 3.4–155.9 ng/mL.

We also checked plasma glucose level and inflammatory reaction markers: C-reactive protein (CRP), IL-6, and ferritin. All parameters were measured at the time of admission (baseline), and their maximal values (max) were fixed.

Blood glucose level was measured using an automatic biochemistry analyzer (Cobas Integra c311 [Roche Diagnostics GmbH, Mannheim, Germany]) and diagnostic kits. Serum IL-6 concentration was determined by enzyme-linked immunosorbent assay on Bio-Rad 680 Microplate Reader equipment (Hercules, CA, USA), using appropriate reagent kits (Vector-Best; Novosibirsk, Russia). Results were processed using Zemfira 4 software for ELISA Bio-Rad analyzer (reference range, 0–7 pg/mL). An automatic biochemistry analyzer (Cobas Integra 400 [Roche Diagnostics GmbH, Mannheim, Germany]) and corresponding diagnostic kits from that manufacturer were used to determine level of CRP by means of turbidimetric method (reference range, 0–5 mg/L). Ferritin level was measured on an Abbott Architect c8000 analyzer (Chicago, IL, USA; reference range, 64–111 µmol/L).

Statistical processing of research results was carried out using the Statistica v. 10 package (StatSoft; Tulsa, OK, USA), with the help of standard methods of variation statistics. Between-group comparison was carried out using the Mann–Whitney criteria for incorrect distribution; results are presented as median (Me) and interquartile range [25%; 75%], as well as mean (M) and standard deviation (SD) for Student criterion in correct distributed parameters. Associations between quantitative parameters were assessed using Spearman's correlation coefficient. To describe relative risk, we calculated the odds ratio (OR), with 95% confidence interval (95% CI) calculated with Fisher's exact method. We explored the association between 25(OH)D level and both COVID-19 severity and fatal outcome using logistic regression (adjusting for age and comorbidities), with results expressed as β coefficients and 95% CI. The criterion for the statistical reliability of the obtained results was $p < 0.05$.

3. Results

As mentioned, vitamin D status was detected in 161 patients. We excluded data from 10 pregnant women, 5 patients with confirmed human immunodeficiency virus infection, and 13 patients taking replacement renal therapy for stage 5 chronic kidney disease. Thus, the final analysis included data from 133 COVID-19 patients (76 men [57%] and 57 women [43%]), aged 21–93 years (mean, 52 ± 14 years).

Based on disease severity, patients were grouped into moderate and severe course. Most hospitalized patients (81%) had a moderate course of disease with CT-confirmed lung damage as CT-2 and CT-3. At the same time, among patients with severe course, 56% had lung damage as CT-4. Severe patients were older and more often had obesity, diabetes mellitus, and cardiovascular diseases, especially coronary artery disease (CAD; Table 1).

Table 1. Patients' characteristics in relation to COVID-19 severity.

Parameter	Severe Course *n* = 25	Moderate Course *n* = 108	*p*
Age, y, M ± SD	57 ± 3	51 ± 1	**0.02**
Sex, m/f, *n* (%)	15(60)/10(40)	61(57)/47(44)	0.75
Obesity, *n* (%)	16 (64)	23 (21)	**0.00**
AH, *n* (%)	15 (60)	46 (43)	0.12
CAD, *n* (%)	11 (44)	25 (23)	**0.04**
DM, *n* (%)	8 (32)	18 (17)	**0.00**
Death, *n* (%)	15 (60)	3 (3)	**0.00**
Volume of lung tissue lesions (CT), *n* (%) 0 1 2 3 4	 0 1 (4) 5 (20) 5 (20) 14 (56)	 7 (7) 19 (18) 41 (38) 32 (30) 9 (8)	**0.00**
25(OH)D, ng/mL, Me [25; 75]	9.7 [6.0; 14.9]	14.6 [10.6; 24.4]	**0.00**
Vitamin D status, *n* (%) Normal Insufficiency Mild deficiency Severe deficiency	 1 (4) 3 (12) 8 (32) 13 (52)	 11 (10) 28 (26) 45 (42) 24 (22)	**0.003**
Bed days, M ± SD	21.0 ± 2.5	17.0 ± 0.9	0.16
Glucose max, mmol/L, Me [25%; 75%]	10.3 [8.4; 18.4]	6.15, 0; 9.7	**0.00**
CRP baseline, mg/L, Me [25%; 75%]	64.7 [36.4; 200.0]	34.7 [15.9; 89.6]	**0.01**

Table 1. *Cont.*

Parameter	Severe Course $n = 25$	Moderate Course $n = 108$	p
CRP max, mg/L, Me [25%; 75%]	265.1 [182.2; 322.0]	60.0 [21.4; 137.3]	**0.00**
IL-6 baseline, pg/mL, Me [25%; 75%]	22.0 [10.8; 75.0]	7.8 [2.4; 20.7]	**0.001**
IL-6 max, pg/mL, Me [25%; 75%]	36.4 [20.1; 282.0]	10.4 [2.8; 25.1]	**0.00**
Ferritin baseline, μg/L, Me [25%; 75%]	895.4 [317.1; 1581.7]	357.3 [172.2; 811.3]	**0.01**
Ferritin max, μg/L, Me [25%; 75%]	1347.6 [835.6; 2197.1]	496.1 [257.4; 1057.4]	**0.00**

M, mean; SD, standard deviation; m, men; f, women; AH, arterial hypertension; CAD, coronary artery disease; DM, diabetes mellitus; CT, computed tomography; CRP, C-reactive protein; IL-6, interleukin 6; $p < 0.05$ values are bolded.

Both baseline and maximal serum CRP, IL-6, and ferritin levels, as well as maximal glucose level in patients with a severe course, were expectedly higher than in patients with a moderate course, characterizing a prominent immune-inflammatory response.

Only 12 patients (9%) had a normal vitamin D status, whereas 91% were insufficient (23%) or deficient (mild deficiency in 40% and severe in 28%). Serum 25(OH)D level in severe-course patients was significantly lower than in moderate-course patients. Moreover, the number of patients with severe vitamin D deficiency [serum 25(OH)D level less than 10 ng/mL] in the severe-course group was larger than that in the moderate-course group (Table 1).

The number of obese patients in the vitamin D deficiency and severe deficiency groups tendered to be larger than in the vitamin D insufficiency group and in the group with normal 25(OH)D level, though the finding had no statistical significance (Table 2). Moreover, prevalence of CAD and DM was significantly higher in patients with vitamin D deficiency and severe deficiency. As noted, vitamin D mild and severe deficiency was associated with severe course of COVID-19. Thus, 35% of severe vitamin D-deficient patients and 15% of mild vitamin D-deficient ones had a severe course of COVID-19.

Table 2. Vitamin D status in 133 COVID-19 patients.

Parameter	Deficiency n =90		Insufficiency $n = 31$	Normal $n = 12$	p
	Severe Deficiency n =37	Mild Deficiency $n = 53$			
Age, y, M ± SD	52 ± 3	53 ± 2	49 ± 2	51 ± 3	0.39 * 0.74
Sex, m/f, n (%)	27 (73)/ 10 (27)	26 (49)/ 27 (51)	14 (45)/ 17 (55)	9 (75)/ 3 (25)	0.17 * **0.02**
Obesity, n (%)	13 (35)	19 (36)	6 (19)	1 (8)	0.06 * 0.36
AH, n (%)	19 (51)	28 (53)	11 (36)	3 (25)	0.09 * 0.43
CAD, n (%)	15 (41)	14 (26)	6 (19)	1 (8)	0.12 * **0.03**
DM, n (%)	7 (19)	14 (26)	5 (16)	0	**0.00** * **0.00**
Severe course, n (%)	13 (35)	8 (15)	3 (10)	1 (8)	0.15 * **0.003**

*, compared with severe deficiency. M, mean; SD, standard deviation; m, men; f, women; AH, arterial hypertension; CAD, coronary artery disease; DM, diabetes mellitus; $p < 0.05$ values are bolded.

Analyzing factors possibly predisposing to death in COVID-19 infection, we found that patients who died were older and significantly more often had obesity, arterial hypertension, or CAD. Moreover, patients who died expectedly had higher blood glucose, CRP, IL-6, and ferritin levels. Vitamin D mild and severe deficiency was strongly associated with death incidence (Table 3).

Table 3. COVID-19 patient characteristics in relation to disease outcome.

Parameter	Death $n = 18$	Discharged $n = 115$	p
Age, y, M ± SD	62 ± 3	50 ± 1	**0.00**
Sex, m/f, n (%)	10 (56)/ 8 (44)	66 (57)/ 49 (43)	0.88
Obesity, n (%)	12 (67)	27 (24)	**0.00**
AH, n (%)	14 (78)	47 (41)	**0.004**
CAD, n (%)	11 (61)	25 (22)	**0.00**
DM, n (%)	6 (33)	20 (17)	**0.00**
Severe course, n (%)	15 (83)	10 (9)	**0.00**
Volume of lung tissue lesions (CT), n (%) 0 1 2 3 4	 0 0 4 (22) 3 (17) 11 (61)	 7 (6) 20 (17) 42 (37) 34 (30) 12 (10)	**0.00**
25(OH)D, ng/mL, Me [25%; 75%] Vitamin D status, n (%) Normal, n (%) Insufficiency, n (%) Mild deficiency, n (%) Severe deficiency, n (%)	9.6 [6.0; 11.5] 1 (6) 0 7 (39) 10 (55)	14.8 [10.1; 24.3] 11 (10) 31 (27) 46 (40) 27 (24)	**0.001** **0.02** **0.005**
Bed days, M ± SD	16 ± 3	18 ± 1	0.27
Glucose baseline, mmol/L	7.0 [6.1; 10.0]	5.9 [5.0; 7.5]	**0.03**
Glucose max, mmol/L	10.8 [7.7; 18.4]	6.3 [5.0; 10.0]	**0.00**
CRP baseline, mg/L	74.6 [36.4; 168.2]	35.5 [16.3; 90.0]	**0.02**
CRP max, mg/L	255.5 [182.2; 308.0]	67.4 [22.2; 140.1]	**0.00**
IL-6 baseline, pg/mL	37.4 [10.8; 87.6]	8.3 [2.4; 20.7]	**0.00**
IL-6 max, pg/mL	37.7 [12.7; 453.5]	11.8 [2.9; 27.6]	**0.001**
Ferritin baseline, µg/L, $n = 131$	965.0 [680.1; 1581.7]	366.5 [172.2; 895.4]	**0.005**
Ferritin max, µg/L	1699.1 [1119.5; 2197.1]	536.1 [260.0; 1051.0]	**0.00**

M, mean; SD, standard deviation; m, men; f, women; AH, arterial hypertension; CAD, coronary artery disease; DM, diabetes mellitus; CT, computed tomography; Me, median, CRP, C-reactive protein; IL-6, interleukin 6; $p < 0.05$ values are bolded.

We observed an inverse relationship between serum 25(OH)D level and CRP max level ($R = -0.21$; $p = 0.02$) and ferritin max level ($R = -0.24$; $p = 0.01$). Serum 25(OH)D level also negatively correlated with glucose max level ($R = -0.25$; $p = 0.04$).

Moreover, a positive correlation existed between glucose max level and baseline inflammatory marker levels: CRP ($R = 0.26$; $p = 0.003$), ferritin ($R = 0.18$; $p = 0.04$), and IL-6 ($R = 0.20$; $p = 0.04$). The correlation was also present for their max values: CRP ($R = 0.44$; $p = 0.001$), ferritin ($R = 0.22$; $p = 0.03$), and IL-6 ($R = 0.26$; $p = 0.006$).

We did not find correlation between 25(OH)D level and CT data in this population.

The threshold for 25(OH)D level associated with increased risk of severe course in this population was 11.7 ng/mL ($AUC_{area} = 0.69$; sensitivity, 71%; and specificity, 68%; $p = 0.003$)

(Figure 1a). Approximately the same 25(OH)D level was associated with increased risk of mortality: 10.9 ng/mL (AUC_{area} = 0.75; sensitivity, 74%; and specificity, 72%; p = 0.001) (Figure 1b). That level corresponds to vitamin D deficiency.

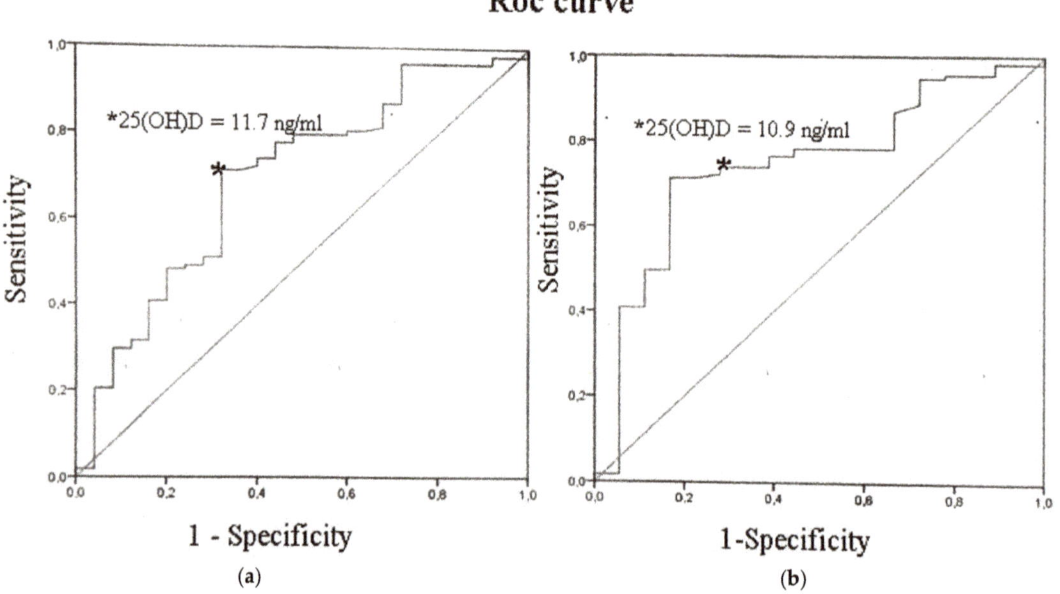

Figure 1. 25-Hydroxyvitamin D level's association with increased risk of (**a**) severe course, p = 0.003; (**b**) mortality, p = 0.001.

We evaluated a possible contribution of vitamin D status and other predictors, such as age, sex, and comorbidities, to the risk of severe course (Table 4) and death (Table 5) in COVID-19 using multivariate ordered logistic regression analysis with both the adjusted and unadjusted models. To adjust for confounding factors, we used two models: model 1 was adjusted for age and sex, whereas model 2 was adjusted for age, sex, and comorbidities. Serum 25(OH)D level < 10.0 ng/mL increased risk of severe coronavirus infection by 3.79 times (95% CI, 1.53–9.39; p = 0.004) and death by 4.07 times (95% CI, 1.46–11.35; p = 0.007). In the unadjusted model, obesity appeared to be a significant predictor for severity (OR = 6.57; 95% CI, 2.57–16.78; p = 0.000) (Table 4) and death (OR = 6.52; 95% CI, 2.23–19.02; p = 0.001) (Table 5), and it significantly correlated with 25(OH)D level (r = –0.18; p = 0.04). Therefore, obesity was not included in the adjusted model. Sex was not a significant risk factor, whereas age and major comorbidities had significant relationships with severity and mortality in the unadjusted models only. By contrast, age was an independent predictor only in model 1 for severity (OR = 1.04; 95% CI, 1.01–1.07; p = 0.04), whereas no significance was evident for model 2. For death, age was a strong independent predictor in both model 1 (OR = 1.09; 95% CI, 1.03–1.16; p = 0.002) and model 2 (OR = 1.07; 95% CI, 1.01–1.15; p = 0.03). Serum 25(OH)D levels < 10 ng/mL in model 1 (OR = 4.09; 95% CI, 1.58–10.67; p = 0.004) and model 2 (OR = 4.17; 95% CI, 1.54–11.27; p = 0.005) were strongly associated with COVID-19 severity. The same pattern concerning severe 25(OH)D deficiency was observed for mortality in model 1 (OR = 5.68; 95% CI, 1.74–18.52; p = 0.004) and in model 2 (OR = 5.79; 95% CI, 1.66–20.22; p = 0.006).

Table 4. Predictors for COVID-19 severity (logistic regression analysis).

Predictor	Unadjusted		Model 1 Adjusted		Model 2 Adjusted	
	OR (95% CI)	p	OR (95% CI)	p	OR (95% CI)	p
Age	1.03 (1.01–1.07)	**0.04**	1.04 (1.01–1.07)	**0.04**	1.03 (0.99–1.08)	0.17
Male	1.16 (0.48–2.80)	0.75	0.77 (0.20–2.02)	0.59	0.85 (0.32–2.29)	0.75
Obesity	6.57 (2.57–16.78)	**0.000**				
AH	2.02 (0.83–4.91)	0.12			0.98 (0.28–3.43)	0.98
CAD	2.61 (1.05–6.46)	**0.04**			1.18 (0.34–4.08)	0.79
DM	2.35 (0.88–6.28)	0.08			2.25 (0.77–6.57)	0.14
25(OH)D < 20 ng/mL	2.97 (0.95–9.27)	0.06	2.72 (0.86–8.59]	0.09	2.48 (0.77–7.99)	0.13
25(OH)D < 10 ng/mL	3.79 (1.53–9.39)	**0.004**	4.09 (1.58–10.67)	**0.004**	4.17 (1.54–11.27)	**0.005**

Model 1 is adjusted for age and sex. Model 2 is adjusted for age, sex, and comorbidities. Data are presented as frequencies (%) and OR (95% CI). OR, odds ratio; CI, confidence interval; AH, arterial hypertension; CAD, coronary artery disease; DM, diabetes mellitus; $p < 0.05$ values are bolded.

Table 5. Predictors for COVID-19 fatal outcome (logistic regression analysis).

Predictor	Unadjusted		Model 1 Adjusted		Model 2 Adjusted	
	OR (95% CI)	p	OR (95% CI)	p	OR (95% CI)	p
Age	1.07 (1.03–1.13)	**0.001**	1.09 (1.03–1.16)	**0.002**	1.07 (1.01–1.15)	**0.03**
Male	0.93 (0.34–2.52)	0.88	0.43 (0.13–1.41)	0.16	0.51 (0.15–1.69)	0.27
Obesity	6.52 (2.23–19.02)	**0.001**				
AH	5.06 (1.57–16.34)	0.07			1.59 (0.35–5.54)	0.56
CAD	5.66 (1.99–16.10)	**0.001**			1.40 (0.35–5.54)	0.63
DM	2.38 (0.79–7.08)	0.12			1.98 (0.57–6.92)	0.28
25(OH)D < 20 ng/mL	9.78 (1.26–76.14)	**0.03**	8.60 (1.07–69.12)	0.05	7.87 (0.96–64.43)	0.06
25(OH)D < 10 ng/mL	4.07 (1.46–11.35)	**0.007**	5.68 (1.74–18.52)	**0.004**	5.79 (1.66–20.22)	**0.006**

Model 1 is adjusted for age and sex. Model 2 is adjusted for age, sex, and comorbidities. Data are presented as frequencies (%) and odds ratio (95% CI). OR, odds ratio; CI, confidence interval; AH, arterial hypertension; CAD, coronary artery disease; DM, diabetes mellitus; $p < 0.05$ values are bolded.

4. Discussion

The available data indicate that vitamin D therapy in people with vitamin D insufficiency and deficiency reduces the likelihood of developing acute respiratory viral infections by 42% [31] and that patients with vitamin D deficiency have a longer, more severe course of disease [32]. Analysis of 25(OH)D levels in COVID-19 patients in China showed a high incidence of vitamin D deficiency in winter and a possible relationship between low vitamin D supply and severity and outcomes of the disease [33,34]. The results of a recent study showed that patients with severe coronavirus infection have the lowest serum 25(OH)D levels [35], in accordance with the data obtained here. Thus, vitamin D deficiency and obesity were most often reported in fatal COVID-19 patients [36,37] The results of the present study show that obesity independently worsens disease prognosis by at least 6 times, requiring an integrated approach in managing such patients. Moreover, severe vitamin D deficiency appears to be a strong independent negative predictor, even when adjusted for age, sex, and comorbidities, that worsens disease prognosis even by 4 times. Though in this study DM did not increase the risk of either severity or mortality of COVID-19 patients, we found associations between serum 25(OH)D level and plasma glucose concentration on one hand and between inflammatory markers and glucose level on the other hand that confirms the relationship between these conditions, and corresponds with previous data [38].

Our results correspond to the data presented in systematic review and meta-analysis of 23 studies that summarized all the existing knowledge concerning the role of vitamin D in COVID-19. According to the presented data, vitamin D deficiency increases the chance of severe COVID-19 development for about five times (OR = 5.1, 95% CI, 2.6–10.3), while there was no significant association between vitamin D status and increased mortality rates (OR = 1.6, 95% CI, 0.5–4.4) [39].

A main factor contributing to COVID-19 severity is believed to be development of the cytokine storm, the uncontrolled release of various inflammatory markers (such as CRP, IL-6, ferritin, tumor necrosis factor α or neutrophil-to-lymphocyte ratio (NLR)). For example, intensive care unit COVID-19 patients have the highest concentrations of IL-1β, IL-6, and IL-6 to IL-10 ratio [40]. Additionally, increased NLRs and decreased eosinophil counts are typical for severe COVID-19 patients, while neutrophils and lymphocytes counts demonstrate respectively positive and negative correlation with COVID-19 severity [41]. Our data confirmed that IL-6, CRP and ferritin levels are higher in severe COVID-19 cases and in patients with fatal outcome and showed negative correlations between these parameters and 25(OH)D.

The cytokine storm can activate intravascular coagulation, forming the basis for multiorgan injury, which is mediated mainly by inflammatory cytokines such as IL-6 [42]. By contrast, coronavirus can directly affect endothelial cells, causing cell death, and induces a cytopathic effect on airway epithelial cells [43]. Additionally, SARS-CoV-2 can affect the alveolar cells by ACE2 binding and suppress surfactant production. This damage might be prevented by vitamin D, as in vitro and in vivo studies have shown that 1.25(OH)2-D induces type II pneumocyte proliferation and surfactant synthesis in the lungs. These data are confirmed in clinical studies by positive correlation between vitamin D status and lung tissue lesions volume according to CT evaluation [44,45]. On the other hand, we did not find similar interlinks in our work.

IL-6 itself can increase the severity of COVID-19 by up-regulating angiotensin-converting enzyme 2 receptor and inducing cathepsin L production in macrophages, thus mediating the cleavage of the S1 subunit of the coronavirus surface spike glycoprotein. The latter is necessary for coronavirus to enter human host cells and to cause all further reactions. Low vitamin D concentration is associated with high IL-6 production, whereas vitamin D supplementation has an anti-inflammatory effect [46].

Taking into account vitamin D's immunomodulatory effects, particularly the inhibition of NF-κB by increasing synthesis of IκBα [47,48], we can assume that vitamin D intake and the consequent achievement of a 25(OH)D concentration of 40–50 ng/mL (100–125 nmol/L) might have a positive effect in patients with coronavirus respiratory infections such as Middle East respiratory syndrome, SARS-CoV, and SARS-CoV-2 [11,49]. This is supported by data from a few studies showing that use of large vitamin D doses in critically ill patients with viral and bacterial pneumonia, under mechanical ventilation, leads to shortening of intensive care unit treatment duration and prognosis improvement [50]. However, further research is required to obtain more reliable information on vitamin D's role in preventing and treating new coronavirus infection.

There are several limitations in the study. First, it is a single-center study with a relatively small sample size, while a larger cohort of COVID-19 patients is preferable to better assess vitamin D status and severity/outcomes of the disease. Secondly, only a part of the immune markers was included in the analysis; we do not have NLRs data for this cohort, so the immune response to SARS-CoV-2 should be characterized in more detail in the future. There were no anthropometric data in patients' medical histories for us to calculate BMI, despite higher BMI being known as a strong predictor for vitamin D deficiency as well as COVID-19 severity. Moreover, this study is not a prospective one and does not provide the information regarding the relationship between vitamin D status dynamic and immune response, as well as the outcomes of COVID-19.

Author Contributions: E.V.S.: Conceptualization, project administration. T.L.K.: Conceptualization, project administration, writing, and editing. A.T.A.: Data analysis and interpretation, writing (review and editing). K.A.G.: Data collection. E.S.B.: Data collection. A.V.S.: writing and editing. M.A.V.: Data collection and review. W.B.G.: Data analysis, conceptualization, writing, and editing. All authors have read and agreed to the published version of the manuscript.

Funding: This study was provided by Almazov National Medical Research Centre and by the special grant of Russian Ministry of Health (no. 121031100284-7).

Institutional Review Board Statement: The study was conducted according to the guidelines of the Declaration of Helsinki and approved by Almazov National Medical Research Centre Local Ethics Committee (0811-20-01C).

Informed Consent Statement: Informed consent was obtained from all subjects involved in the study.

Data Availability Statement: The data generated and analyzed during this study are included in this published article and its supplementary information files. Additional information is available from the corresponding author on reasonable request.

Acknowledgments: We thank study participants for supplying samples to advance our understanding of vitamin D's role in COVID-19 infection. We acknowledge health care professionals from Botkin Clinical Infectious Hospital for assistance in collecting data.

Conflicts of Interest: Sunlight, Nutrition, and Health Research Center (W.B.G.) receives funding from Bio-Tech Pharmacal Inc. (Fayetteville, AR, USA). The other authors declare no conflict of interest.

References

1. Karonova, T.L.; Vashukova, M.A.; Gusev, D.A.; Golovatuk, K.A.; Grineva, E.N. Vitamin D deficiency as a factor for immunity stimulation and lower risk of acute respiratory infections and COVID-19. *Arter. Hypertens.* **2020**, *26*, 295–303. [CrossRef]
2. Zemb, P.; Bergman, P.; Camargo, C.A., Jr.; Cavalier, E.; Cormier, C.; Courbebaisse, M.; Hollis, B.; Joulia, F.; Minisola, S.; Pilz, S.; et al. Vitamin D deficiency and COVID-19 pandemic. *Glob. Antimicrob. Resist.* **2020**, *22*, 133–134. [CrossRef] [PubMed]
3. Baeke, F.; Korf, H.; Overbergh, L.; van Etten, E.; Verstuyf, A.; Gysemans, C.; Mathieu, C. Human T lymphocytes are direct targets of 1,25-dihydroxyvitamin D3 in the immune system. *J. Steroid Biochem. Mol. Biol.* **2010**, *121*, 221–227. [CrossRef]
4. Hewison, M.; Freeman, L.; Hughes, S.V.; Evans, K.N.; Bland, R.; Eliopoulos, A.G.; Kilby, M.D.; Moss, P.A.; Chakraverty, R. Differential regulation of vitamin D receptor and its ligand in human monocyte-derived dendritic cells. *J. Immunol.* **2003**, *170*, 5382–5390. [CrossRef]
5. Ginde, A.A.; Liu, M.C.; Camargo, C.A., Jr. Demographic differences and trends of vitamin D insufficiency in the US population, 1988–2004. *Arch. Intern. Med.* **2009**, *169*, 626–632. [CrossRef] [PubMed]
6. Rondanelli, M.; Miccono, A.; Lamburghini, S.; Avanzato, I.; Riva, A.; Allegrini, P.; Faliva, M.A.; Peroni, G.; Nichetti, M.; Perna, S. Self-Care for Common Colds: The Pivotal Role of Vitamin D, Vitamin C, Zinc, and Echinacea in Three Main Immune Interactive Clusters (Physical Barriers, Innate and Adaptive Immunity) Involved during an Episode of Common Colds-Practical Advice on Dosages and on the Time to Take These Nutrients/Botanicals in order to Prevent or Treat Common Colds. *Evid. Based Complement. Altern. Med.* **2018**, *2018*, 5813095. [CrossRef]
7. White, J.H. Regulation of intracrine production of 1,25-dihydroxyvitamin D and its role in innate immune defense against infection. *Arch. Biochem. Biophys.* **2012**, *523*, 58–63. [CrossRef]
8. Liu, P.T.; Stenger, S.; Li, H.; Wenzel, L.; Tan, B.H.; Krutzik, S.R.; Ochoa, M.T.; Schauber, J.; Wu, K.; Meinken, C.; et al. Toll-like receptor triggering of a vitamin D-mediated human antimicrobial response. *Science* **2006**, *311*, 1770–1773. [CrossRef]
9. Adams, J.S.; Ren, S.; Liu, P.T.; Chun, R.F.; Lagishetty, V.; Gombart, A.F.; Borregaard, N.; Modlin, R.L.; Hewison, M. Vitamin d-directed rheostatic regulation of monocyte antibacterial responses. *J. Immunol.* **2009**, *182*, 4289–4295. [CrossRef]
10. Laaksi, I. Vitamin D and respiratory infection in adults. *Proc. Nutr. Soc.* **2012**, *71*, 90–97. [CrossRef]
11. Campbell, P.A.; Wu-Young, M.; Lee, R.C. Rapid response to Elisabeth Mahase E: Covid-19: What treatments are being investigated? *BMJ* **2020**, *368*, m1252. [CrossRef]
12. Lemire, J.M.; Adams, J.S.; Kermani-Arab, V.; Bakke, A.C.; Sakai, R.; Jordan, S.C. 1,25-Dihydroxyvitamin D3 suppresses human T helper/inducer lymphocyte activity in vitro. *J. Immunol.* **1985**, *134*, 3032–3035.
13. Cantorna, M.T.; Snyder, L.; Lin, Y.D.; Yang, L. Vitamin D and 1,25(OH)2D regulation of T cells. *Nutrients* **2015**, *7*, 3011–3021. [CrossRef]
14. Jeffery, L.E.; Burke, F.; Mura, M.; Zheng, Y.; Qureshi, O.S.; Hewison, M.; Walker, L.S.; Lammas, D.A.; Raza, K.; Sansom, D.M. 1,25-Dihydroxyvitamin D3 and IL-2 combine to inhibit T cell production of inflammatory cytokines and promote development of regulatory T cells expressing CTLA-4 and FoxP3. *J. Immunol.* **2009**, *183*, 5458–5467. [CrossRef] [PubMed]
15. Cutolo, M.; Paolino, S.; Smith, V. Evidences for a protective role of vitamin D in COVID-19. *RMD Open* **2020**, *6*, e001454. [CrossRef] [PubMed]

16. Yang, J.; Zhang, H.; Xu, J. Effect of Vitamin D on ACE2 and Vitamin D receptor expression in rats with LPS-induced acute lung injury. *Chin. J. Emerg. Med.* **2016**, *25*, 1284–1289. [CrossRef]
17. Vankadari, N.; Wilce, J.A. Emerging WuHan (COVID-19) coronavirus: Glycan shield and structure prediction of spike glycoprotein and its interaction with human CD26. *Emerg. Microbes Infect.* **2020**, *9*, 601–604. [CrossRef] [PubMed]
18. Vasarhelyi, B.; Satori, A.; Olajos, F.; Szabo, A.; Beko, G. Low vitamin D levels among patients at Semmelweis University: Retrospective analysis during a one-year period. *Orv. Hetil.* **2011**, *152*, 1272–1277. [CrossRef] [PubMed]
19. Huang, C.; Wang, Y.; Li, X.; Ren, L.; Zhao, J.; Hu, Y.; Zhang, L.; Fan, G.; Xu, J.; Gu, X.; et al. Clinical features of patients infected with 2019 novel coronavirus in Wuhan, China. *Lancet* **2020**. [CrossRef]
20. Zhonghua, L.Z.B.X.Z.Z. The epidemiological characteristics of an outbreak of 2019 novel coronavirus diseases (COVID-19) in China. *CMA.J.* **2020**, *41*, 145–151. [CrossRef]
21. White, C.; Nafilyan, V. *Coronavirus (COVID-19) Related Deaths by Ethnic Group, England and Wales: 2 March 2020 to 10 April 2020*; Office for National Statistics: Newport, UK, 2020.
22. Hope-Simpson, R.E. The role of season in the epidemiology of influenza. *J. Hyg.* **1981**, *86*, 35–47. [CrossRef] [PubMed]
23. Cannell, J.J.; Vieth, R.; Umhau, J.C.; Holick, M.F.; Grant, W.B.; Madronich, S.; Garland, C.F.; Giovannucci, E. Epidemic influenza and vitamin D. *Epidemiol. Infect.* **2006**, *134*, 1129–1140. [CrossRef]
24. Lavie, C.J.; Sanchis-Gomar, F.; Henry, B.M.; Lippi, G. COVID-19 and obesity: Links and risks. *Expert Rev. Endocrinol. Metab.* **2020**, *15*, 215–216. [CrossRef]
25. Hussain, A.; Bhowmik, B.; do Vale Moreira, N.C. COVID-19 and diabetes: Knowledge in progress. *Diabetes Res. Clin. Pract.* **2020**, *162*, 108142. [CrossRef] [PubMed]
26. Grant, W.B.; Lahore, H.; McDonnell, S.L.; Baggerly, C.A.; Franch, C.B.; Aliano, J.L.; Bhattoa, H.P. Evidence That Vitamin D Supplementation Could Reduce Risk of Influenza and COVID-19 Infections and Deaths. *Nutrients* **2020**, *12*, 988. [CrossRef]
27. D'Avolio, A.; Avataneo, V.; Manca, A.; Cusato, J.; De Nicolo, A.; Lucchini, R.; Keller, F.; Cantu, M. 25-Hydroxyvitamin D Concentrations Are Lower in Patients with Positive PCR for SARS-CoV-2. *Nutrients* **2020**, *12*, 1359. [CrossRef]
28. Karonova, T.; Andreeva, A.; Nikitina, I.; Belyaeva, O.; Mokhova, E.; Galkina, O.; Vasilyeva, E.; Grineva, E. Prevalence of Vitamin D deficiency in the North-West region of Russia: A cross-sectional study. *J. Steroid Biochem. Mol. Biol.* **2016**, *164*, 230–234. [CrossRef] [PubMed]
29. Holick, M.F.; Binkley, N.C.; Bischoff-Ferrari, H.A.; Gordon, C.M.; Hanley, D.A.; Heaney, R.P.; Murad, M.H.; Weaver, C.M.; Endocrine, S. Evaluation, treatment, and prevention of vitamin D deficiency: An Endocrine Society clinical practice guideline. *J. Clin. Endocrinol. Metab.* **2011**, *96*, 1911–1930. [CrossRef]
30. Pigarova, E.A.; Rozhinskaya, L.Y.; Belaya, J.E.; Dzeranova, L.K.; Karonova, T.L.; Ilyin, A.V.; Melnichenko, G.A.; Dedov, I.I. Russian Association of endocrinologists recommendations for diagnosis, treatment and prevention of vitamin D deficiency in adults. *Probl. Endocrinol.* **2016**, *62*, 60–84. [CrossRef]
31. Martineau, A.R.; Jolliffe, D.A.; Hooper, R.L.; Greenberg, L.; Aloia, J.F.; Bergman, P.; Dubnov-Raz, G.; Esposito, S.; Ganmaa, D.; Ginde, A.A.; et al. Vitamin D supplementation to prevent acute respiratory tract infections: Systematic review and meta-analysis of individual participant data. *BMJ* **2017**, *356*, i6583. [CrossRef]
32. Pham, H.; Rahman, A.; Majidi, A.; Waterhouse, M.; Neale, R.E. Acute Respiratory Tract Infection and 25-Hydroxyvitamin D Concentration: A Systematic Review and Meta-Analysis. *Int. J. Environ. Res. Public Health* **2019**, *16*, 3020. [CrossRef]
33. Wang, D.; Hu, B.; Hu, C.; Zhu, F.; Liu, X.; Zhang, J.; Wang, B.; Xiang, H.; Cheng, Z.; Xiong, Y.; et al. Clinical characteristics of 138 hospitalized patients with 2019 novel coronavirus-infected pneumonia in Wuhan, China. *JAMA* **2020**, *323*, 1061. [CrossRef]
34. Zhou, F.; Yu, T.; Du, R.; Fan, G.; Liu, Y.; Liu, Z.; Xiang, J.; Wang, Y.; Song, B.; Gu, X.; et al. Clinical course and risk factors for mortality of adult inpatients with COVID-19 in Wuhan, China: A retrospective cohort study. *Lancet* **2020**, *395*, 1045–1062. [CrossRef]
35. Infante, M.; Buoso, A.; Pieri, M.; Lupisella, S.; Nuccetelli, M.; Bernardini, S.; Fabbri, A.; Iannetta, M.; Andreoni, M.; Colizzi, V.; et al. Low vitamin D status at admission as a risk factor for poor survival in hospitalized patients with COVID-19: An italian retrospective study. *J. Am. Coll. Nutr.* **2021**, *18*, 1–16. [CrossRef] [PubMed]
36. Chua, M.W.J.; Zheng, S. Obesity and COVID-19: The clash of two pandemics. *Obes. Res. Clin. Pract.* **2020**, *14*, 380–382. [CrossRef]
37. Carter, S.J.; Baranauskas, M.N.; Fly, A.D. Considerations for obesity, vitamin D, and physical activity amid the COVID-19 pandemic. *Obesity* **2020**, *28*, 1176–1177. [CrossRef] [PubMed]
38. Kumar, A.; Arora, A.; Sharma, P.; Anikhindi, S.A.; Bansal, N.; Singla, V.; Khare, S.; Srivastava, A. Is diabetes mellitus associated with mortality and severity of COVID-19? A meta-analysis. *Diabetes Metab. Syndr. Clin. Res. Rev.* **2020**, *14*, 535–545. [CrossRef]
39. Ghasemian, R.; Shamshirian, A.; Heydari, K.; Malekan, M.; Alizadeh-Navaei, R.; Ebrahimzadeh, M.A.; Ebrahimi Warkiani, M.; Jafarpour, H.; Razavi Bazaz, S.; Rezaei Shahmirzadi, A.; et al. The role of vitamin D in the age of COVID-19: A systematic review and meta-analysis. *Int. J. Clin. Pract.* **2021**, *29*, e14675. [CrossRef]
40. McElvaney, O.J.; McEvoy, N.L.; McElvaney, O.F.; Carroll, T.P.; Murphy, M.P.; Dunlea, D.M.; Ní Choileáin, O.; Clarke, J.; O'Connor, E.; Hogan, G.; et al. Characterization of the Inflammatory Response to Severe COVID-19 Illness. *Am. J. Respir. Crit. Care Med.* **2020**, *202*, 812–821. [CrossRef]
41. Zneng, F.; Chen, R.; Yao, R.; Huang, Y.; Tan, X.; Liu, J.; Li, N.; Xie, Y. Dynamic changes in the immune response correlate with disease severity and outcomes during infection with SARS-CoV-2. *Infect. Dis. Ther.* **2021**, *10*, 1391–1405. [CrossRef] [PubMed]

42. Merad, M.; Martin, J.C. Pathological inflammation in patients with COVID-19: A key role for monocytes and macrophages. *Nat. Rev. Immunol.* **2020**, *20*, 355–362. [CrossRef] [PubMed]
43. Park, W.B.; Kwon, N.J.; Choi, S.J.; Kang, C.K.; Choe, P.G.; Kim, J.Y.; Yun, J.; Lee, G.W.; Seong, M.W.; Kim, N.J.; et al. Virus Isolation from the First Patient with SARS-CoV-2 in Korea. *J. Korean Med. Sci.* **2020**, *35*, e84. [CrossRef]
44. Sulli, A.; Gotelli, E.; Casabella, A.; Paolino, S.; Pizzorni, C.; Alessandri, E.; Grosso, M.; Ferone, D.; Smith, V.; Cutolo, M. Vitamin D and lung outcomes in elderly COVID-19 patients. *Nutrients* **2021**, *13*, 717. [CrossRef]
45. Fakhoury, H.M.A.; Kvietys, P.R.; Shakir, I.; Shams, H.; Grant, W.B.; Alkattan, K. Lung-centric inflammation of COVID-19: Potential modulation by vitamin D. *Nutrients* **2021**, *13*, 2216. [CrossRef] [PubMed]
46. Mercola, J.; Grant, W.B.; Wagner, C.L. Evidence Regarding Vitamin D and Risk of COVID-19 and Its Severity. *Nutrients* **2020**, *12*, 3361. [CrossRef] [PubMed]
47. Beard, J.A.; Bearden, A.; Striker, R. Vitamin D and the anti-viral state. *J. Clin. Virol.* **2011**, *50*, 194–200. [CrossRef]
48. Cohen-Lahav, M.; Shany, S.; Tobvin, D.; Chaimovitz, C.; Douvdevani, A. Vitamin D decreases NFkappaB activity by increasing IkappaBalpha levels. *Nephrol. Dial. Transplant.* **2006**, *21*, 889–897. [CrossRef]
49. DeDiego, M.L.; Nieto-Torres, J.L.; Regla-Nava, J.A.; Jimenez-Guardeno, J.M.; Fernandez-Delgado, R.; Fett, C.; Castano-Rodriguez, C.; Perlman, S.; Enjuanes, L. Inhibition of NF-kappaB-mediated inflammation in severe acute respiratory syndrome coronavirus-infected mice increases survival. *J. Virol.* **2014**, *88*, 913–924. [CrossRef]
50. Han, J.E.; Jones, J.L.; Tangpricha, V.; Brown, M.A.; Brown, L.A.S.; Hao, L.; Hebbar, G.; Lee, M.J.; Liu, S.; Ziegler, T.R.; et al. High Dose Vitamin D Administration in Ventilated Intensive Care Unit Patients: A Pilot Double Blind Randomized Controlled Trial. *J. Clin. Transl. Endocrinol.* **2016**, *4*, 59–65. [CrossRef] [PubMed]

MDPI

Article

Effects of a 2-Week 5000 IU versus 1000 IU Vitamin D3 Supplementation on Recovery of Symptoms in Patients with Mild to Moderate Covid-19: A Randomized Clinical Trial

Shaun Sabico [1], Mushira A. Enani [2], Eman Sheshah [3], Naji J. Aljohani [1,4], Dara A. Aldisi [5], Naif H. Alotaibi [6], Naemah Alshingetti [7], Suliman Y. Alomar [8], Abdullah M. Alnaami [1], Osama E. Amer [1], Syed D. Hussain [1] and Nasser M. Al-Daghri [1,*]

1 Chair for Biomarkers of Chronic Diseases, Biochemistry Department, College of Science, King Saud University, Riyadh 11451, Saudi Arabia; ssabico@ksu.edu.sa (S.S.); najij@hotmail.com (N.J.A.); aalnaami@ksu.edu.sa (A.M.A.); oamer1@ksu.edu.sa (O.E.A.); shussain@ksu.edu.sa (S.D.H.)
2 Infectious Diseases Section, King Fahad Medical City, Riyadh 59046, Saudi Arabia; menani@kfmc.med.sa
3 Diabetes Care Center, King Salman Bin Abdulaziz Hospital, Riyadh 12769, Saudi Arabia; eman_shesha@hotmail.com
4 Obesity, Endocrine and Metabolism Center, Department of Medicine, King Fahad Medical City, Riyadh 59046, Saudi Arabia
5 Department of Community Health Sciences, College of Applied Medical Sciences, King Saud University, Riyadh 11451, Saudi Arabia; daldisi@ksu.edu.sa
6 Department of Medicine, College of Medicine, King Saud University, Riyadh 12372, Saudi Arabia; dr-alotaibi@hotmail.com
7 Obstetrics and Gynaecology Department, King Salman Bin Abdulaziz Hospital, Riyadh 11564, Saudi Arabia; alshingetti@yahoo.com
8 Doping Research Chair, Department of Zoology, College of Science, King Saud University, Riyadh 11495, Saudi Arabia; syalomar@ksu.edu.sa
* Correspondence: ndaghri@ksu.edu.sa

Citation: Sabico, S.; Enani, M.A.; Sheshah, E.; Aljohani, N.J.; Aldisi, D.A.; Alotaibi, N.H.; Alshingetti, N.; Alomar, S.Y.; Alnaami, A.M.; Amer, O.E.; et al. Effects of a 2-Week 5000 IU versus 1000 IU Vitamin D3 Supplementation on Recovery of Symptoms in Patients with Mild to Moderate Covid-19: A Randomized Clinical Trial. *Nutrients* **2021**, *13*, 2170. https://doi.org/10.3390/nu13072170

Academic Editors: William B. Grant and Ronan Lordan

Received: 27 May 2021
Accepted: 21 June 2021
Published: 24 June 2021

Abstract: Objective: Vitamin D deficiency has been associated with an increased risk of COVID-19 severity. This multi-center randomized clinical trial aims to determine the effects of 5000 IU versus 1000 IU daily oral vitamin D3 supplementation in the recovery of symptoms and other clinical parameters among mild to moderate COVID-19 patients with sub-optimal vitamin D status. Study Design and Setting: A total of 69 reverse transcriptase polymerase chain reaction (RT-PCR) SARS-CoV-2 positive adults who were hospitalized for mild to moderate COVID-19 disease were allocated to receive once daily for 2 weeks either 5000 IU oral vitamin D3 (n = 36, 21 males; 15 females) or 1000 IU oral vitamin D3 (standard control) (n = 33, 13 males; 20 females). Anthropometrics were measured and blood samples were taken pre- and post-supplementation. Fasting blood glucose, lipids, serum 25(OH)D, and inflammatory markers were measured. COVID-19 symptoms were noted on admission and monitored until full recovery. Results: Vitamin D supplementation for 2 weeks caused a significant increase in serum 25(OH)D levels in the 5000 IU group only (adjusted $p = 0.003$). Within-group comparisons also showed a significant decrease in BMI and IL-6 levels overtime in both groups (p-values < 0.05) but was not clinically significant in between-group comparisons. Kaplan–Meier survival analysis revealed that the 5000 IU group had a significantly shorter time to recovery (days) than the 1000 IU group in resolving cough, even after adjusting for age, sex, baseline BMI, and D-dimer (6.2 ± 0.8 versus 9.1 ± 0.8; $p = 0.039$), and ageusia (loss of taste) (11.4 ± 1.0 versus 16.9 ± 1.7; $p = 0.035$). Conclusion: A 5000 IU daily oral vitamin D3 supplementation for 2 weeks reduces the time to recovery for cough and gustatory sensory loss among patients with sub-optimal vitamin D status and mild to moderate COVID-19 symptoms. The use of 5000 IU vitamin D3 as an adjuvant therapy for COVID-19 patients with suboptimal vitamin D status, even for a short duration, is recommended.

Keywords: COVID-19; vitamin D; clinical trial; saudi; vitamin D insufficiency

1. Introduction

The apocalyptic and exponential spread of the coronavirus disease 2019 (COVID-19) has so far claimed almost 4 million human lives globally since it was declared a pandemic in 2020 [1], bringing the entire world to a full stop as it struggled to quickly understand and control the highly contagious severe acute respiratory syndrome coronavirus-2 (SARS-CoV-2), the causative pathogen of COVID-19 [2]. As the months progressed and the strict national lockdowns were eased, it was observed that a large majority of the SARS-CoV-2 carriers were asymptomatic and that the natural course of COVID-19 among infected people eventually led to full recovery, especially if the individual had no pre-existing health conditions [3]. In parallel, advances in COVID-19 management started to increase everything from empirical antivirals and repurposed drugs [4] to the emergency use of potentially efficacious COVID-19 vaccines [5].

Indeed, much has been accomplished by the global medical and academic communities in understanding the etiology and appropriate therapy for COVID-19, especially given the short span of time. In the initial months of the pandemic, a preventive and promising adjuvant therapy was favored given its established role in the prevention of asthmatic exacerbations, viral respiratory infections, pneumonia, and overall mortality in high-risk populations such as the elderly. This well-known supplement is vitamin D [6–9]. Consequently, accumulating evidence has suggested associations between low levels of vitamin D and the severity of COVID-19 outcome [10–13]. Among the well-established theories of this association is the biophysical and structural evidence that SARS-CoV-2's point of cellular entry is the angiotensin converting enzyme 2 (ACE2) receptor protein, which is found in abundance on the surfaces of respiratory cells and is the same point of entry observed in SARS-CoV-1 [14]. Vitamin D heightens the expression of the ACE2 receptor protein, which balances the pathways that are known to be disrupted by coronaviruses, ACE/ACE2 and angiotensin II (ANG)/ANG 1-7 [15,16]. Another interesting theory is that vitamin D is a negative acute phase reactant in most acute and chronic inflammatory conditions [17], which also explains why vitamin D deficiency is common in states that harbor low-grade systemic inflammation such as diabetes, hypertension, heart disease, and aging [18].

Given that both COVID-19 and vitamin D deficiency are global pandemics, and the consistent significant associations between low vitamin D status and many pathologic extra-skeletal conditions including respiratory diseases, clinical trials are thus warranted to provide robust evidence as to whether vitamin D status optimization through supplementation can be preventive and/or therapeutic against coronavirus epidemics. Such empirical investigations are crucial for accurate and up-to-date management as to the true value of vitamin D in the on-going COVID-19 pandemic [19].

As it has already been documented that severe vitamin D deficiency is a predictor of mortality among SA residents [13,20] and that the vitamin D status of confirmed SARS-CoV-2 positive patients are significantly lower than those who tested negative [16], it is the appropriate strategy to move the field forward by conducting intervention trials. To fill this gap, the present randomized, open-label clinical trial aims to determine the beneficial effects of a 2-week, daily 5000 IU versus standard therapy (1000 IU) vitamin D3 supplementation on the recovery times of symptoms among patients having mild to moderate COVID-19 with sub-optimal vitamin D status being treated in tertiary care hospitals in Riyadh, Saudi Arabia.

2. Materials and Methods

2.1. Study Design and Participants

The present study is a multi-center, randomized clinical trial conducted from 29 July–22 September 2020. In retrospect, the study period coincided with a marked reduction in the daily confirmed COVID-19 cases nationwide (1643 confirmed cases on 29 July 2020 down to 561 cases on 22 September 2020) [21]. Participating centers were all tertiary care hospitals in Riyadh, Kingdom of Saudi Arabia (KSA), and included King Fahad Medical City (KFMC),

King Salman Hospital (KSH), and King Saud University Medical City (KSUMC). Male and female adult participants aged 20–75 years old who had an RT-PCR confirmed SARS-CoV-2 positive diagnosis (not more than 3 days prior to inclusion) and were presenting with mild to moderate symptoms and who consented voluntarily (written and verbal) were enrolled in the trial. As per the definition of the Saudi Ministry of Health (MoH) protocol for RT-PCR-confirmed COVID-19 cases, a mild-moderate category meant that the patient required no O2 on presentation, had no evidence of pneumonia but had clinical symptoms (e.g., fever), the management of which was supportive care [22]. Criteria for hospital admission required a confirmed/suspected COVID-19 patient who was symptomatic with evidence of pneumonia, above 65 years, ARDS, the presence of comorbidities and other illnesses that require admission, amongst others (a full list is provided by the MoH hospital admission criteria version 1.1) [23]. Severe COVID-19 cases (those that required intensive care (e.g., respiratory rate $\geq$30/min, oxygen saturation $\leq$93%, presence of bilateral lung infiltrates >50% of the lung field)), children and pregnant women, and those whose baseline 25(OH)D were above 75 nmol/L were excluded. Individuals who were SARS-CoV-2 negative and/or SARS-CoV-2 positive but asymptomatic (for home isolation) were also not included. Ethical approval was granted by the Institutional Review Board (IRB) of KFMC, Riyadh, KSA (IRB Log No. 20-282). The study was also registered in the Saudi Clinical Trial Registry (SCTR No. 20061006; Protocol No. H-01-R-012, 7 July 2020) [24]. Figure 1 shows the flowchart of participants. The CONSORT reporting guidelines were used as a checklist for the present randomized trial [25].

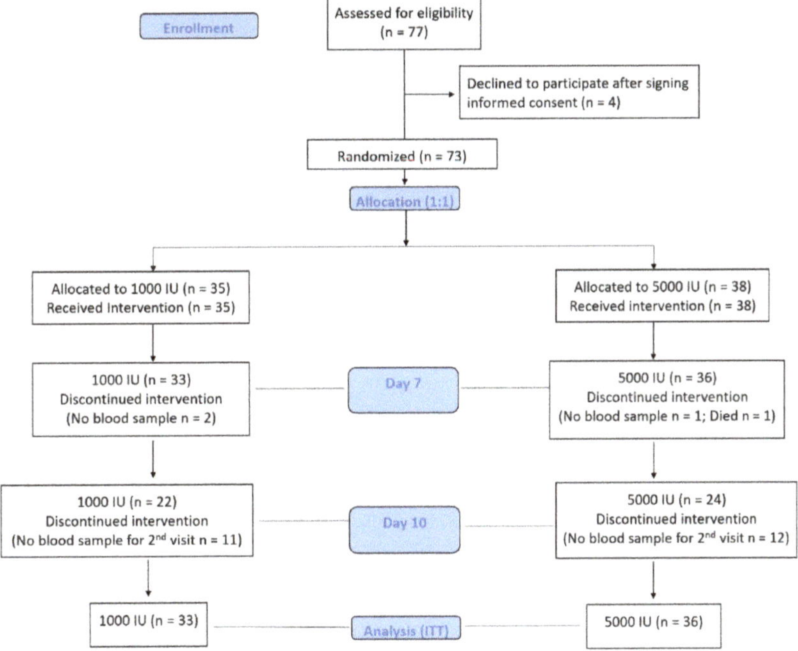

Figure 1. Flowchart of participants.

2.2. Randomization

Patients were allocated (1:1) to receive either standard vitamin D therapy (1000 IU (control)) or 5000 IU vitamin D3 for 14 days. Randomization of the study was done at the KFMC Pharmacy, which also provided the Investigational Drug Service (IDS) clearance and the site for dispensing the supplements. The randomization scheme was computer-generated using four permuted blocks of equal size for the two treatment groups.

2.3. Study Protocol

Patients in the 5000 IU group were given Ultra-D® 5000 IU containing 125 μg cholecalciferol (vitamin D3) (Synergy Pharma, Dubai, UAE) while patients in the 1000 IU group were given Vita-D® 1000 IU containing 25 μg cholecalciferol (Synergy Pharma, Dubai, UAE). Both supplements were taken orally daily for 2 weeks. The supplements provided were different in color in both packaging and tablet with unit labels stamped in both the tablet and blister, making blinding of the trial impossible. To monitor compliance, participants were given blisters containing 7 tablets at the baseline visit and were asked to return after one week (Day 7) with any unused tablets for a fresh refill and to monitor symptoms. All participants were advised to continue supplementation until Day 14, even if deisolated/discharged earlier. All participants with pre-existing conditions were advised to continue medications for those pre-existing conditions. Anthropometrics and blood collection were done at baseline (Day 0) and Day 7 or on the discharge day. The monitoring of the primary outcomes (existing symptoms) noted at baseline (Day 0) were followed up on Day 7 or on discharge day and 30 days after discharge and/or the last vitamin dose through a mobile phone call by a data collector who was blind to the treatment received by patients. The primary outcome was the number of days to resolve symptoms. Secondary outcomes include changes in the metabolic profile. Other outcomes such as days to discharge, ICU admission as well as mortality were noted. For the purpose of this study, a recovered case (discharged) was based on the guidelines set by the MoH for symptomatic patients, defined as '10 days after onset of symptoms, plus at least 3 days without symptoms (fever and respiratory symptoms) or 3 days without symptoms and one negative RT-PCR test' [21].

2.4. Data Collection

A general questionnaire was administered to all participants, which included demographics, baseline symptoms, medical history, supplements taken as well as baseline anthropometrics. Anthropometrics included height (m), weight (kg), waist (cm), and hip (cm) measurements. Body mass index (BMI) was calculated (kg/m^2).

All blood sample analyses were sent and carried out in the Biosafety Level 2-facility (BSL-2) with Biological Safety Cabinet Class II (BSC-II), College of American Pathologists (CAP) accredited virology laboratory of KSUMC, Riyadh, SA. Laboratory investigations included complete blood count (including prothrombin time, activated partial thromboplastin time (APTT), international normalized ratio (INR), and bicarbonate), liver profile (bilirubin, bilirubin direct, alkaline phosphatase (ALP), alanine transferase (ALT) and lactate dehydrogenase (LDH)), renal profile (creatinine and urea), inflammatory markers (D-dimer and ferritin), and fasting blood glucose and lipid profile (triglycerides, low-density lipoprotein (LDL-) and high-density lipoprotein (HDL)-cholesterol)), all of which were measured routinely. Interleukin-6 (IL-6) was measured using the Milliplex® MAP Human High Sensitivity T Cell Panel kit (Cat: HSTCMAG) (Millipore Corporation, Billerica, MA, USA) on the FlexMAP 3D System (Luminex Corporation, Austin, TX, USA). The standard curve range for IL-6 is 0.18–750pg/mL, with an inter- and intra-assay coefficient of variation (CVs) of <15% and <10%, respectively. C-reactive protein (CRP) was measured using Maglumi CRP chemiluminescent immunoassays (CLIA) (Shenzhen New Industries Biomedical Engineering Co., Ltd. (SNIBE) Diagnostics, Shenzen, China), with an inter- and intra-assay CVs of <15% and <10%, respectively, and a standard curve range of 0–10,000 μg/mL. Serum 25(OH)D was assessed using the CDC-approved CLIA assays (Maglumi 25OHD, SNIBE Diagnostics, Shenzen, China) as certified by the Vitamin D Standardization-Certification Program (VDSCP) [26], with an assay range of 7.5 nmol/L to 375 nmol/L. Both CRP and 25(OH)D were assessed using a fully automated CLIA analyzer (Maglumi 1000) (SNIBE Diagnostics, Shenzhen, China). Vitamin D deficiency [25(OH)D < 50nmol/L] and vitamin D sufficiency [25(OH)D $\geq$ 75 nmol/L] were defined based on national and regional recommendations (25, 25). The use of 1000 IU as a control was also based on the standard management of vitamin D deficiency in the GCC region [27,28].

2.5. Data Analysis and Sample Size Calculation

Data were entered and analyzed using SPSS version 21.0 (IBM, Chicago, IL, USA). Statistical analysis was performed using intention-to-treat (ITT) analysis, where missing data were managed using the last observation carried forward (LOCF) method. Results were presented as mean $\pm$ standard deviation for the continuous normal variables and mean $\pm$ standard error (SE) for the continuous non-normal variables. Categorical variables were presented as frequencies (N) and percentages (%). Comparisons between vitamin D doses and other categorical variables were tested using the chi-square test of independence. An independent sample T-test was used to compare clinical variables. Mixed method analysis of covariance (ANCOVA) was used to determine within and between group comparisons overtime, adjusting for baseline covariates age, sex, and BMI. Lastly, Kaplan–Meier survival analysis was done to determine the differences in the recovery time of symptoms, adjusted for age, sex, baseline BMI, and D-dimer. p-value < 0.05 was considered significant.

The sample size was taken from published literature [29], reporting a 73% reduction in clinically verified infection (non-SARS-CoV2) among vitamin D deficient patients using vitamin D supplementation. With odds of 0.27 and 80% power, the total required sample size for analysis at a 95% confidence interval (CI) was $n = 26$ ($n = 13$ per arm). A total of 60 cases would thus be recruited to anticipate dropouts ($n = 30$ per arm). A post-hoc power analysis indicated that this study achieved a power of 0.95, with an average difference of 2.9 days between the two doses of vitamin D to resolve cough symptoms, with a standard deviation of 2.8.

3. Results

3.1. Baseline Characteristics of Participants

A total of 77 participants ($n = 57$ in-patients from KFMC and $n = 20$ outpatients from KSH) were assessed for eligibility (not shown in tables). Table 1 shows the baseline clinical characteristics of the participants overall and after stratification according to vitamin D dose. A total of 69 COVID-19 patients (33 males and 36 females) (mean BMI of 30.7 kg/m^2 $\pm$ 7.8) participated in the present study. The 5000 IU group was significantly younger compared to the 1000 IU group ($p = 0.03$). In contrast, the 1000 IU group had significantly higher BMI than the 5000 IU group ($p = 0.02$). The rest of the baseline anthropometrics and vital signs were not significantly different from one another.

With regard to medical history, hypertension was observed in more than half of all of the participants and was the most common pre-existing condition (55%) followed by type 2 diabetes mellitus (51%), obesity (33%), hyperlipidemia (13%), chronic kidney disease (CKD) (7%), cardiovascular disease (6%), and asthma (4%). No significant differences were found between groups. The rest of the medical history is found in Table 1. The intake of supplements, particularly vitamin C, was noted in 47% of patients. None of the participants claimed to be taking vitamin D supplements prior to COVID-19 diagnosis.

Among the symptoms, fever (77%), dyspnea (71%) muscle pain (59%), and cough (51%) affected more than half of the participants, followed by headache (45%), joint pain (33%), and nausea (25%). Vomiting and sore throat were the least common symptoms (both at 19.2%). No significant differences in the symptoms were seen in both groups. Finally, the clinical conditions of 5 (1%) participants eventually deteriorated and required intensive care. One patient died. The median days to discharge were 7 (CI 5–9). No significant differences were observed in the outcomes of both groups (Table 1). Worthy of note was that vitamin D deficiency was observed in 40 cases (55%), with no difference between the groups ($p = 0.1$), while the rest had vitamin D insufficiency (not shown in table). Other baseline clinical and serologic characteristics of the participants are provided in Supplementary Table S1.

Table 1. Baseline Descriptive Characteristics and Symptoms on Admission.

Parameters	All	1000 IU	5000 IU	*p*-Value
n	69	33	36	
Anthropometrics/Vital Signs				
Age	49.8 ± 14.3	53.5 ± 12.3	46.3 ± 15.2	0.03
BMI	30.7 ± 7.8	32.0 ± 6.5	28.2 ± 7.1	0.02
Male/Female	34/35	13/20	21/15	0.12
WHR	0.91 ± 0.11	0.91 ± 0.11	0.90 ± 0.14	0.45
Systolic BP (mmHg)	128.2 ± 17.2	128.3 ± 20.7	128.1 ± 13.4	0.96
Diastolic BP (mmHg)	74.0 ± 13.7	72.8 ± 16.5	75.1 ± 10.6	0.47
Temperature (°C)	37.5 ± 0.9	37.3 ± 0.9	37.7 ± 0.9	0.06
Pulse Rate	93.9 ± 17.2	93.2 ± 17.4	94.5 ± 17.4	0.76
Respiratory Rate	23.9 ± 4.7	24.7 ± 5.0	23.2 ± 4.2	0.19
Medical History (%)				
Hypertension	38 (5)	18 (54)	20 (56)	0.61
T2DM	35 (51)	17 (52)	18 (50)	0.76
Obesity	23 (33)	12 (36)	11 (31)	0.54
Hyperlipidaemia	9 (13)	4 (12)	5 (14)	1.0
CKD	5 (7)	4 (12)	1 (3)	0.19
Cardiovascular Disease	4 (6)	3 (9)	1 (3)	0.34
Asthma	3 (4)	2 (6)	1 (3)	0.60
Rheumatoid	2 (3)	1 (3)	1 (3)	1.0
Thyroid	2 (3)	1 (3)	1 (3)	1.0
Epilepsy	1 (1)	1 (3)	–	1.0
Supplements (%)				
Vitamin C	34 (47)	14 (40)	20 (53)	0.28
Symptoms (%)				
Fever	56 (77)	24 (69)	32 (84)	0.18
Dyspnea	52 (71)	26 (74)	26 (68)	0.58
Fatigue	43 (59)	22 (63)	21 (55)	0.51
Cough	37 (51)	21 (60)	16 (42)	0.28
Headache	33 (45)	13 (37)	20 (53)	0.17
Joint pain	24 (33)	12 (34)	12 (32)	0.85
Nausea	18 (25)	9 (26)	9 (24)	0.31
Diarrhea	16 (22)	8 (23)	8 (21)	0.17
Sore throat	14 (19)	5 (14)	9 (24)	0.17
Vomiting	14 (19)	8 (23)	6 (16)	0.42
Outcomes (N)				
ICU Admission	5	3	2	1.0
Mortality	1	–	1	–
Days to Discharge	7 (5–9)	7 (0–10)	6 (5–8)	0.14

Note: Data presented as N (%) for frequencies and mean ± SD for continuous variables.

3.2. Primary Endpoints

The average days to resolve symptoms in both groups are shown in Table 2. Unadjusted Kaplan–Meier survival analysis was used to determine the differences in recovery times and revealed that the number of days to resolve cough was significantly shorter in the 5000 IU group than the 1000 IU group (6.2 ± 0.8 versus 9.1 ± 0.8; unadjusted $p = 0.007$) (Figure 2A). The same shorter period was observed for ageusia (loss of taste), again in favor of the 5000 IU group (11.4 ± 1.0 versus 16.9 ± 1.7; unadjusted $p = 0.035$) (Figure 2B). None of the other symptom recovery times were significantly different in either groups (Table 2). The significance for cough decreased but persisted even after adjusting for age, sex, baseline BMI, and D-dimer ($p = 0.039$), while the same significance was observed for ageusia ($p = 0.035$) (not mentioned in the figure).

Table 2. Average Days to Resolve Covid-19 Symptoms according to Vitamin D Dose.

Symptoms	1000 IU	5000 IU	*p*-Value
Fever	9.9 ± 1.7	8.5 ± 0.9	0.97
Dyspnea	11.2 ± 1.6	8.9 ± 1.1	0.24
Fatigue	8.9 ± 0.5	7.7 ± 0.8	0.27
Cough	9.1 ± 0.8	6.2 ± 0.8	0.007
Headache	10.6 ± 0.9	8.7 ± 0.8	0.24
GI symptoms	9.7 ± 1.2	7.6 ± 0.7	0.89
Sore throat	9.5 ± 0.6	12.5 ± 0.7	0.15
Body Aches	9.2 ± 0.9	9.6 ± 0.9	0.68
Chills	17.6 ± 1.2	11.2 ± 1.1	0.14
Anosmia	16.3 ± 1.7	11.2 ± 1.1	0.14
Ageusia	16.9 ± 1.7	11.4 ± 1.0	0.035

Note: Data presented as estimated mean ± SE obtained from Kaplan–Meier survival analysis; *p*-value < 0.05 considered significant.

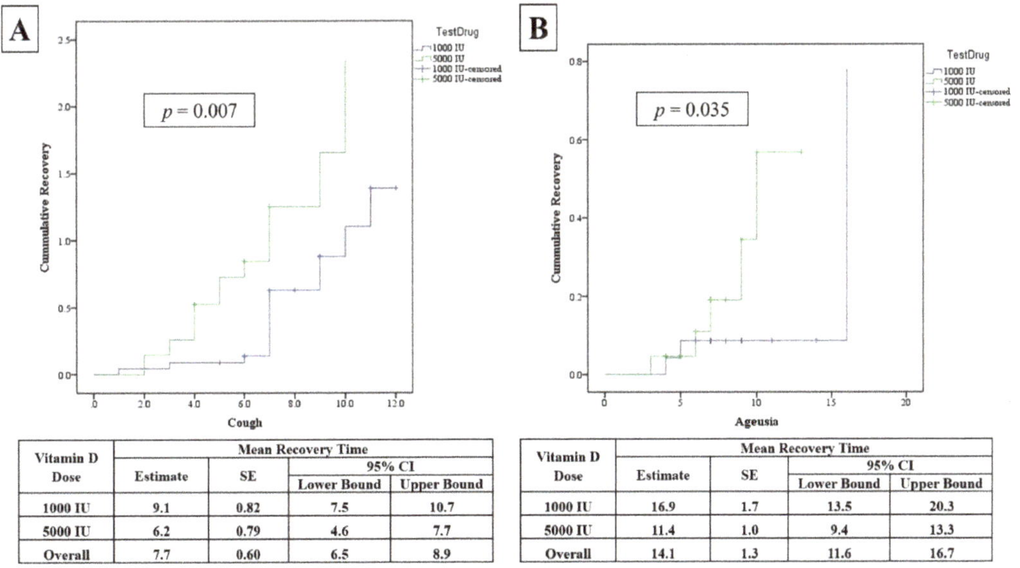

Vitamin D Dose	Mean Recovery Time			
	Estimate	SE	95% CI	
			Lower Bound	Upper Bound
1000 IU	9.1	0.82	7.5	10.7
5000 IU	6.2	0.79	4.6	7.7
Overall	7.7	0.60	6.5	8.9

Vitamin D Dose	Mean Recovery Time			
	Estimate	SE	95% CI	
			Lower Bound	Upper Bound
1000 IU	16.9	1.7	13.5	20.3
5000 IU	11.4	1.0	9.4	13.3
Overall	14.1	1.3	11.6	16.7

Figure 2. Unadjusted Kaplan–Meier Plot showing the recovery times for cough (**A**) and ageusia (**B**) according to vitamin D dose.

3.3. *Secondary Endpoints: Clinical Characteristics Overtime*

No adverse events with respect to treatment were reported in either arm. Table 3 shows that within group comparisons, there was a significant decrease in BMI overtime in both the 1000 IU and 5000 IU groups ($p < 0.05$). Furthermore, in both groups, a significant increase was also observed in WBC count, monocyte, ALT, and a significant decrease was seen in levels of IL-6 in ($p < 0.05$) post-intervention. In the 1000 IU group alone, there was a significant increase in hematocrit ($p = 0.04$) and lymphocyte ($p = 0.03$), with a parallel significant decrease in prothrombin time ($p = 0.05$) and ferritin ($p = 0.004$) over time. On the other hand, in the 5000 IU group, there was a significant increase in neutrophil ($p = 0.03$) and urea ($p < 0.001$). Levels of 25(OH)D significantly increased only in the 5000 IU group ($p = 0.001$), and this significance persisted even after the adjustment for covariates ($p = 0.003$) (Figure 3). No significant changes in lipids and glucose were seen in either group post-supplementation. A. Unadjusted between-group comparisons revealed a clinically significant decrease in BMI in favor of the 1000 IU group ($p = 0.035$). This significance was lost after adjustments for baseline BMI, sex, and age ($p = 0.08$). Between-group comparisons revealed no clinically significant differences between the groups with the exception of D-dimer, which was notably higher in the 1000 IU group (Table 3).

Table 3. Pre and Post Clinical parameters according to Vitamin D supplementation.

Parameters	1000 IU (n = 33)			5000 IU (n = 36)			Between Group p-Value
	Pre-	Post	p-Value	Pre-	Post	p-Value	
Anthropometrics							
BMI (kg/m^2)	32.0 ± 6.5	31.6 ± 6.0	0.04	28.2 ± 7.1	27.9 ± 5.4	0.049	0.08
WHR	0.91 ± 0.11	0.91 ± 0.1	0.84	0.9 ± 0.14	0.9 ± 0.1	0.65	0.73
Complete Blood Count							
Hemoglobin (g/L)	12.7 ± 1.8	13.2 ± 2.2	0.17	13.0 ± 2.8	13.4 ± 2.4	0.03	0.88
Hematocrit (%)	38.5 ± 5.5	40.2 ± 7.2	0.04	40.3 ± 5.7	40.5 ± 6.4	0.66	0.51
RBC count	4.6 ± 0.6	4.8 ± 0.9	0.18	4.8 ± 0.5	4.8 ± 0.7	0.53	0.43
WBC count #	8.5 ± 1.0	9.4 ± 0.9	0.03	6.9 ± 0.4	9.5 ± 0.8	0.001	0.74
Platelet count #	269 ± 29	403 ± 24	<0.001	241 ± 16	380 ± 27	<0.001	0.53
Lymphocyte #	1.0 ± 0.1	1.7 ± 0.2	0.03	2.4 ± 1.1	1.5 ± 0.2	0.95	0.37
Monocyte #	0.5 ± 0.1	0.6 ± 0.1	0.01	0.4 ± 0.0	0.6 ± 0.1	<0.001	0.37
Eosinophil #	0.3 ± 0.1	0.2 ± 0.1	0.85	0.1 ± 0.0	0.1 ± 0.0	0.35	0.30
Neutrophil #	6.2 ± 0.7	6.3 ± 0.5	0.56	5.3 ± 0.5	7.1 ± 0.8	0.03	0.80
Prothrombin Time	13.6 ± 1.6	13.0 ± 1.3	0.05	13.1 ± 1.3	12.9 ± 1.7	0.79	0.76
APTT	32.7 ± 4.8	33.8 ± 7.9	0.85	31.9 ± 4.7	33.8 ± 11.4	0.24	0.74
INR	1.2 ± 0.1	1.1 ± 0.1	0.06	1.1 ± 0.1	1.1 ± 0.1	0.80	0.78
Bicarbonate (mEq/L)	20.8 ± 3.6	22.5 ± 3.1	0.36	21.8 ± 2.7	21.9 ± 6.1	0.51	0.79
Liver Profile							
Bilirubin #	7.1 ± 1.2	6.2 ± 0.7	0.65	9.1 ± 1.2	8.8 ± 0.7	0.86	0.06
Bilirubin (direct) #	4.1 ± 0.4	3.9 ± 0.5	0.55	5.3 ± 0.6	4.5 ± 0.3	0.10	0.12
ALP (U/L) #	97.5 ± 16.2	85.9 ± 13.5	0.22	88.5 ± 11.0	106.4 ± 18.6	0.48	0.67
ALT (U/L) #	62.1 ± 17.6	84.7 ± 20.8	0.02	65.3 ± 14.6	114.9 ± 33.5	0.002	0.73
LDH (U/L) #	564 ± 56	484 ± 40	0.32	487 ± 36	410 ± 28	0.16	0.32
Renal Profile							
Creatinine (μmol/L)	71.6 ± 16.2	70.9 ± 12.1	0.68	67.0 ± 19.1	66.8 ± 6.3	0.50	0.46

Table 3. *Cont.*

Parameters	1000 IU (n = 33)			5000 IU (n = 36)			Between Group p-Value
	Pre-	Post	p-Value	Pre-	Post	p-Value	
Urea (mg/dl) #	9.1 ± 1.8	8.6 ± 1.7	0.89	5.1 ± 0.5	8.0 ± 1.6	<0.001	0.14
Lipid Profile							
Triglycerides (mmol/L) #	1.5 ± 0.1	2.0 ± 0.2	0.48	1.4 ± 0.1	2.0 ± 0.2	0.36	0.52
Total Cholesterol (mmol/L)	4.0 ± 1.4	4.4 ± 1.4	0.86	4.0 ± 0.9	4.5 ± 1.4	0.97	0.75
HDL-Cholesterol (mmol/L)	1.0 ± 0.2	1.1 ± 0.4	0.39	1.0 ± 0.3	1.1 ± 0.4	0.52	0.48
LDL-Cholesterol (mmol/L)	2.4 ± 1.2	2.4 ± 1.1	0.30	2.3 ± 0.8	2.4 ± 1.1	0.81	0.58
Inflammatory Markers							
D-Dimer (µg/mL) #	3.4 ± 2.0	1.9 ± 0.5	0.26	0.6 ± 0.1	1.3 ± 0.6	0.08	0.02
Ferritin (µg/mL) #	784 ± 112	526 ± 76	0.004	733 ± 153	519 ± 96	0.19	0.69
CRP (mg/L) #	47.9 ± 6.8	33.1 ± 7.1	0.10	33.7 ± 5.7	34.2 ± 6.4	0.58	0.25
IL-6 (pg/mL) #	23.9 ± 5.9	19.2 ± 5.6	0.03	18.6 ± 4.6	10.5 ± 2.9	0.01	0.83
Glycemic Profile							
Fasting Glucose (mmol/L) #	10.3 ± 1.1	11.2 ± 1.2	0.38	10.4 ± 1.1	11.4 ± 1.0	0.13	0.91
Vitamin D							
25(OH)D (nmol/L) (75–250) #	63.0 ± 2.5	59.9 ± 3.9	0.66	53.4 ± 2.9	62.5 ± 3.4	0.001	0.67

Note: Data presented as mean ± SD for normal variables while mean ± SE for non-normal variables (#); adjusted p-values obtained from mixed methods ANCOVA, adjusted for age, sex, and BMI; significant at $p < 0.05$.

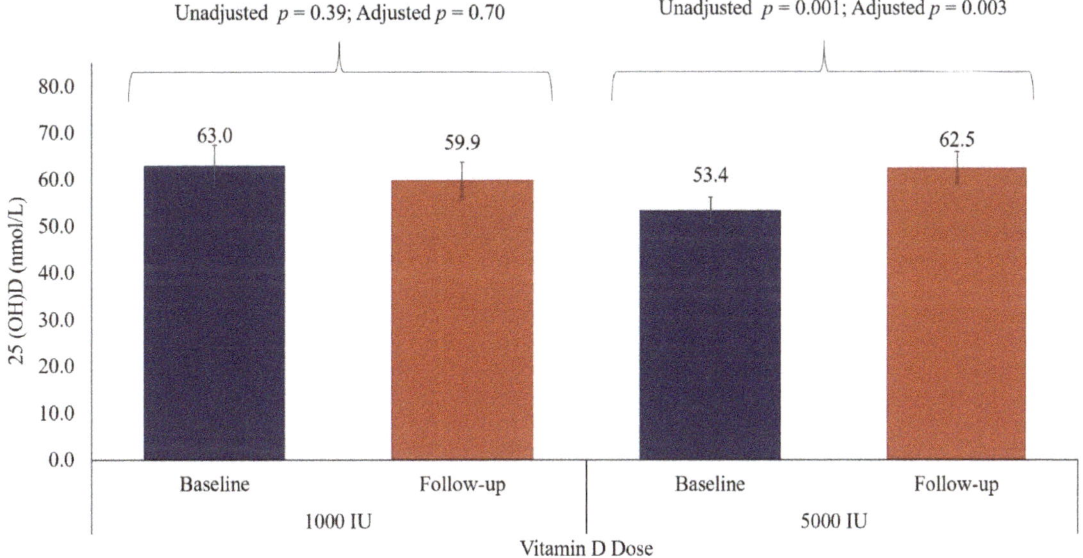

Figure 3. Mean 25(OH) D concentration before and after supplementation.

4. Discussion

To the best of our knowledge, the present study is the first clinical trial for vitamin D and COVID-19 conducted in the Middle East, a region with one of the highest prevalences of vitamin D deficiency in the world, especially in Saudi Arabia (SA) [13,16,20], which consequently, is one of the hardest hit by COVID-19 within the Gulf Cooperation Council

(GCC) countries [1]. The goal of the present randomized clinical trial is primarily to determine whether a short-term 5000 IU vitamin D3 supplementation can reduce recovery times of COVID-19 symptoms among mostly in-patients with mild-moderate symptoms. From this trial, it was observed that 5000 IU oral vitamin D3 taken daily for 2 weeks can substantially reduce the days of recovery from cough and ageusia, and this was clinically significant compared to those who took the standard dose for vitamin D deficiency management. It is worth highlighting that the circulating 25(OH)D levels of almost all of the participants at baseline were either in the insufficiency or mild deficiency range, and that 5000 IU vitamin D3 administered for 2 weeks is safe and tolerable, given the acceptable upper safety dose is 4000IU [30]. In a recent case-control study done in KSA, the majority of the 150 hospitalized patients who screened positive for SARS-CoV-2 had severe manifestations of COVID-19 (80% had radiographically confirmed lung infiltrates) and had a much lower vitamin D status (75% had 25(OH)D < 50 nmol) compared to their non-COVID-19 counterparts (n = 72), who also experienced severe symptoms but tested negative for SARS-CoV-2 [20]. In comparison, the present participants whose COVID-19 conditions were under the mild to moderate category also had sub-optimal but relatively higher 25(OH)D levels than both groups in the mentioned study. While causality cannot be derived from these observations, the inverse association of 25(OH)D to the severity of COVID-19 outcomes is evident and as such, the possibility of benefitting from vitamin D supplementations needs to be tested.

Preliminary trials on the use of vitamin D supplementation against COVID-19 are limited but accumulating. In a pilot study done in Spain, early high dose vitamin D3 prevented ICU admission among COVID-19 patients in combination with the best available standard care for severe cases [31]. In a case-series of COVID-19 patients who received 50,000 IU daily for 5 days, a marked reduction in recovery time and inflammatory markers were observed compared to those who received 1000 IU [32]. A recent quasi-experimental study also showed that among the frail elderly with COVID-19, those who received boluses of 50,000 IU per month or 80,000–100,000 IU per 2–3 months were associated with less severe and improved survival (OR = 0.08 (0.01: 0.81), p = 0.03) [33]. Among the negative trials, a single high dose vitamin D (200,000 IU) given to severe COVID-19 patients (n = 114) did not reduce the hospital stay and severity of outcome compared to the placebo group (n = 118) (Hazard Ratio 1.12) [34]. The trials mentioned have mostly focused on severe cases and mega-doses of vitamin D compared to the present study, which focused on mild cases and lower daily vitamin D doses. In a large-scale meta-analysis conducted involving almost 11,000 participants in 25 clinical trials on the prevention of acute respiratory infections, the protective effects were the greatest among those vitamin D deficient individuals who received daily or weekly doses as opposed to boluses [9]. The dose used in the present study is somewhat similar to a previous RCT, which demonstrated, albeit during a longer term (12 months), that the supplementation of 4000 IU Vitamin D3 prevented acute respiratory infections by as much as 36% (Relative Risk 0.64, 95 % CI 0.43–0.94) based on a cohort of 140 adults with increased risk of acute respiratory infections (>4 infections/year) [35].

As mentioned previously, the extra-skeletal roles of vitamin D are well-established, not only in respiratory infections but in the regulation of the innate immune system overall. Observations from past coronavirus pandemics such as SARS-CoV-1 demonstrated that coronaviruses inhibit type 1 interferon (IFN) receptors, which inversely affect innate immunity [36]. When unbound, the vitamin D receptor (VDR) deteriorates the beneficial antiviral effects of IFN through the removal of a key transcription factor (STAT1) in IFN signaling. This inverse association between VDR and STAT1 implicates that the unbinding of STAT1 through the increased circulation of biologically active forms of vitamin D (calcitriol) (e.g., supplementation) heightens the type 1 IFN response, consequently improving the innate immune system [37]. Another theory by which elevating the circulating 25(OH)D can enhance coronavirus degradation is the acidification of endolysosomes, cellular organelles in charge of the release of SARS-CoV-2 in the cytosol, thereby stimulating autophagy [38,39]. These mechanisms, together with the ones mentioned previously, may partially explain how vitamin D supplementation can alleviate COVID-19 symptoms, which in the present study includes cough and ageusia. Ageusia is of interest as not much has been published on the role of vitamin D in the reversal of this symptom. Loss of taste and smell however are common in respiratory viral and bacterial infections [40]. Wang and colleagues observed that Toll-like receptor (TLR) and interferon (IFN) pathways were found to be present in taste tissue, and these pathways are activated in response to inflammation (e.g., respiratory infection), which inadvertently interferes normal taste transduction [41]. Vitamin D may restore gustatory function via the suppression of these pathways in the presence of infection, consequently downregulating the inflammatory response [42]. This mechanism is reversed among healthy individuals, where vitamin D may stimulate TLR expression, but in preparation for pathogen exposure [43]. Another explanation can be due to the neuroprotective effects of vitamin D [44], which includes the regulation of the neurotrophins responsible for the development of the gustatory taste system [45].

In the present study, it was apparent that both vitamin D groups had significant reductions in BMI following COVID-19 diagnosis. This observed clinical weight loss can be secondary to the loss of olfactory and gustatory sensations associated with appetite and may have nothing to do with the vitamin D. Unintentional weight loss was observed as one of the collaterals of COVID-19 [46]. Given that most participants in the present study were either overweight or obese, this consequence maybe considered positive for the present cohort, but it also suggests that nutritional therapy may be needed for full recovery of COVID-19 patients following hospital admission and/or isolation [47].

Lastly, while circulating calcium and the parathyroid hormone were not assessed in the present trial, it is important to mention their influence on COVID-19 severity. Calcium in particular plays important roles in virus entry and gene expression [48], with hypocalcemia being commonly observed as a common biochemical abnormality among patients with severe COVID-19 manifestations [49,50], which, in combination with vitamin D deficiency, contributes to a unique osteo-metabolic phenotype [51]. Therefore, vitamin D correction, which controls the entire body's calcium homeostasis, may further benefit COVID-19 patients with suboptimal 25(OH)D levels by maintaining calcium balance, consequently decreasing risk of COVID-19 severity.

Strengths and Limitations

The results of the present clinical trial should be interpreted with full consideration of its limitations. Risk of bias is apparent given the study's open-label design since blinding was impossible. To minimize this, the assessment of symptoms at follow-up were collected over the phone by a blinded data collector. The beneficial effects of 5000 IU vitamin D3 supplementation in this case applies only to mild and moderate COVID-19 cases with sub-optimal vitamin D status (mild deficiency to insufficiency), and whether the same dose and duration will also apply to severe COVID-19 cases with worse vitamin D status needs to be investigated in future clinical trials. The duration of intervention was primarily based on MoH guidelines in terms of deisolation/discharge of COVID-19 cases presenting with mild to moderate symptoms, as it was difficult to monitor the participants physically given the existing COVID-19 restrictions imposed during the study period. While the study had no placebo, the use of 1000 IU is standard and served as a control since it would be deemed unethical not to provide vitamin D supplements if participants were known to have suboptimal vitamin D status. Finally, baseline differences in the parameters were evident despite randomization, as is true for most clinical trials. While age and BMI were used as covariates in all models, these necessary adjustments added stringency to the analysis given its small sample size. Nevertheless, the findings are robust and well powered. The present clinical trial is one of the first interventions globally and the first in the Middle East to use vitamin D as a short-term adjuvant therapy in improving mild to moderate COVID-19 symptoms among patients with sub-optimal vitamin D levels. Prospective cohort studies are needed to determine whether these beneficial effects ultimately extend to prevention of SARS-CoV-2 infection.

5. Conclusions

In summary, a 2-week oral supplementation of 5000 IU vitamin D3 was superior to 1000 IU in resolving cough and gustatory sensory loss among COVID-19 patients with sub-optimal vitamin D presenting with mild to moderate symptoms. The present findings add to the growing body of evidence on the beneficial effects of vitamin D supplementation against COVID-19, particularly among those with suboptimal levels.

Supplementary Materials: The following are available online at https://www.mdpi.com/article/10.3390/nu13072170/s1, Table S1: Baseline Serological Characteristics of Groups.

Author Contributions: The authors' responsibilities were as follows: S.S., M.A.E., and N.M.A.-D. conceived and designed the experiments; M.A.E., E.S., N.J.A., and N.H.A. recruited the patients and performed the data collection; D.A.A., S.Y.A., A.M.A., and O.E.A. performed and validated the sample collection and analysis; S.S. and S.D.H. analyzed the data; D.A.A., N.A., and S.Y.A. contributed reagents/materials/analysis tools; S.S. wrote and revised the paper. N.M.A.-D. takes responsibility for the integrity of the data and the accuracy of the data analysis and is the guarantor. All authors have read and agreed to the published version of the manuscript.

Funding: The study was funded by the Deanship of Scientific Research, Chair for Biomarkers of Chronic Diseases at King Saud University, Riyadh, KSA. Vitamin D supplements used in the intervention were provided by Synergy Pharma (Dubai, UAE).

Institutional Review Board Statement: The study was conducted according to the guidelines of the Declaration of Helsinki, and approved by the Institutional Review Board of King Fahad Medical City, Riyadh, Saudi Arabia (IRB Log No. 20-282, 28 July 2020).

Informed Consent Statement: Informed consent was obtained from all subjects involved in the study.

Data Availability Statement: The datasets used and/or analyzed during the current study are available from the corresponding author upon reasonable request.

Acknowledgments: The authors are grateful to Synergy Pharma (Dubai, UAE) for providing the vitamin D3 supplements used in this trial. Synergy Pharma had no role in the design of the study; in the collection, analyses, or interpretation of the data; in the writing of the manuscript; and in the decision to publish the results. The authors are also grateful to Fahad Al-Salhi for coordinating with the team and patients, Kholoud Alharthi for data collection, and Muhrah AlMowalad, the senior pharmacist in charge of IDS at KFMC, for the randomization and dispensing of the vitamin D tablets.

Conflicts of Interest: All authors declare no conflict of interest.

Abbreviations

ACE2—Angiotensin converting enzyme; ANCOVA—Analysis of Covariance; BMI—Body mass index; BSC-II—Biological Safety Cabinet Class II; BSL 2—Biosafety Level 2; CAP—College of American Pathologists; CLIA—Chemiluminescent Immunoassay; COVID-19—Coronavirus disease-19; CRP—C-Reactive Protein; GCC—Gulf Cooperation Council; HDL—High density lipoprotein; IFN—Interferon; IL-6—Interleukin 6; IRB—Institutional Review Board; ITT—Intention to treat; KFMC—King Fahad Medical City; KSH—King Salman Hospital; KSUMC—King Saud University Medical City; LDL—Low density lipoprotein; LOCF—Last Observation Carried Forward; RT-PCR—Reverse Transcriptase Polymerase Chain Reaction; SA - Saudi Arabia; SARS-CoV-2—Severe Acute Respiratory Syndrome Coronavirus-2; SD—Standard Deviation; SE – Standard Error; SCTR—Saudi Clinical Trial Registry; SNIBE—Shenzhen New Industries Biomedical Engineering; UAE—United Arab Emirates; VDR—Vitamin D Receptor; VDSCP—Vitamin D Standardization-Certification Program.

References

1. Dong, E.; Du, H.; Gardner, L. An Interactive Web-Based Dashboard to Track COVID-19 in Real Time. *Lancet Infect. Dis.* **2020**, *20*, 533–534. [CrossRef]
2. Coronaviridae Study Group of the International Committee on Taxonomy of Viruses. The species Severe acute respiratory syndrome-related coronavirus: Classifying 2019-nCoV and naming it SARS-CoV-2. *Nat. Microbiol.* **2020**, *5*, 536–544. [CrossRef] [PubMed]

3. Oran, D.P.; Topol, E.J. Prevalence of Asymptomatic SARS-CoV-2 Infection. *Ann. Intern. Med.* **2020**, *173*, 362–367. [CrossRef] [PubMed]
4. Sanders, J.M.; Monogue, M.L.; Jodlowski, T.Z.; Cutrell, J.B. Pharmacologic Treatments for Coronavirus Disease 2019 (COVID-19). *JAMA* **2020**, *323*, 1824–1836. [CrossRef] [PubMed]
5. Lancet, T. COVID-19 Vaccines: No Time for Complacency. *Lancet* **2020**, *396*, 1607. [CrossRef]
6. Jolliffe, D.A.; Greenberg, L.; Hooper, R.L.; Griffiths, C.J.; Camargo, C.A., Jr.; Kerley, C.P.; Jensen, M.E.; Mauger, D.; Stelmach, I.; Urashima, M.; et al. Vitamin D supplementation to prevent asthma exacerbations: A systematic review and meta-analysis of individual participant data. *Lancet Respir. Med.* **2017**, *5*, 881–890. [CrossRef]
7. Bjelakovic, G.; Gluud, L.L.; Nikolova, D.; Whitfield, K.; Wetterslev, J.; Simonetti, R.G.; Bjelakovic, M.; Gluud, C. Vitamin D Supplementation for Prevention of Mortality in Adults. *Cochrane Database Syst. Rev.* **2014**, CD007470. [CrossRef]
8. Sarhan, T.S.; Elrifai, A. Serum Level of Vitamin D as a Predictor for Severity and Outcome of Pneumonia. *Clin. Nutr.* **2021**, *40*, 2389–2393. [CrossRef]
9. Martineau, A.R.; Jolliffe, D.A.; Hooper, R.L.; Greenberg, L.; Aloia, J.F.; Bergman, P.; Dubnov-Raz, G.; Esposito, S.; Ganmaa, D.; Ginde, A.A.; et al. Vitamin D supplementation to prevent acute respiratory tract infections: Systematic review and meta-analysis of individual participant data. *BMJ* **2017**, *356*, i6583. [CrossRef]
10. Ma, H.; Zhou, T.; Heianza, Y.; Qi, L. Habitual Use of Vitamin D Supplements and Risk of Coronavirus Disease 2019 (COVID-19) Infection: A Prospective Study in UK Biobank. *Am. J. Clin. Nutr.* **2021**, *113*, 1275–1281. [CrossRef]
11. Radujkovic, A.; Hippchen, T.; Tiwari-Heckler, S.; Dreher, S.; Boxberger, M.; Merle, U. Vitamin D Deficiency and Outcome of COVID-19 Patients. *Nutrients* **2020**, *12*, 2757. [CrossRef]
12. Mercola, J.; Grant, W.B.; Wagner, C.L. Evidence Regarding Vitamin D and Risk of COVID-19 and Its Severity. *Nutrients* **2020**, *12*, 3361. [CrossRef]
13. Alguwaihes, A.M.; Al-Sofiani, M.E.; Megdad, M.; Albader, S.S.; Alsari, M.H.; Alelayan, A.; Alzahrani, S.H.; Sabico, S.; Al-Daghri, N.M.; Jammah, A.A. Diabetes and Covid-19 among hospitalized patients in Saudi Arabia: A single-centre retrospective study. *Cardiovasc. Diabetol.* **2020**, *19*, 205. [CrossRef] [PubMed]
14. Wrapp, D.; Wang, N.; Corbett, K.S.; Goldsmith, J.A.; Hsieh, C.-L.; Abiona, O.; Graham, B.S.; McLellan, J.S. Cryo-EM structure of the 2019-nCoV spike in the prefusion conformation. *Science* **2020**, *367*, 1260–1263. [CrossRef]
15. Musavi, H.; Abazari, O.; Barartabar, Z.; Kalaki-Jouybari, F.; Hemmati-Dinarvand, M.; Esmaeili, P.; Mahjoub, S. The Benefits of Vitamin D in the COVID-19 Pandemic: Biochemical and Immunological Mechanisms. *Arch. Physiol. Biochem.* **2020**, 1–9. [CrossRef] [PubMed]
16. Al-Daghri, N.M.; Amer, O.E.; Alotaibi, N.H.; Aldisi, D.A.; Enani, M.A.; Sheshah, E.; Aljohani, N.J.; Alshingetti, N.; Alomar, S.Y.; Alfawaz, H.; et al. Vitamin D status of Arab Gulf residents screened for SARS-CoV-2 and its association with COVID-19 infection: A multi-centre case-control study. *J. Transl. Med.* **2021**, *19*, 166. [CrossRef]
17. Waldron, J.L.; Ashby, H.L.; Cornes, M.P.; Bechervaise, J.; Razavi, C.; Thomas, O.L.; Chugh, S.; Deshpande, S.; Ford, C.; Gama, R. Vitamin D: A Negative Acute Phase Reactant. *J. Clin. Pathol.* **2013**, *66*, 620–622. [CrossRef] [PubMed]
18. Goncalves de Carvalho, C.M.; Ribeiro, S.M. Aging, low-grade systemic inflammation and vitamin D: A mini-review. *Eur. J. Clin Nutr.* **2017**, *71*, 434–440. [CrossRef]
19. Bassatne, A.; Basbous, M.; Chakhtoura, M.; El Zein, O.; Rahme, M.; El-Hajj Fuleihan, G. The link between COVID-19 and Vitamin D (VIVID): A systematic review and meta-analysis. *Metabolism* **2021**, *119*, 154753. [CrossRef] [PubMed]
20. Alguwaihes, A.M.; Sabico, S.; Hasanato, R.; Al-Sofiani, M.E.; Megdad, M.; Albader, S.S.; Alsari, M.H.; Alelayan, A.; Alyusuf, E.Y.; Alzahrani, S.H.; et al. Severe vitamin D deficiency is not related to SARS-CoV-2 infection but may increase mortality risk in hospitalized adults: A retrospective case-control study in an Arab Gulf country. *Aging Clin. Exp. Res.* **2021**, *33*, 1415–1422. [CrossRef]
21. Ministry of Health. Saudi Arabia COVID-19 Dashboard. Available online: https://covid19.moh.gov.sa (accessed on 8 December 2020).
22. Saudi MoH Protocol for Patients Suspected of/Confirmed with COVID-19 (Version 2.1 July 2020). Available online: https://www.moh.gov.sa/en/Ministry/MediaCenter/Publications/Pages/covid19.aspx (accessed on 9 December 2020).
23. Saudi MoH Hospital Admission Criteria for COVID-19 Patients (May 2020). Available online: https://www.moh.gov.sa/Ministry/MediaCenter/Publications/Documents/admission-criteria.pdf (accessed on 5 March 2021).
24. Saudi Clinical Trials Registry. Available online: https://old.sfda.gov.sa/en/drug/clinical_trials/pages/registered-clinical.aspx (accessed on 24 May 2021).
25. Schulz, K.F.; Altman, D.G.; Moher, D. CONSORT 2010 statement: Updated guidelines for reporting parallel group randomised trials. *J. Pharmacol. Pharmacother.* **2010**, *1*, 100–107. [CrossRef]
26. VDSCP Vitamin D Certified Assays. Certifications from 2020 (Updated September 2020). Available online: https://www.cdc.gov/labstandards/pdf/hs/CDC_Certified_Vitamin_D_Procedures-508.pdf (accessed on 8 December 2020).
27. Al-Saleh, Y.; Sulimani, R.; Sabico, S.; Raef, H.; Fouda, M.; Alshahrani, F.; Al Shaker, M.; Al Wahabi, B.; Sadat-Ali, M.; Al Rayes, H.; et al. 2015 Guidelines for Osteoporosis in Saudi Arabia: Recommendations from the Saudi Osteoporosis Society. *Ann. Saudi Med.* **2015**, *35*, 1–12. [CrossRef]

28. Al Saleh, Y.; Beshyah, S.A.; Hussein, W.; Almadani, A.; Hassoun, A.; Al Mamari, A.; Ba-Essa, E.; Al-Dhafiri, E.; Hassanein, M.; Fouda, M.A.; et al. Diagnosis and management of vitamin D deficiency in the Gulf Cooperative Council (GCC) countries: An expert consensus summary statement from the GCC vitamin D advisory board. *Arch. Osteoporos.* **2020**, *15*, 35. [CrossRef]
29. Simpson, S.; van der Mei, I.; Stewart, N.; Blizzard, L.; Tettey, P.; Taylor, B. Weekly cholecalciferol supplementation results in significant reductions in infection risk among the vitamin D deficient: Results from the CIPRIS pilot RCT. *BMC Nutr.* **2015**, *1*, 7. [CrossRef]
30. Rizzoli, R. Vitamin D supplementation: Upper limit for safety revisited? *Aging Clin. Exp. Res.* **2021**, *33*, 19–24. [CrossRef]
31. Entrenas Castillo, M.; Entrenas Costa, L.M.; Vaquero Barrios, J.M.; Alcala Diaz, J.F.; Lopez Miranda, J.; Bouillon, R.; Quesada Gomez, J.M. Effect of calcifediol treatment and best available therapy versus best available therapy on intensive care unit admission and mortality among patients hospitalized for COVID-19: A pilot randomized clinical study. *J. Steroid. Biochem. Mol. Biol.* **2020**, *203*, 105751. [CrossRef]
32. Ohaegbulam, K.C.; Swalih, M.; Patel, P.; Smith, M.A.; Perrin, R. Vitamin D Supplementation in COVID-19 Patients: A Clinical Case Series. *Am. J. Ther.* **2020**, *27*, e485–e490. [CrossRef] [PubMed]
33. Annweiler, G.; Corvaisier, M.; Gautier, J.; Dubee, V.; Legrand, E.; Sacco, G.; Annweiler, C. Vitamin D Supplementation Associated to Better Survival in Hospitalized Frail Elderly COVID-19 Patients: The GERIA-COVID Quasi-Experimental Study. *Nutrients* **2020**, *12*, 3377. [CrossRef] [PubMed]
34. Murai, I.H.; Fernandes, A.L.; Sales, L.P.; Pinto, A.J.; Goessler, K.F.; Duran, C.S.C.; Silva, C.B.R.; Franco, A.S.; Macedo, M.B.; Dalmolin, H.H.H.; et al. Effect of a Single High Dose of Vitamin D3 on Hospital Length of Stay in Patients with Moderate to Severe COVID-19: A Randomized Clinical Trial. *JAMA* **2021**, *325*, 1053–1060. [CrossRef] [PubMed]
35. Bergman, P.; Norlin, A.C.; Hansen, S.; Bjorkhem-Bergman, L. Vitamin D supplementation improves well-being in patients with frequent respiratory tract infections: A post hoc analysis of a randomized, placebo-controlled trial. *BMC Res. Notes* **2015**, *8*, 498. [CrossRef] [PubMed]
36. Minakshi, R.; Padhan, K.; Rani, M.; Khan, N.; Ahmad, F.; Jameel, S. The SARS Coronavirus 3a protein causes endoplasmic reticulum stress and induces ligand-independent downregulation of the type 1 interferon receptor. *PLoS ONE* **2009**, *4*, e8342. [CrossRef]
37. Jakovac, H. COVID-19 and vitamin D-Is there a link and an opportunity for intervention? *Am. J. Physiol. Endocrinol. Metab.* **2020**, *318*, e589. [CrossRef] [PubMed]
38. Khan, N.; Chen, X.; Geiger, J.D. Role of Endolysosomes in Severe Acute Respiratory Syndrome Coronavirus-2 Infection and Coronavirus Disease 2019 Pathogenesis: Implications for Potential Treatments. *Front. Pharmacol.* **2020**, *11*, 595888. [CrossRef] [PubMed]
39. Geiger, J.D.; Khan, N.; Murugan, M.; Boison, D. Possible Role of Adenosine in COVID-19 Pathogenesis and Therapeutic Opportunities. *Front. Pharmacol.* **2020**, *11*, 594487. [CrossRef]
40. Rawson, N.E.; Huang, L. Symposium overview: Impact of oronasal inflammation on taste and smell. *Ann. N. Y. Acad. Sci.* **2009**, *1170*, 581–584. [CrossRef]
41. Wang, H.; Zhou, M.; Brand, J.; Huang, L. Inflammation and taste disorders: Mechanisms in taste buds. *Ann. N. Y. Acad. Sci.* **2009**, *1170*, 596–603. [CrossRef]
42. Do, J.E.; Kwon, S.Y.; Park, S.; Lee, E.S. Effects of vitamin D on expression of Toll-like receptors of monocytes from patients with Behcet's disease. *Rheumatology* **2008**, *47*, 840–848. [CrossRef]
43. Ojaimi, S.; Skinner, N.A.; Strauss, B.J.; Sundararajan, V.; Woolley, I.; Visvanathan, K. Vitamin D deficiency impacts on expression of toll-like receptor-2 and cytokine profile: A pilot study. *J. Transl. Med.* **2013**, *11*, 176. [CrossRef] [PubMed]
44. Xu, Y.; Baylink, D.J.; Chen, C.S.; Reeves, M.E.; Xiao, J.; Lacy, C.; Lau, E.; Cao, H. The importance of vitamin d metabolism as a potential prophylactic, immunoregulatory and neuroprotective treatment for COVID-19. *J. Transl. Med.* **2020**, *18*, 322. [CrossRef] [PubMed]
45. Meng, L.; Jiang, X.; Ji, R. Role of neurotrophin in the taste system following gustatory nerve injury. *Metab. Brain. Dis.* **2015**, *30*, 605–613. [CrossRef]
46. Di Filippo, L.; De Lorenzo, R.; D'Amico, M.; Sofia, V.; Roveri, L.; Mele, R.; Saibene, A.; Rovere-Querini, P.; Conte, C. COVID-19 is associated with clinically significant weight loss and risk of malnutrition, independent of hospitalisation: A post-hoc analysis of a prospective cohort study. *Clin. Nutr.* **2021**, *40*, 2420–2426. [CrossRef] [PubMed]
47. Pironi, L.; Sasdelli, A.S.; Ravaioli, F.; Baracco, B.; Battaiola, C.; Bocedi, G.; Brodosi, L.; Leoni, L.; Mari, G.A.; Musio, A. Malnutrition and nutritional therapy in patients with SARS-CoV-2 disease. *Clin. Nutr.* **2021**, *40*, 1330–1337. [CrossRef]
48. Zhou, Y.; Frey, T.K.; Yang, J.J. Viral calciomics: Interplays between Ca^{2+} and virus. *Cell Calcium* **2009**, *46*, 1–17. [CrossRef]
49. Di Filippo, L.; Doga, M.; Frara, S.; Giustina, A. Hypocalcemia in COVID-19: Prevalence, clinical significance and therapeutic implications. *Rev. Endocr. Metab. Disord.* **2021**, *13*, 1–10. [CrossRef]
50. Martha, J.W.; Wibowo, A.; Pranata, R. Hypocalcemia is associated with severe COVID-19: A systematic review and meta-analysis. *Diabetes Metab. Syndr.* **2021**, *15*, 337–342. [CrossRef] [PubMed]
51. Di Filippo, L.; Frara, S.; Giustina, A. The emerging osteo-metabolic phenotype of COVID-19: Clinical and pathophysiological aspects. *Nat. Rev. Endocrinol.* **2021**, 1–2. [CrossRef]

 *nutrients*

Article

The Association between Vitamin D and Zinc Status and the Progression of Clinical Symptoms among Outpatients Infected with SARS-CoV-2 and Potentially Non-Infected Participants: A Cross-Sectional Study

Sahar Golabi [1], Maryam Adelipour [2], Sara Mobarak [3], Maghsud Piri [4], Maryam Seyedtabib [5], Reza Bagheri [6], Katsuhiko Suzuki [7], Damoon Ashtary-Larky [8], Fatemeh Maghsoudi [9] and Mahshid Naghashpour [10,*]

1 Department of Medical Physiology, School of Medicine, Abadan University of Medical Sciences, Abadan 6313833177, Iran; s.golabi@abadanums.ac.ir
2 Department of Clinical Biochemistry, School of Medicine, Ahvaz Jundishapur University of Medical Sciences, Ahvaz 6135715794, Iran; adelipour-m@ajums.ac.ir
3 Department of Infectious Diseases, School of Medicine, Abadan University of Medical Sciences, Abadan 6313833177, Iran; s.mobarak@abadanums.ac.ir
4 Vice Chancellor for Health, Abadan University of Medical Sciences, Abadan 6313833177, Iran; maghsudpiri@gmail.com
5 Department of Biostatistics and Epidemiology, School of Public Health, Ahvaz Jundishapur University of Medical Sciences, Ahvaz 6135715794, Iran; m.stabib3@gmail.com
6 Department of Exercise Physiology, University of Isfahan, Isfahan 8174673441, Iran; will.fivb@yahoo.com
7 Faculty of Sport Sciences, Waseda University, 2-579-15 Mikajima, Tokorozawa 359-1192, Japan; katsu.suzu@waseda.jp
8 Nutrition and Metabolic Diseases Research Center, Ahvaz Jundishapur University of Medical Sciences, Ahvaz 6135715794, Iran; damoon_ashtary@yahoo.com
9 Department of Public Health, School of Health, Abadan University of Medical Sciences, Abadan 6313833177, Iran; fatemehmagh627@gmail.com
10 Department of Nutrition, School of Medicine, Abadan University of Medical Sciences, Abadan 6313833177, Iran
* Correspondence: m.naghashpour@abadanums.ac.ir; Tel.: +98-9166157338

Citation: Golabi, S.; Adelipour, M.; Mobarak, S.; Piri, M.; Seyedtabib, M.; Bagheri, R.; Suzuki, K.; Ashtary-Larky, D.; Maghsoudi, F.; Naghashpour, M. The Association between Vitamin D and Zinc Status and the Progression of Clinical Symptoms among Outpatients Infected with SARS-CoV-2 and Potentially Non-Infected Participants: A Cross-Sectional Study. *Nutrients* **2021**, *13*, 3368. https://doi.org/10.3390/nu13103368

Academic Editors: William B. Grant, Ronan Lordan and Lutz Schomburg

Received: 2 September 2021
Accepted: 21 September 2021
Published: 25 September 2021

Abstract: Vitamin D and zinc are important components of nutritional immunity. This study compared the serum concentrations of 25-hydroxyvitamin D (25(OH)D) and zinc in COVID-19 outpatients with those of potentially non-infected participants. The association of clinical symptoms with vitamin D and zinc status was also examined. A checklist and laboratory examination were applied to collect data in a cross-sectional study conducted on 53 infected outpatients with COVID-19 and 53 potentially non-infected participants. Serum concentration of 25(OH)D were not significantly lower in patients with moderate illness (19 $\pm$ 12 ng/mL) than patients with asymptomatic or mild illness (29 $\pm$ 18 ng/mL), with a trend noted for a lower serum concentration of 25(OH)D in moderate than asymptomatic or mild illness patients (p = 0.054). Infected patients (101 $\pm$ 18 µg/dL) showed a lower serum concentration of zinc than potentially non-infected participants (114 $\pm$ 13 µg/dL) (p = 0.01). Patients with normal (odds ratio (OR), 0.19; $p \leq 0.001$) and insufficient (OR, 0.3; p = 0.007) vitamin D status at the second to seventh days of disease had decreased OR of general symptoms compared to patients with vitamin D deficiency. This study revealed the importance of 25(OH)D measurement to predict the progression of general and pulmonary symptoms and showed that infected patients had significantly lower zinc concentrations than potentially non-infected participants.

Keywords: clinical symptoms; vitamin D status; zinc status; sunlight exposure; COVID-19

1. Introduction

Novel coronavirus disease 2019 (COVID-19), caused by severe acute respiratory syndrome coronavirus 2 (SARS-CoV-2), is a worldwide pandemic that originally emerged in Wuhan, China [1]. As of 3 April 2020, Iran has been among the countries with the highest burden of the COVID-19 outbreak [2]. COVID-19 is characterized by the symptoms of viral pneumonia, such as fever, fatigue, dry cough, and lymphopenia. Patients have reported comorbidities such as diabetes, cardiovascular disease, liver disease, kidney disease, and malignant tumors [3]. This disease also affects physical activity, sedentary action, and psychological emotion [4].

While therapeutic options are still under investigation, and some vaccines have been approved, cost-effective ways to reduce the probability of or even prevent infection and the shift from mild symptoms to more serious detrimental disease are highly worthwhile [5].

An appropriate diet and good nutritional status are essential for an optimal immune response to prevent infections. On the other hand, a poor diet and deficiency of these nutrients will increase the disease burden. Evidence proposes that nutrients are involved in the development of COVID-19 [6].

Vitamin D3 is a pre-pro-hormone that begins its biosynthesis pathway with the solar UVB irradiation of 7-dehydrocholesterol on bare skin exposed to strong sunlight and exhibits multifaceted effects beyond calcium and bone metabolism. Vitamin D is essential to balance immune responses [7]. Since vitamin D receptors are expressed on immune cells (B, T, and antigen-presenting cells), which can synthesize the active metabolite of vitamin D, this vitamin can act in an autocrine manner in a local immunological environment. Iran is a country with a high prevalence of vitamin D deficiency among various age groups, with the more apparent prevalence of this deficiency in Tehran, the capital of Iran. According to a systematic review and meta-analysis, the prevalence rate of vitamin D deficiency among the Iranian population is wide-ranging from 2.5% to 98% in various studies and regions [8]. Vitamin D deficiency is associated with increased autoimmunity and increased susceptibility to infection [9].

Epidemiological evidence suggests a significant association between vitamin D deficiency and an increased incidence of several infectious diseases, viral respiratory tract infections [10], and influenza [11]. A recent epidemiologic study reported a strong significant relationship between the serum concentration of vitamin D and the number of deaths per million people from COVID-19 across 20 European countries [12]. Previous findings have shown that individuals with vitamin D deficiency have a higher risk of contracting a severe COVID-19 disease [13].

It is well-known that zinc is a critical mineral in many biological processes due to its functions as a cofactor, signaling molecule, and structural element [14]. Furthermore, zinc has an important role in the regulation of the immune system by regulating the proliferation, differentiation, maturation, and functioning of leukocytes and lymphocytes [15]. Zinc also plays a signaling role in the modulation of inflammatory responses [16]. It is also a component of nutritional immunity [17]. Previously published data demonstrate that zinc status is associated with the prevalence of respiratory tract infections in children and adults [18,19]. It is also thought that zinc has the potential to support COVID-19 therapy due to its immunomodulatory roles and direct antiviral effects [20].

Moreover, adequate dietary intake of zinc and vitamin D is essential for suitable immunocompetence and resistance to viral infections [21]. In addition, an ecological study demonstrates that intake levels of vitamin D are inversely accompanied by higher COVID-19 incidence and/or mortality, especially in populations that are genetically predisposed to low micronutrient status [22]. Moreover, it is suggested that nutrition intervention acquiring an adequate status of some vitamins and minerals including vitamin D and zinc might protect against COVID-19 and alleviate the course of the disease [21,22]. However, dietary recommendations alone are not enough to ensure the adequacy of these nutrients [21]. As a result, nutritional data assessing nutrients are essential for immune system function [22].

So far, data on the association between vitamin D and zinc status and the progression of symptoms during the clinical course among COVID-19 outpatients are limited. The high prevalence of vitamin D and zinc deficiency in the elderly, smokers, patients with chronic diseases, and obese individuals suggests that vitamin D and zinc play a role as therapeutic agents against COVID-19. Here, we evaluated the role of the nutritional status of vitamin D and zinc in the perspective of COVID-19 and the progression of symptoms during the clinical course of the disease. Therefore, we compared the demographics, baseline comorbidities, and serum concentrations of vitamin D and zinc at the second to seventh days of disease between infected outpatients with COVID-19 and potentially non-infected participants from an academic health care setting in southwestern Iran. The association between serum concentrations of vitamin D and zinc at the second to seventh days of disease and the progression of symptoms during the clinical course of the disease was also determined.

2. Materials and Methods

2.1. Study Design

To examine the potential association between vitamin D and zinc status and the disease progression of COVID-19 among the clients of a health care setting, we designed a health service center-based cross-sectional and descriptive–analytical study aimed to compare infected outpatients with COVID-19 and potentially non-infected participants in terms of demographics, baseline comorbidities, and serum concentrations of vitamin D and zinc at the second to seventh days of disease. In addition, patients who tested positive for COVID-19 by reverse transcription–polymerase chain reaction (RT-PCR) were followed up from day 1 to day 28 after the onset of symptoms to evaluate the effect of vitamin D and zinc status at the second to seventh days of disease on the symptom progression during the clinical course of COVID-19. This research was approved by the Ethics Committee of Abadan University of Medical Sciences (Ethics code: IR.ABADANUMS.REC.1399.073). The criterion for entering the infected and potentially non-infected participants was a positive or negative RT-PCR result. Infected patients: patients with a laboratory-confirmed COVID-19 diagnosis based on the RT-PCR test. Potentially non-infected participants: individuals whose COVID-19 had not been confirmed based on the RT-PCR test with no history of positive RT-PCR test during the COVID-19 pandemic or recovering COVID recently, no clinical signs associated with COVID-19, including high fever, and high-risk occupations, including medical staff.

2.2. Setting

Sixteen-hour COVID-19 health service centers operate under the supervision of Abadan University of Medical Sciences. These centers were activated following the outbreak of the COVID-19 pandemic and work on an outpatient basis due to the need to provide health services for the citizens of Abadan (located in southwestern Iran). Outpatients with COVID-19 and potentially non-infected participants referred to centers from 6 June 2020, to 12 August 2020, were recruited in the study.

2.3. Study Population and Sample

The population of this study comprised clients referred to the 16-hour outpatient centers mentioned in the previous section. All participants provided written informed consent during recruitment for study participation and repeat contact. All clinical investigations were conducted according to the ethical standards of the World Medical Association's Declaration of Helsinki. Infected patients were at the second to seventh days of COVID-19 disease.

Age- and sex-matched potentially non-infected participants with negative RT-PCR test results were recruited from the same 16-hour health service center by telephone and underwent screening by a study team member.

All infected patients and potentially non-infected participants underwent respiratory sampling, including nasal and pharyngeal swabs, bronchoalveolar lavage fluid, sputum, or bronchial aspirates, in one of the 16-hour outpatient centers to evaluate COVID-19. An RT-PCR kit (COVITECH, Tehran, Iran) was used to qualitatively detect the presence or absence of *SARS-CoV-2* infection, which is currently used in the Iranian health centers to diagnose COVID-19 disease. Cut-off Ct value < 36 was considered as a positive result.

We used an open-source calculator to calculate the minimum sample size required based on the probability of a type I error of alpha = 0.5 and type II error of beta = 0.2 (power = 80%). According to this calculation, at least 53 cases and 53 controls were needed. Individuals with a clear RT-PCR result (either positive or negative) meeting the essential criteria to enter the study were selected by a simple sampling method, so that every client had an equal probability of admission and inclusion in the study. We used the demographic factors of age and sex as factors to ensure that our potentially non-infected participants were matched with our infected patients. As such, a potentially non-infected participant with a specific age and sex was included in the study for each infected patient of the same age and sex.

2.4. Inclusion and Exclusion Criteria

Participants ≥11 years of age of both sexes were included in the study. Moreover, to be included, participants needed to have a clear RT-PCR result (positive or negative) and be willing to participate in the study. They also needed to have the ability to understand the relevant information and complete the informed consent form. Pregnant and lactating women, participants with uncertain RT-PCR test results, and patients with sickle cell anemia or thalassemia were excluded [23].

2.5. Variables

The variable presented a positive result for the specific test for COVID-19 detection. Moreover, to identify the stage of COVID disease, infected patients were categorized according to disease severity and prognosis using Center for Disease Control and Prevention (CDC) criteria, which include the following. (1) *Asymptomatic or presymptomatic disease*: individuals who presented positive results for the RT-PCR test but showed no symptoms of COVID-19. (2) *Mild illness*: individuals who had any of the symptoms of COVID-19 (e.g., fever, headache, cough, sore throat, muscle pain, malaise, vomiting, nausea, diarrhea, and smell and taste disorders) but did not have dyspnea, shortness of breath, or abnormal chest imaging. (3) *Moderate illness*: individuals who indicated evidence of lower respiratory disease during clinical assessment or imaging and oxygen saturation (SpO_2) of ≥94% in room air at sea level. (4) *Severe illness*: individuals who had an SpO_2 of <94% in room air at sea level, a pressure of oxygen to fraction of inspired oxygen (PaO_2/FiO_2) of <300 mm Hg, a respiratory frequency of >30 breaths/min, or lung infiltrates at >50%. (5) *Critical illness*: individuals with septic shock, respiratory failure, and/or multiple organ dysfunction [24].

In the present study, no infected patients with severe or critical diseases were found among the participants. Furthermore, *asymptomatic* and *mild* categories were defined as "mild and no sign" in the data analysis.

Primary outcomes were based on clinical and laboratory examinations, as well as exposure to sunlight; secondary outcomes were related to clinical symptoms. Additionally, demographic evidence (age, sex, marital status, education level, and smoking habits), comorbidities, body mass index (BMI), and taking nutritional supplements were potential confounders.

2.6. Data Sources and Measurements

After the RT-PCR test results were determined, a checklist was given to all infected and potentially non-infected participants so that they could provide information on demographic and anthropometric characteristics, signs and symptoms, current smoking status, and any comorbidities or other conditions that have been linked to the disease (e.g., cardiovascular disease, diabetes mellitus, chronic obstructive pulmonary disease and other lung diseases, cancers, chronic kidney disease, obesity, taking nutritional supplements, and smoking) [25]. In addition, sunlight exposure was quantified through a questionnaire as a proxy measure for vitamin D status [26].

Clinical examinations including respiratory rate (RR), pulse rate (PR), and SpO_2 levels were measured by a pulse oximeter at the time of admission on day 1 (second to seventh days of disease) in the health service center.

Laboratory examination including serum concentrations of a total of 25-hydroxyvitamin D (25(OH)D) and zinc was conducted on the admission day. After informed consent had been obtained, around 5 mL of blood was collected following 8 h of fasting. Biochemical analysis was performed on the serum sample after separation, and the serum concentrations of zinc were measured with a fully automated analyzer (Miura, ISE Co., Italy) using a kit for the quantitative determination of the zinc according to the manufacturer's protocol (PaadCo Co., Iran) following a direct colorimetric method. The reference values used for serum concentration of zinc were 68–107 μg/dL. To assess the whole-body vitamin D status of the participants, serum concentrations of a total of 25(OH)D were measured retrospectively in serum samples collected in gel tubes at the time of admission. Serum concentrations of 25(OH)D were quantified using a commercially available immunoassay (Vitamin D 96 ELISA Kit. Ideal, Ideal Tashkhis Ateieh, Tehran, Iran). The mean inter-assay coefficients of variation (CVs) for the 25(OH)D and zinc concentrations were 8.3% and 5.9%, respectively. Intra-assay CVs were not conducted for 25(OH)D and zinc measurements.

However, all experiments were performed in a clinical laboratory having a quality control certificate from Iran Health Reference Laboratory. The procedure of 25(OH)D measurement in the serum has been illustrated in Video S1 (Supplementary Materials).

We used further stratification for the serum concentrations of 25(OH)D and categorized infected patients and potentially non-infected participants in terms of serum concentrations of 25(OH)D to normal, insufficient, and deficient vitamin D status, so that the cut-off point of 25(OH)D 12–20 ng/mL (30–50 nM) was defined as vitamin D insufficiency, and <12 ng/mL (equivalent to <30 nM) as vitamin D deficiency. Additionally, a cut-off point of >20 ng/mL (>50 nM) was defined as normal. This categorization was according to the criteria of the Institute of Medicine (US) Committee to Review Dietary Reference Intakes for Vitamin D and Calcium Dietary Reference Intakes for Calcium and Vitamin D [27]. This categorization was used to compare the vitamin D status between infected patients and potentially non-infected participants.

The data of clinical symptoms were collected from both asymptomatic and symptomatic COVID-19-infected patients to evaluate the disease progression by recording self-reported health information weekly. The recorded information included the symptoms and pre-existing medical conditions obtained on day 1 at the sampling site and then on days 7, 14, 21, and 28 of the first symptoms observed by telephone contact. It was assumed that individuals with negative RT-PCR results, no clinical signs or symptoms of COVID-19, and no high-risk occupations (e.g., medical staff, taxi drivers) were not infected. Commonly presented clinical symptoms of COVID-19 fell into four categories: (1) general (fatigue, fever, night sweats, asthenia, flushing, chills, hypothermia, runny nose, sore throat), (2) pulmonary (chest pain, shortness of breath, dyspnea, cough), (3) gastrointestinal (anorexia, abdominal cramps, diarrhea, vomiting, nausea, constipation, bloating), and (4) neurologic (headache, muscle pain, joint pain, ear pain, new smell and taste disorders such as anosmia and dysgeusia) [28].

2.7. Statistical Analysis

We matched the data of the infected patients with those of the potentially non-infected individuals of the same sex and age. Testing of data for normal distribution was carried out using the Kolmogorov–Smirnov test. Characteristics of the infected patients and potentially non-infected participants were compared using the χ^2 test for discrete variables and the independent sample *t*-test for continuous variables. A generalized estimating equation (GEE) regression model with a logistic link function and an exchangeable correlation structure for each individual was employed to assess the odds ratio (OR) and 95% confidence interval (95% CI) of the disease symptoms on days 1, 7, 14, 21, and 28 after the onset of the first symptoms. GEE model was restricted to patients infected with SARS-CoV-2. The model was adjusted for potential confounding variables including age, sex, marital status, education levels, and BMI.

All descriptive analyses and the GEE modeling were performed using IBM SPSS Statistics (version 26). In all tests, $p < 0.05$ was considered to indicate statistical significance.

3. Results

3.1. Participants' Characteristics

As illustrated in Figure 1, a total of 1181 potentially eligible clients were admitted to the health centers. Among them, 1169 clients with a confirmed RT-PCR test result (691 clients with positive and 478 clients with negative RT-PCR test results) visited the health service center from 6 June 2020 to 12 August 2020 and their eligibility was confirmed. Following the simple randomization by telephone call and matching in terms of age and sex, a total of 108 individuals (54 infected patients and 54 potentially non-infected participants) contributed to the study and blood sampling. One infected patient was excluded from the study following the diagnosis of pregnancy. Additionally, one potentially non-infected participant was excluded due to unwillingness to continue the study. Ultimately, 53 infected patients (male = 68%; mean age = 41 years) and 53 age- and sex-matched potentially non-infected participants (male = 72%; mean age = 40 years) completed the follow-up and analysis.

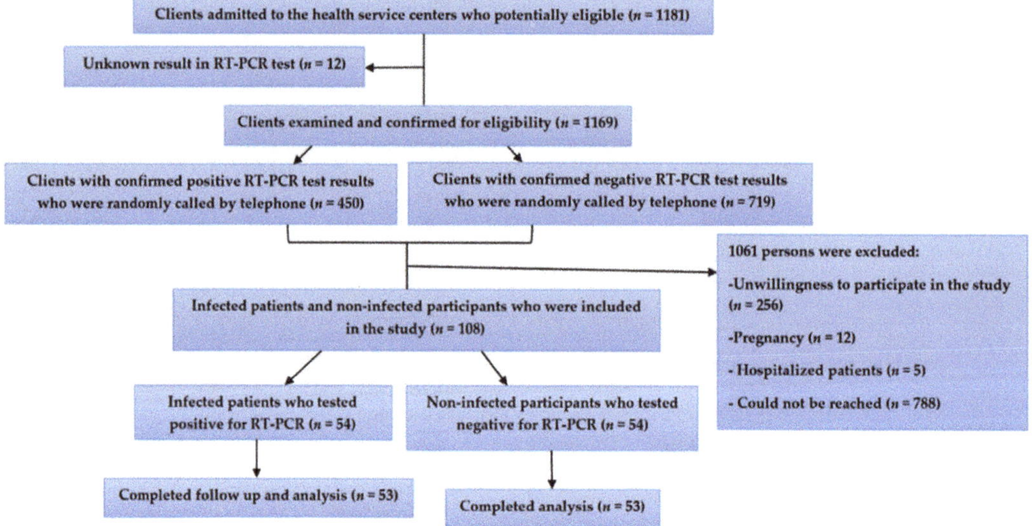

Figure 1. Flow diagram of the study design.

The infected patients' and potentially non-infected participants' characteristics in the study are given in Table 1. There was no significant difference in mean age among infected patients and potentially non-infected participants. Participants were predominantly male and had no significant differences in terms of their marital status, education level, cigarette smoking status, comorbidities, and BMI. Additionally, respiratory rate (RR) was significantly higher in infected patients than in potentially non-infected participants ($p = 0.001$). Moreover, SpO_2 was significantly lower among infected patients than in potentially non-infected participants ($p = 0.03$). Furthermore, 28 (53%) infected patients and 25 (47%) potentially non-infected participants took vitamin D supplements monthly. Additionally, three (6%) infected patients took zinc supplements, whereas no potentially non-infected participants did. However, there were no significant differences between the two study groups in terms of taking vitamin D and zinc supplements (data not shown in the table).

Table 1. Demographic, clinical, comorbidity, and anthropometric characteristics of patients infected with SARS-CoV-2 and matched controls at the second to seventh days of disease [1].

Characteristics	Infected Patients (n = 53)	Potentially Non-Infected Participants (n = 53)	p-Value
Age (year)	41 ± 13	40 ± 14	0.609
Sex			
Male, n (%)	36 (68)	38 (72)	0.672
Married status, n (%)			
Single	12 (23)	12 (23)	0.592
Married	41 (77)	41 (77)	
Education levels, n (%)			
Illiterate	2 (4)	1 (2)	
Under diploma	15 (29)	19 (37)	0.754
Diploma	13 (25)	10 (19)	
College education	22 (42)	22 (42)	
Cigarette smoking, n (%)			
No	40 (76)	45 (85)	0.223
Yes	13 (25)	8 (15)	
RR (number/min)	14 ± 0.2	13 ± 0.3	0.001
PR (number/min)	91 ± 3	87 ± 2	0.271
SpO_2 (%)	97 ± 1.4	97 ± 1.2	0.032
Duration of disease (day) [2]	7 ± 2	-	
Comorbidities, n (%)			
Chronic pulmonary diseases	2 (4)	0 (0)	0.153
Hypertension	10 (19)	5 (9)	0.164
Diabetes mellitus	6 (11)	4 (8)	0.506
Obesity	13 (25)	21 (40)	0.096
Malnutrition	1 (2)	0 (0)	0.315
Cancer	2 (4)	0 (0)	0.153
Liver disease	5 (9)	3 (6)	0.462
Chronic neurological diseases	2 (4)	1 (2)	0.547
Chronic hematologic diseases	2 (4)	0 (0)	0.157
Renal diseases	3 (6)	4 (8)	0.696
Chronic heart disease	4 (8)	2 (4)	0.414
HIV	2 (4)	0 (0)	0.153
Asthma and allergy	6 (11)	5 (9)	0.750
Others [3]	8 (15)	16 (30)	0.063
BMI (Kg/m^2)	27 ± 5	28 ± 4	0.663

[1] Independent sample t-test was conducted to analyze continuous variables, and the results were stated as mean ± standard deviation. Categorical variables were analyzed by chi-square test, and the results were presented as number (%). [2] Duration of disease indicates the number of days since the onset of the patient's first clinical symptoms obtained by asking the infected patients and recording in the questionnaire. [3] Others including autoimmune disease, hemoglobinopathies, migraine, digestive system problems, hypothyroidism, hyperthyroidism, hyperlipidemia, endometrioses, neck, and back disk. BMI, body mass index; RR, respiratory rate; PR, pulse rate; SpO_2, oxygen saturation.

3.2. Vitamin D Status and Sunlight Exposure of Infected Patients and Potentially Non-Infected Participants

The laboratory measurements were generally performed 7 ± 2 days after the RT-PCR test, and a statistically significant difference in days away was not found within the infected patients and potentially non-infected participants.

As represented in Table 2, we did not inspect the statistical significance in either 25(OH)D concentration or vitamin D status category between infected patients (26 ng/mL) compared with the potentially non-infected participants (29 ng/mL). More than a quarter of the potentially non-infected participants (i.e., 14 (26%) individuals) had vitamin D insufficiency (12–20 ng/mL); three (6%) individuals were deficient (<12 ng/mL), and 36 (68%) individuals had normal vitamin D (≤15 nmol/L).

Table 2. Vitamin D status and characteristic items were used to measure individual sunlight exposure among patients infected with SARS-CoV-2 and potentially non-infected participants [1].

Components of Individual UV Exposure and Modifying Factors	Infected Patients (n = 53)	Potentially Non-Infected Participants (n = 53)	p-Value
25(OH)D (ng/mL)	26 ± 17	29 ± 16.	0.424
25(OH)D status, n (%)			
Vitamin D deficiency	10 (19)	3 (6)	
Vitamin D insufficiency	9 (17)	14 (26)	0.086
Normal vitamin D	34 (64)	36 (68)	
Daily sun exposure (minute)	78 ± 104	87 ± 60	0.585
How much time did you spend outdoors between the hours of 9 and 11 a.m.? (hour)	1.3 ±0.9	1.3 ±0.9	0.855
How much time did you spend outdoors between the hours of 11 a.m. and 1 p.m.? (hour)	1.2 ± 0.9	1.1 ± 1	0.810
How much time did you spend outdoors between the hours of 7 and 9 a.m.? (hour)	1.1 ± 1	1.1 ± 1	0.888
How much time did you spend outdoors between the hours of 1 and 3 p.m.? (hour)	0.8 ± 0.9	0.7 ± 0.9	0.594
How much time did you spend outdoors between the hours of 3 and 5 p.m.? (hour)	0.7 ± 0.9	0.6 ± 0.9	0.676
How much time did you spend outdoors between the hours of 5 and 7 p.m.? (hour)	0.8 ± 0.9	1 ± 0.9	0.328
What percent of this time did you spend under shade (e.g., tree or beach shade)? (%)	78 ± 22	63 ± 32	0.006
What percent of time did you wear a brimmed hat? (%)	18 ± 37	11 ± 31	0.330
What percent of time did you wear long sleeves? Long pants? (%)	84 ± 34	76 ± 39	0.264
What percent of time did you wear sunscreen? (%)	9 ± 26	2 ± 10	0.090

[1] χ^2 test for discrete and the independent sample t-test for continuous variables were applied to analyze data. The results have been shown with mean ± standard deviation for continuous and number (%) for discrete data. UV, ultraviolet.

The comparison of the 25(OH)D concentration between infected patients with *asymptomatic and mild illness* and patients with *moderate illness* is illustrated in Figure 2. We observed a marginally significant difference in terms of 25(OH)D concentration between patients with *moderate illness* (19 ± 12 ng/mL) compared to patients with *asymptomatic and mild illness* (29 ± 18 ng/mL) (p = 0.054).

Typical questions used to assess sunlight exposure are also listed in Table 2. The comparison of the components of sunlight exposure between infected patients and potentially non-infected participants revealed that the percentage of time spent in the shade was significantly higher in patients than in potentially non-infected participants. However, the other components did not show any significant differences between the two study groups.

3.3. Zinc Status of the Infected Patients and Potentially Non-Infected Participants

As shown in Figure 3, infected patients showed a significantly lower serum concentration of zinc than potentially non-infected participants (101 ± 18 μg/dL in infected patients vs. 114 ± 13 μg/dL in potentially non-infected participants) (p = 0.013).

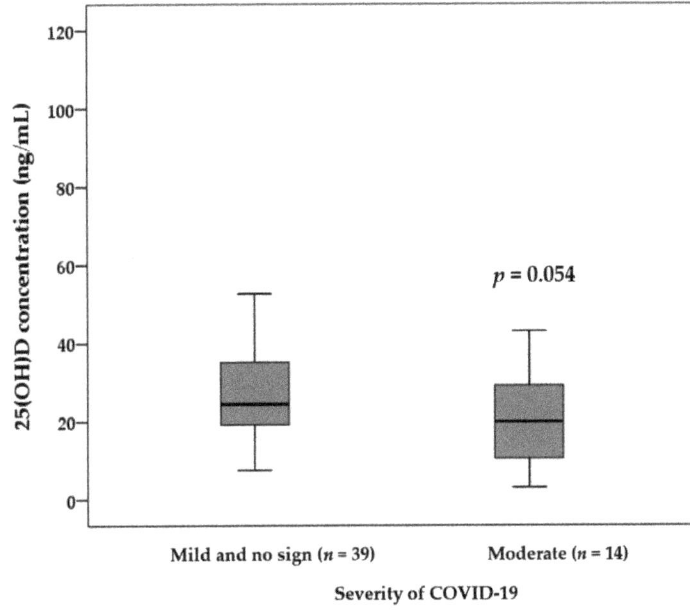

Figure 2. The comparison of the serum concentration of 25(OH)D between infected patients with different severity of COVID-19. An independent sample *t*-test was applied to analyze data. Patients with moderate COVID-19 showed a trend noted for a lower serum concentration of 25(OH)D than *mild and no sign illness* patients.

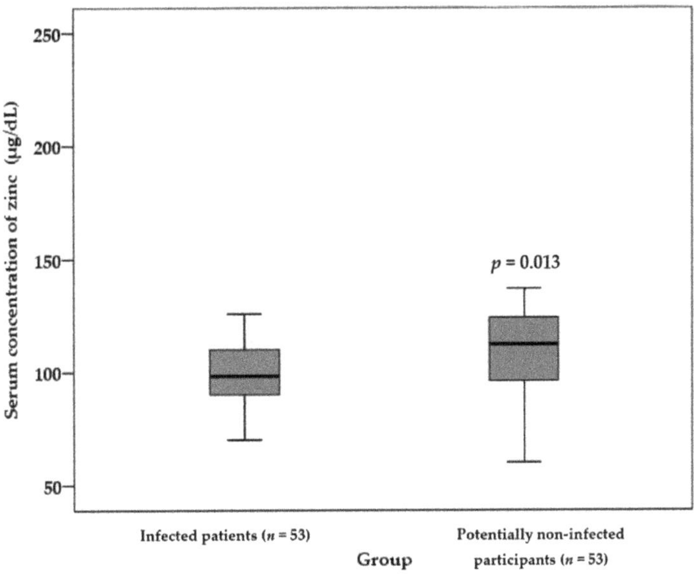

Figure 3. The comparison of the serum concentration of zinc between infected patients and potentially non-infected participants. An independent sample *t*-test was applied to analyze data. The serum concentration of zinc was significantly lower among infected patients than potentially non-infected participants.

The comparison of the serum concentration of zinc between infected patients with *asymptomatic and mild illness* and patients with *moderate illness* is illustrated in Figure 4. The results demonstrated a lower serum concentration of zinc in patients with moderate illness (97 ± 17 μg/dL) compared to those with *asymptomatic and mild illness* (102 ± 18 μg/dL). However, this difference was not statistically significant (p = 0.412).

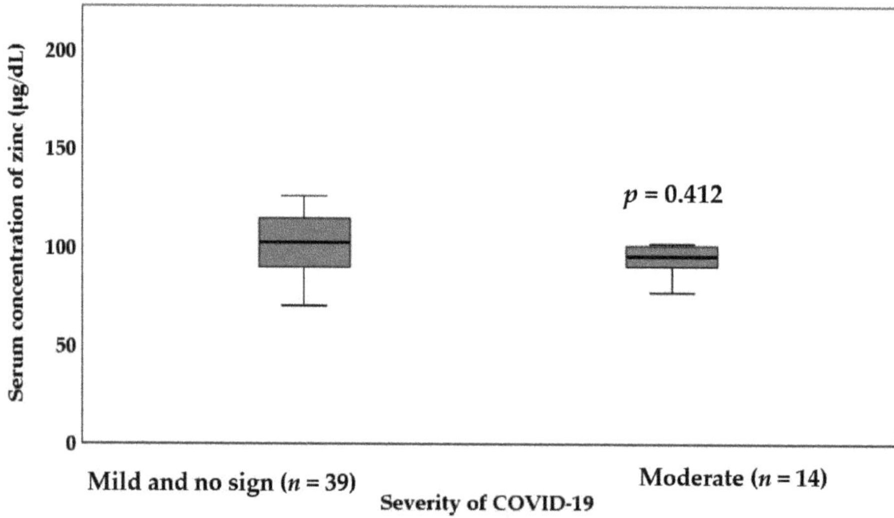

Figure 4. The comparison of the serum concentration of zinc between infected patients with different severity of COVID-19. An independent sample t-test was applied to analyze data. We did not find any significant difference between infected patients with *asymptomatic and mild illness* and patients with *moderate illness* in terms of serum concentration of zinc.

3.4. Symptom Follow-Up, Outcomes, and Associations with Demographic, BMI and Laboratory Parameters

Table 3 shows the changes in clinical symptoms among infected patients from day 1 to 28 of the disease. The most common general, pulmonary, gastrointestinal, and neurologic symptoms at days 1 and 7 were fatigue, cough, anorexia, and smell disorder, respectively. Additionally, fatigue, cough, bloating, and smell disorder were the most common symptoms at days 14 and 21 of the disease. In addition, fatigue and sore throat, cough, constipation, and muscle pain were among the most common general, pulmonary, gastrointestinal, and neurologic symptoms on day 28 of the disease. Finally, all clinical symptoms except constipation and hypothermia showed a decreasing trend from days 1 to 28 of the disease.

Table 4 shows the results of the GEE model for the longitudinal relationship between vitamin D and zinc status at the second to seventh days of disease and clinical symptoms adjusted for age, sex, marital status, education levels, and BMI among infected patients with COVID-19.

The results revealed that the odds ratio of general symptoms of COVID-19 was three times higher among males than females (OR = 3.06; 95% CI, 1.13–8.33; p = 0.03). However, the odds ratio of neurologic symptoms in males was 0.41 times that of females (OR = 0.41; 95% CI, 0.17–0.98; p = 0.045).

Furthermore, the patients who had normal vitamin D status were less likely to experience general symptoms and pulmonary symptoms than patients with vitamin D deficiency with ORs of 0.10 (95% CI, 0.04–0.24; $p \leq 0.001$) and 0.27 (95% CI, 0.07–0.99; p = 0.05), respectively.

Table 3. Changes in clinical symptoms of COVID-19 among infected patients from day 1 to 28 of follow-up [1].

Clinical Symptoms / Days of Follow-Up	Day 1	Day 7	Day 14	Day 21	Day 28
General					
Fatigue	32 (60)	16 (30)	11 (21)	7 (13)	4 (8)
Fever	12 (23)	3 (6)	1 (2)	1 (2)	0(0)
Night sweats	25 (47)	10 (19)	8 (15)	2 (4)	0 (0)
Flushing	1 (2)	1 (2)	1 (2)	1 (2)	1 (2)
Chills	5 (9)	6 (11)	3 (6)	1 (2)	1 (2)
Hypothermia	0(0)	2 (4)	2 (4)	0 (0)	1 (2)
Runny nose	4 (8)	2 (4)	2 (4)	2 (4)	1 (2)
Sore throat	14 (26)	7 (13)	4 (8)	1 (2)	4 (8)
Pulmonary					
Chest pain	5 (9)	8 (15)	4 (8)	4 (8)	4 (8)
Shortness of breath	14 (26)	8 (15)	6 (11)	4 (8)	1 (2)
Cough	23 (43)	18 (34)	15 (28)	9 (17)	5 (10)
Gastrointestinal					
Anorexia	24 (45)	12 (23)	3 (6)	2 (4)	4 (8)
Abdominal cramps	10 (19)	9 (17)	4 (8)	3 (6)	1 (2)
Diarrhea	19 (36)	8 (15)	(0)	5 (9)	1 (2)
Vomiting	3 (6)	1 (2)	1 (2)	1 (2)	1 (2)
Nausea	11 (21)	3 (6)	2 (4)	1 (2)	1 (2)
Constipation	5 (9)	4 (8)	3 (6)	2 (4)	6 (11)
Bloating	8 (15)	7 (13)	6 (11)	3 (6)	1 (2)
Neurologic					
Headache	18 (34)	9 (17)	4 (8)	3 (6)	1 (2)
Muscle pain	11 (21)	7 (13)	6 (11)	3 (6)	5 (10)
Joint pain	13 (25)	4 (8)	4 (8)	2 (4)	1 (2)
Ear pain	5 (9)	5 (9)	3 (6)	1 (2)	2 (4)
Smell disorders	33 (62)	15 (28)	9 (17)	8 (15)	5 (9)
Taste disorder	26 (49)	11 (21)	5 (9)	2 (4)	3 (6)

[1] Descriptive statistic were conducted to analyze data. Data were shown as *n (%)*.

Table 4. Estimates of observed symptom progression of COVID-19 and the association with demographic, BMI, and laboratory parameters among infected patients (n = 53) [1].

Parameters	Symptom Categories			
	General	Pulmonary	Gastrointestinal	Neurologic
Age (year)	1.02 (0.99–1.05)	1.05 (1.00–1.11)	0.97 (0.92–1.02)	0.97 (0.93–1.02)
Sex				
Female	1 (ref)	1 (ref)	1 (ref)	1 (ref)
Male	3.06 (1.13–8.33) [2]	1.11 (0.33–3.68)	1.71 (0.62–4.75)	0.41 (0.17–0.98) [2]
Married status				
Single	1 (ref)	1 (ref)	1 (ref)	1 (ref)
Married	0.91 (0.23–3.54)	1.32 (0.39–4.42)	2.56 (0.87–7.32)	1.97 (0.60–6.69)
Education levels				
Illiterate	1 (ref)	1 (ref)	1 (ref)	1 (ref)
Under diploma	0.70 (0.08–5.96)	18.45 (0.73–463.77)	0.28 (0.03–3.08)	0.10 (0.01–1.38)
Diploma	0.54 (0.06–5.10)	7.45 (0.21–212.65)	0.29 (0.02–4.12)	0.26 (0.02–3.35)
College education	1.24 (0.17–9.09)	12.80 (0.46–354.25)	0.64 (0.06–7.05)	0.38 (0.03–4.69)
Category of vitamin D status				
Deficiency	1 (ref)	1 (ref)	1 (ref)	1 (ref)
Insufficiency	0.19 (0.06–0.65) [4]	0.63 (0.10–3.87)	0.52 (0.10–2.81)	1.09 (0.32–3.72)
Normal	0.10 (0.04–0.24) [3]	0.27 (0.07–0.99) [2]	0.39 (0.12–1.21)	0.50 (0.21–1.19)
Serum concentration of zinc (μg/dL)	0.98 (0.96–1.01)	1.00 (0.98–1.02)	0.99 (0.97–1.01)	1.02 (0.99–1.05)
BMI (Kg/m^2)	1.06 (0.97–1.17)	0.98 (0.90–1.07)	1.04 (0.95–1.15)	1.08 (0.97–1.19)
Time (Day)	0.88 (0.84–0.93) [4]	0.91 (0.88–0.95) [4]	0.91 (0.87–0.94) [4]	0.89 (0.86–0.93) [4]

[1] Odds of common clinical signs and symptoms of COVID-19 followed in days 1 to 28 of disease (95% CI). General estimation equation (GEE) was applied to analyze data. [2] $p < 0.05$, [3] $p < 0.01$, [4] $p < 0.001$.

Additionally, the odds ratio of general symptoms of COVID-19 in patients with insufficient vitamin D was 0.2 times that of patients with vitamin D deficiency (OR = 0.19; 95% CI, 0.06–0.65; $p = 0.008$).

In the present study, all symptoms showed a decreasing trend over time. However, the marital status, education, age, BMI, and serum concentration of zinc variables were not significantly associated with clinical symptoms ($p \geq 0.05$).

4. Discussion

This is likely the first study to characterize the association of vitamin D and zinc status with the severity and progression of symptoms of COVID-19. In this study, 53 outpatients with laboratory-confirmed COVID-19 disease and 53 potentially non-infected participants for whom the disease was excluded by RT-PCR were included to compare the vitamin D and zinc status in the body. In addition, the associations between the serum concentrations of 25(OH)D and zinc at the second to seventh days of disease and the progression of clinical symptoms among infected patients were evaluated by the GEE model adjusted for age, sex, marital status, education levels, and BMI. Our findings showed that in terms of vitamin D status, although the serum concentrations of 25(OH)D of infected patients and potentially non-infected individuals were statistically similar, a trend was noted for a lower serum concentration of 25(OH)D in moderate than asymptomatic or mild illness patients.

One caveat to consider is that the patients with normal vitamin D status were less likely to experience general and pulmonary symptoms than patients with vitamin D deficiency. Additionally, patients with inadequate vitamin D status were less likely to report general symptoms of COVID-19 than patients with vitamin D deficiency. In other words, a normal vitamin D status at the second to seventh days of disease reduced the odds of general and pulmonary symptoms during the disease. Based on the results of the comparison between infected patients and potentially non-infected participants in terms of vitamin D status, it can be inferred that vitamin D status affects the severity of COVID-19 and the progression of symptoms during the clinical course of the disease.

Similarly, a recent study determined that the diagnosis of vitamin D deficiency could be useful in evaluating COVID-19 patients' potential risk of disease development and severity [29]. A cross-sectional study in Qom City, Iran, in patients with COVID-19 reported a significant association between a hospital stay and a lower serum concentration of vitamin D. However, the correlation between vitamin D status and death rate (or the time interval to obtain a normal oxygen level) was not significant [30]. This may be due to the role of vitamin D in the immune system, from its receptors on the majority of immune cells to increase the production of anti-inflammatory cytokines versus pro-inflammatory cytokines or even the production of an antimicrobial peptide against enwrapped coronaviruses. Additionally, vitamin D upregulates angiotensin-converting enzyme 2 (ACE2) expression, which in the lungs has shown a protective effect against acute lung injury [31].

In addition, a case–control study confirmed that the serum concentration of vitamin D deficiency is associated with more severe lung involvement, longer disease duration, and the severity of radiologic pulmonary involvement as evaluated by computed tomography. In particular, serum concentration of 25(OH)D were significantly lower in COVID-19 patients with either multiple lung consolidations or diffuse/severe interstitial lung involvement than in those with mild involvement [32].

Moreover, a systematic review and meta-analysis concluded that a lower serum concentration of 25(OH)D accompanies severe presentation and mortality relating to COVID-19 disease [33]. A recent study described the relationship between vitamin D status and complications and mortality from COVID-19 in 46 countries. The results showed that the serum concentration of 25(OH)D in each country had a significant relationship with the number of deaths, the risk of being infected with SARS-CoV-2, and the severity of the disease [34]. Additionally, a short report in 20 European countries indicated that the serum concentration of 25(OH)D was also extremely low in elderly populations, especially in Spain, Italy, and Switzerland. It was also the most vulnerable population group in terms of COVID-19. This study concluded that vitamin D supplementation is recommended to protect against SARS-CoV-2 infection [12].

Although our results showed a trend for a lower serum concentration of 25(OH)D among moderate than asymptomatic or mild illness patients, these findings conflict with those of some previous studies showing strong protective effects of vitamin D. Moreover, a cross-sectional study conducted on biobank samples of participants from England, Scotland, and Wales showed that the serum concentration of 25(OH)D was associated with COVID-19 risk; however, this association disappeared after controlling for confounding factors [35]. These controversies suggest that further studies are needed to evaluate the protective effects of normal vitamin D status in COVID-19 patients.

Moreover, the evaluation of sunlight exposure and modifying factors among infected patients and potentially non-infected participants showed that the percentage of time spent in the shade was significantly greater in patients than in potentially non-infected participants. However, daily sun exposure, time spent outdoors, and time spent wearing a brimmed hat, long sleeves, and sunscreen did not indicate any significant differences between the two study groups. This result could support our finding that there is no significant difference between the two groups in terms of the serum concentration of vitamin D.

The protective effect of the serum concentration of 25(OH)D on the severity of COVID-19 and the progression of symptoms during the clinical course of the disease might underlie some mechanisms: vitamin D has beneficial effects on protective immunity in part due to its effects on the innate immune system and β-cell function [7,9]. Immune cells express vitamin D receptors (VDRs). It is known that macrophages identify lipopolysaccharide (LPS), a surrogate for bacterial infection, through Toll-like receptors (TLRs). TLR binding increases the expression of both VDRs and 1-α-hydroxylase [36,37]. This results in the binding of the 1,25 D-VDR-RXR heterodimer to vitamin D response elements (VDREs), leading to the translocation of the complex into the cell nucleus, where it modifies the expression of hundreds of genes, including those involved in cytokine production [38]. The complex also induces the production of antimicrobial peptides, including cathelicidin and beta-defensin 4 [39]. These peptides co-localize within phagosomes with injected bacteria, as they disturb bacterial cell membranes and exhibit strong anti-microbacterial activity. The transcription of cathelicidin is dependent on sufficient 25(OH)D [37].

The administration of vitamin D in a dose of 5000 IU/kg has been shown to reduce the replication of rotavirus both in vitro and in vivo [40]. Vitamin D administration can also reduce the production of T helper type 1 (Th1) cell cytokines, such as interferon-γ and tumor necrosis factor-α (TNF-α), and the expression of pro-inflammatory cytokines by macrophages. It can also increase anti-inflammatory cytokine levels [41,42]. The induction of cytokine storms is also reduced by vitamin D. However, vitamin D supplementation did seem to non-significantly increase the risk of in-hospital mortality among COVID-19 patients addressing the maintenance of serum concentrations of vitamin D and zinc in a normal range to prevent the incidence or progression of clinical symptoms among COVID-19 patients [43].

The innate immune system generates both pro-inflammatory and anti-inflammatory cytokines in patients suffering from COVID-19 [1]. Binding to dipeptidyl peptidase-4 receptor (DPP-4/CD26) is one of the molecular virulence mechanisms employed by a coronavirus. It has been demonstrated recently that human DPP-4/CD26 interacts with the S1 domain of the SARS-CoV-2 spike glycoprotein [44]. In this context, vitamin D deficiency has been shown to remarkably reduce the expression of the DPP4/CD26 receptor in vivo [45]. Vitamin D is a strong inducer of autophagy [46] and inhibits HIV replication in macrophages via vitamin D-mediated induction of cathelicidin, perhaps by enhancing autophagy and phagosomal maturation [47].

In the present study, patients infected with SARS-CoV-2 had significantly lower serum concentrations of zinc than potentially non-infected individuals. However, the serum concentration of zinc was not different among COVID-19 patients with *mild or asymptomatic illnesses* compared to participants who had *moderate* COVID-19. Moreover, the serum concentration of zinc at the second to seventh days of disease showed no significant association with common clinical symptoms of COVID-19 in four categories during the period of day 1 to day 28 after the disease onset.

Our finding is consistent with that of a prospective observational study conducted on COVID-19 inpatients at the time of hospitalization, which reported that the serum concentration of zinc was significantly lower in patients compared to healthy controls [48]. Additionally, a recent study in Turkey reported that in the first trimester of pregnancy, the serum concentration of zinc was significantly lower in pregnant women with COVID-19 compared to controls [49]. Moreover, a single-center study carried out on hospitalized patients with COVID-19 found that the serum concentration of zinc was significantly lower in patients who died than those who were admitted to an intensive care unit (ICU) or non-ICU and survived. However, contrary to our finding, the serum concentration of zinc at the time of admission could affect clinical outcomes in COVID-19 patients [50].

Additionally, as mentioned in a recently published review study, zinc may have beneficial effects including a decreased susceptibility to infection in the current and future pandemics [51]. In contrast to our study results, a review study revealed that a pre-existing severe zinc deficiency predisposes patients to a stronger progression of SARS-CoV-2 infections, and even a mild zinc deficiency should be corrected to prevent a more severe viral infection [5]. However, in the present study, no significant difference was observed between the serum concentration of zinc and the severity of COVID-19 disease, which may be due to insufficient sample size.

In terms of the effect of zinc during the COVID-19 pandemic, it is believed that zinc is a potential supportive treatment in therapy against COVID-19 disease due to its positive effects on the immune response [20].

Previous studies strongly revealed that zinc status is a critical factor that can influence antiviral immunity [52]. A meta-analysis of mostly high-quality studies by Aggarwal et al. [53] showed that the risk of lower respiratory tract infections or pneumonia and diarrhea or dysentery could be reduced in children after zinc administration. Additionally, a retrospective review reported that zinc supplementation at a total dosage of 2–2.5 mg/kg/day improved COVID-19 symptoms after 7 days of treatment. However, this study had some limitations, including the absence of blinding and a control group [54]. Moreover, an uncontrolled case series reported that the administration of a high dose of zinc salt oral lozenges for four consecutive outpatients with clinical characteristics of and/or laboratory-confirmed COVID-19 led to a significant improvement in symptomatic COVID-19 measures after one day of high-dose therapy, suggesting that zinc therapy played a role in clinical recovery [55].

It is thought that the supportive effects of zinc in patients with COVID-19 exist because of its immunomodulatory effects and several direct and indirect effects against a wide variety of viral species, predominantly RNA viruses [56,57]. It has been previously shown that the zinc cation (especially in combination with ionophore pyrithione) can inhibit the RNA polymerase of the SARS-CoV-2 virus, and this evidence makes zinc a potential therapeutic agent for patients with COVID-19 in combination with antiviral medications [57–59]. Accordingly, zinc can inhibit the elongation step of RNA transcription [57]. Zinc can induce its antiviral effects by suppressing RNA-dependent RNA polymerase (RDRP) and blocking the further replication of viral RNA as demonstrated for SARS-CoV-1 [60]. In addition, there is some evidence that suggests zinc can reduce ACE2 activity [31], which is the receptor for SARS-CoV-2 [61]. The modulation of antiviral immunity by zinc can also limit SARS-CoV-2 infection through the upregulation of interferon-alpha (IFN-α) production through the Janus kinase signal transducer and activator of transcription-1 (JAK/STAT1) signaling pathway in leukocytes [62] and increasing its antiviral activity [63]. In addition to its immunomodulatory effects, zinc, as an antiviral agent, exerts its beneficial roles and potential applications in the management of COVID-19, possibly by the enhancement of total antioxidant capacity [64].

Moreover, zinc has anti-inflammatory effects by blocking the inhibitor of nuclear factor kappa B (IκB) kinase (IKK) activity and subsequent nuclear factor kappa B (NF-κB) signaling, resulting in the downregulation of pro-inflammatory cytokine production [65,66]. On the other hand, a viral infection-related inflammatory response resulting in the overproduction of pro-inflammatory cytokines and cytokine storm is known to play a significant role in COVID-19 pathogenesis and patient outcomes [67]. Additionally, the coexistence of noncommunicable chronic diseases (NCDs) in COVID-19 patients may strengthen the inflammatory pathology and increase the risk for adverse outcomes and mortality [68]. In turn, inflammation can be under- or overestimated micronutrient deficiencies. Besides, zinc is a negative acute-phase reactant; therefore, inflammation accompanies serum hypozincemia [15,69]. Accordingly, the adjustment of zinc concentrations for inflammation is necessary when evaluating the zinc status among the population [69,70]. Several methods have been suggested to adjust for the effect of inflammation on the zinc status; however, to our knowledge, none have been examined in adults in whom chronic inflammation is common [70]. Additionally, there is no established agreement on how to control for the effect of inflammation on the serum concentration of zinc, which has a consequence for precise estimates of zinc status at the population level [69].

It is necessary to mention that our study covered a wide age range of participants, from children to the elderly population, who are among the high-risk groups for zinc deficiency. In addition, COVID-19 symptoms may exacerbate zinc deficiency, which is a threat to current high-risk groups [51]. Therefore, the cross-sectional nature of this study does not allow us to determine the causality relationship between zinc status and the progression of COVID-19 disease.

Adequate dietary intake of zinc and vitamin D could be considered as a possible solution to compensate for the low status of vitamin D and zinc, which to some extent may be effective on immunocompetence. However, vulnerable sections of populations may need supplements besides dietary advice to secure adequacy for these nutrients. In the case of low vitamin D status (<50 nmol/L), vitamin D supplementation (40 μg D3/day) is considered as an approach for the prevention of a destructive course of the inflammation induced by COVID-19. Moreover, a dietary zinc intake $\leq$ 25 mg/day was recommended as a preventive dose for COVID-19 on a long-term basis [21]. In addition, foods rich in zinc and zinc supplements could serve as adjuvants in combination with vaccines for the treatment of COVID-19 [64].

Our study has several limitations. First, only 53 laboratory-confirmed COVID-19 outpatients and 53 potentially non-infected participants were involved. Consequently, the small sample size has led to a cautious interpretation of the results. Second, as many of these findings are non-specific, they might overlap with other potentially coexisting deficiencies and illnesses. Our limited nutritional assessment suggests that other nutritional deficiencies might also affect the clinical signs and symptoms of COVID-19. Additional research in this area is needed. Third, recall bias is possible because data of clinical symptoms were self-reported. Fourth, the present study had a longitudinal component where the symptom progression of COVID-19 was observed, but there were no observations before the positive RT-PCR result.

However, the strengths of our study were the longitudinal nature and the follow-ups with the infected participants for one month, which helped us determine the relationship between the nutritional status of vitamin D and zinc at the second to seventh days of disease and the progression of clinical symptoms and recovery time. Additionally, we observed the differences between study groups after age and sex matching, as there is the belief that the difference between infected patients and potentially non-infected individuals might be affected by sex and age structure.

5. Conclusions

The results of the present study underline that although serum concentrations of 25(OH)D in infected patients and potentially non-infected participants were statistically similar, the role of vitamin D in the severity of COVID-19 was marginally significant. In addition, the severity of vitamin D deficiency is associated with the progression of general and pulmonary symptoms, indicating the importance of the evaluation of the vitamin D status at the onset of the disease as a relatively easy option to predict disease severity and the progression of COVID-19 symptoms.

In terms of the zinc status, the results of the present study underline that patients with COVID-19 can have a lower serum concentration of zinc. However, the serum concentration of zinc was not different among COVID-19 patients with *mild or asymptomatic illness* when compared to participants who had *moderate* COVID-19. Moreover, serum concentrations of zinc at the second to seventh days of disease were not associated with the progression of symptoms among the COVID-19 patients. In other words, the serum concentration of zinc of the outpatients might not affect disease severity or the progression of symptoms.

Accordingly, serum concentrations of 25(OH)D and zinc should be examined in all inpatients and outpatients with COVID-19 and at different stages of the disease to maintain or promptly increase concentrations of 25(OH)D and zinc in the optimal range. Further studies are needed to confirm our findings.

Supplementary Materials: The following are available online at https://zenodo.org/record/5266352, Video S1: The procedure of 25(OH)D measurements in the serum.

Author Contributions: Conceptualization, S.G. and M.N.; methodology, M.A.; formal analysis, F.M. and M.S.; investigation, S.M.; resources, M.P.; writing—original draft preparation, M.N.; writing—review and editing, S.G., R.B., M.S., K.S. and D.A.-L.; supervision, M.N. and S.M.; funding acquisition, M.N. All authors have read and agreed to the published version of the manuscript.

Funding: This research was funded by the Vice-Chancellor for research, ABADAN UNIVERSITY OF MEDICAL SCIENCES, ABADAN, IRAN, grant number 99U-826.

Institutional Review Board Statement: This study was conducted according to the guidelines of the Declaration of Helsinki, and approved by the Ethics Committee of ABADAN UNIVERSITY OF MEDICAL SCIENCES, ABADAN, IRAN (protocol code IR.ABADANUMS.REC.1399.073 on 22 August 2020).

Informed Consent Statement: Informed consent was obtained from all participants involved in the study.

Data Availability Statement: The data presented in this study are available on request from the corresponding author.

Acknowledgments: The authors would like to thank the personnel of Imam-Khomeini 16-hour corona health service centers for providing the COVID-19 patient and potentially non-infected samples. We also appreciate the Salamat Laboratory, Abadan, Iran, for analyzing the laboratory tests. Finally, the authors acknowledge the COVID-19 patients and potentially non-infected participants who were involved with the study.

Conflicts of Interest: The authors declare no conflict of interest. The funders had no role in the design of the study; in the collection, analyses, or interpretation of data; in the writing of the manuscript, or in the decision to publish the results.

References

1. Huang, C.; Wang, Y.; Li, X.; Ren, L.; Zhao, J.; Hu, Y.; Zhang, L.; Fan, G.; Xu, J.; Gu, X.; et al. Clinical features of patients infected with 2019 novel coronavirus in Wuhan, China. *Lancet* **2020**, *395*, 497–506. [CrossRef]
2. Raoofi, A.; Takian, A.; Sari, A.A.; Olyaeemanesh, A.; Haghighi, H.; Aarabi, M. COVID-19 Pandemic and Comparative Health Policy Learning in Iran. *Arch. Iran. Med.* **2020**, *23*, 220–234. [CrossRef] [PubMed]
3. Guan, W.-J.; Liang, W.-H.; Zhao, Y.; Liang, H.-R.; Chen, Z.-S.; Li, Y.-M.; Liu, X.-Q.; Chen, R.-C.; Tang, C.-L.; Wang, T.; et al. Comorbidity and its impact on 1590 patients with COVID-19 in China: A nationwide analysis. *Eur. Respir. J.* **2020**, *55*, 2000547. [CrossRef] [PubMed]
4. Ugbolue, U.; Duclos, M.; Urzeala, C.; Berthon, M.; Kulik, K.; Bota, A.; Thivel, D.; Bagheri, R.; Gu, Y.; Baker, J.; et al. An Assessment of the Novel COVISTRESS Questionnaire: COVID-19 Impact on Physical Activity, Sedentary Action and Psychological Emotion. *J. Clin. Med.* **2020**, *9*, 3352. [CrossRef] [PubMed]

5. Wessels, I.; Rolles, B.; Slusarenko, A.J.; Rink, L. Zinc deficiency as a possible risk factor for increased susceptibility and severe progression of Corona Virus Disease 19. *Br. J. Nutr.* **2021**, 1–19. [CrossRef] [PubMed]
6. Im, J.H.; Je, Y.S.; Baek, J.; Chung, M.-H.; Kwon, H.Y.; Lee, J.-S. Nutritional status of patients with COVID-19. *Int. J. Infect. Dis.* **2020**, *100*, 390–393. [CrossRef]
7. Ashtary-Larky, D.; Kheirollah, A.; Bagheri, R.; Ghaffari, M.A.; Mard, S.A.; Hashemi, S.J.; Mir, I.; Wong, A. A single injection of vitamin D3 improves insulin sensitivity and β-cell function but not muscle damage or the inflammatory and cardiovascular responses to an acute bout of resistance exercise in vitamin D-deficient resistance-trained males. *Br. J. Nutr.* **2019**, *123*, 394–401. [CrossRef]
8. Tabrizi, R.; Moosazadeh, M.; Akbari, M.; Dabbaghmanesh, M.H.; Mohamadkhani, M.; Asemi, Z.; Heydari, S.T.; Akbari, M.; Lankarani, K.B. High Prevalence of Vitamin D Deficiency among Iranian Population: A Systematic Review and Meta-Analysis. *Iran. J. Med. Sci.* **2018**, *43*, 125–139.
9. Aranow, C. Vitamin D and the Immune System. *J. Investig. Med.* **2011**, *59*, 881–886. [CrossRef]
10. Ginde, A.A.; Mansbach, J.M.; Camargo, C.A. Association Between Serum 25-Hydroxyvitamin D Level and Upper Respiratory Tract Infection in the Third National Health and Nutrition Examination Survey. *Arch. Intern. Med.* **2009**, *169*, 384–390. [CrossRef]
11. Cannell, J.J.; Vieth, R.; Umhau, J.C.; Holick, M.F.; Grant, W.B.; Madronich, S.; Garland, C.F.; Giovannucci, E. Epidemic influenza and vitamin D. *Epidemiol. Infect.* **2006**, *134*, 1129–1140. [CrossRef]
12. Ilie, P.C.; Stefanescu, S.; Smith, L. The role of vitamin D in the prevention of coronavirus disease 2019 infection and mortality. *Aging Clin. Exp. Res.* **2020**, *32*, 1195–1198. [CrossRef]
13. Ebadi, M.; Montano-Loza, A.J. Perspective: Improving vitamin D status in the management of COVID-19. *Eur. J. Clin. Nutr.* **2020**, *74*, 856–859. [CrossRef]
14. Prasad, A.S. Discovery of Zinc for Human Health and Biomarkers of Zinc deficiency. In *Molecular, Genetic, and Nutritional Aspects of Major and Trace Minerals*; Elsevier: Amsterdam, The Netherlands, 2017; pp. 241–260.
15. Wessels, I.; Maywald, M.; Rink, L. Zinc as a Gatekeeper of Immune Function. *Nutrients* **2017**, *9*, 1286. [CrossRef]
16. Maywald, M.; Wessels, I.; Rink, L. Zinc Signals and Immunity. *Int. J. Mol. Sci.* **2017**, *18*, 2222. [CrossRef] [PubMed]
17. Haase, H.; Rink, L. Multiple impacts of zinc on immune function. *Metallomics* **2014**, *6*, 1175–1180. [CrossRef] [PubMed]
18. Walker, C.L.F.; Rudan, I.; Liu, L.; Nair, H.; Theodoratou, E.; Bhutta, Z.; O'Brien, K.; Campbell, H.; Black, R. Global burden of childhood pneumonia and diarrhoea. *Lancet* **2013**, *381*, 1405–1416. [CrossRef]
19. Aftanas, L.; Eyu, B.; Varenik, V.; Grabeklis, A.; Kiselev, M.; Lakarova, E.; Nechiporenko, S.P.; Nikolaev, V.A.; Skalny, A.V.; Skalnaya, M.G. Element status of population of Central Federal Region. In *Element Status of Population of Russia. Part II.*; Skalny, A.V., Kiselev, M.F., Eds.; ELBI-SPb: Saint Petersburg, Russia, 2011; p. 430.
20. Zhang, L.; Liu, Y. Potential interventions for novel coronavirus in China: A systematic review. *J. Med. Virol.* **2020**, *92*, 479–490. [CrossRef]
21. Alexander, J.; Tinkov, A.; Strand, T.A.; Alehagen, U.; Skalny, A.; Aaseth, J. Early Nutritional Interventions with Zinc, Selenium and Vitamin D for Raising Anti-Viral Resistance Against Progressive COVID-19. *Nutrients* **2020**, *12*, 2358. [CrossRef]
22. Galmés, S.; Serra, F.; Palou, A. Current State of Evidence: Influence of Nutritional and Nutrigenetic Factors on Immunity in the COVID-19 Pandemic Framework. *Nutrients* **2020**, *12*, 2738. [CrossRef]
23. Gay, J.D.L.; Grimes, J.D. Idiopathic hypoparathyroidism with impaired vitamin B12 absorption and neuropathy. *Can. Med. Assoc. J.* **1972**, *107*, 54–58.
24. CDC Coronavirus Disease 2019 (COVID-19). Available online: https://www.cdc.gov/coronavirus/2019-ncov/need-extra-precautions/people-with-medical-conditions.html (accessed on 18 September 2020).
25. COVID-19: Clinical Features. Available online: https://www.uptodate.com/contents/covid-19-clinical-features#H2858229650 (accessed on 10 June 2021).
26. McCarty, C. Sunlight exposure assessment: Can we accurately assess vitamin D exposure from sunlight questionnaires? *Am. J. Clin. Nutr.* **2008**, *87*, 1097S–1101S. [CrossRef] [PubMed]
27. Ross, A.C.; Taylor, C.L.; Yaktine, A.L.; Del Valle, H.B. *Institute of Medicine (US) Committee to Review Dietary Reference Intakes for Vitamin D and Calcium Dietary Reference Intakes for Calcium and Vitamin D*; National Academies Press: Washington, DC, USA, 2011.
28. Mehra, P.; Wolford, L.M. Serum nutrient deficiencies in the patient with complex temporomandibular joint problems. *Bayl. Univ. Med. Cent. Proc.* **2008**, *21*, 243–247. [CrossRef] [PubMed]
29. Munshi, R.; Hussein, M.H.; Toraih, E.A.; Elshazli, R.M.; Jardak, C.; Sultana, N.; Youssef, M.R.; Omar, M.; Attia, A.S.; Fawzy, M.S.; et al. Vitamin D insufficiency as a potential culprit in critical COVID-19 patients. *J. Med. Virol.* **2020**, *93*, 733–740. [CrossRef]
30. Nasiri, M.; Khodadadi, J.; Molaei, S. Does vitamin D serum level affect prognosis of COVID-19 patients? *Int. J. Infect. Dis.* **2021**, *107*, 264–267. [CrossRef]
31. Zhang, H.; Penninger, J.M.; Li, Y.; Zhong, N.; Slutsky, A.S. Angiotensin-converting enzyme 2 (ACE2) as a SARS-CoV-2 receptor: Molecular mechanisms and potential therapeutic target. *Intensive Care. Med.* **2020**, *46*, 586–590. [CrossRef]
32. Sulli, A.; Gotelli, E.; Casabella, A.; Paolino, S.; Pizzorni, C.; Alessandri, E.; Grosso, M.; Ferone, D.; Smith, V.; Cutolo, M. Vitamin D and Lung Outcomes in Elderly COVID-19 Patients. *Nutrients* **2021**, *13*, 717. [CrossRef]
33. Akbar, M.R.; Wibowo, A.; Pranata, R.; Setiabudiawan, B. Low Serum 25-hydroxyvitamin D (Vitamin D) Level Is Associated with Susceptibility to COVID-19, Severity, and Mortality: A Systematic Review and Meta-Analysis. *Front. Nutr.* **2021**, *8*. [CrossRef]

34. Mariani, J.; Giménez, V.M.M.; Bergam, I.; Tajer, C.; Antonietti, L.; Inserra, F.; Ferder, L.; Manucha, W. Association Between Vitamin D Deficiency and COVID-19 Incidence, Complications, and Mortality in 46 Countries: An Ecological Study. *Health Secur.* **2021**, *19*, 302–308. [CrossRef]
35. Hastie, C.E.; Mackay, D.F.; Ho, F.; Celis-Morales, C.A.; Katikireddi, S.V.; Niedzwiedz, C.L.; Jani, B.D.; Welsh, P.; Mair, F.S.; Gray, S.R.; et al. Vitamin D concentrations and COVID-19 infection in UK Biobank. Diabetes. *Metab. Syndr.* **2020**, *14*, 561–565. [CrossRef] [PubMed]
36. Wang, T.T.; Nestel, F.P.; Bourdeau, V.; Nagai, Y.; Wang, Q.; Liao, J.; Tavera-Mendoza, L.; Lin, R.; Hanrahan, J.W.; Mader, S.; et al. Cutting edge: 1,25-dihydroxyvitamin D3 is a direct inducer of antimicrobial peptide gene expression. *J. Immunol.* **2004**, *173*, 2909–2912. [CrossRef]
37. Liu, P.T.; Stenger, S.; Li, H.; Wenzel, L.; Tan, B.H.; Krutzik, S.R.; Ochoa, M.T.; Schauber, J.; Wu, K.; Meinken, C.; et al. Toll-Like Receptor Triggering of a Vitamin D-Mediated Human Antimicrobial Response. *Science* **2006**, *311*, 1770–1773. [CrossRef] [PubMed]
38. Takeuti, F.A.C.; Guimaraes, F.D.S.F.; Guimaraes, P.D.S.F. Applications of vitamin D in sepsis prevention. *Discov. Med.* **2018**, *25*, 291–297. [PubMed]
39. Gombart, A.F.; Borregaard, N.; Koeffler, H.P. Human cathelicidin antimicrobial peptide (CAMP) gene is a direct target of the vitamin D receptor and is strongly up-regulated in myeloid cells by 1,25-dihydroxyvitamin D3. *FASEB J.* **2005**, *19*, 1067–1077. [CrossRef] [PubMed]
40. Zhao, Y.; Ran, Z.; Jiang, Q.; Hu, N.; Yu, B.; Zhu, L.; Shen, L.; Zhang, S.; Chen, L.; Chen, H.; et al. Vitamin D Alleviates Rotavirus Infection through a Microrna-155-5p Mediated Regulation of the TBK1/IRF3 Signaling Pathway In Vivo and In Vitro. *Int. J. Mol. Sci.* **2019**, *20*, 3562. [CrossRef] [PubMed]
41. Sharifi, A.; Vahedi, H.; Nedjat, S.; Rafiei, H.; Hosseinzadeh-Attar, M.J. Effect of single-dose injection of vitamin D on immune cytokines in ulcerative colitis patients: A randomized placebo-controlled trial. *APMIS* **2019**, *127*, 681–687. [CrossRef] [PubMed]
42. Gombart, A.F.; Pierre, A.; Maggini, S. A Review of Micronutrients and the Immune System–Working in Harmony to Reduce the Risk of Infection. *Nutrients* **2020**, *12*, 236. [CrossRef]
43. Cereda, E.; Bogliolo, L.; Lobascio, F.; Barichella, M.; Zecchinelli, A.L.; Pezzoli, G.; Caccialanza, R. Vitamin D supplementation and outcomes in coronavirus disease 2019 (COVID-19) patients from the outbreak area of Lombardy, Italy. *Nutrition* **2020**, *82*, 111055. [CrossRef] [PubMed]
44. Vankadari, N.; Wilce, J.A. Emerging WuHan (COVID-19) coronavirus: Glycan shield and structure prediction of spike glycoprotein and its interaction with human CD26. *Emerg. Microbes Infect.* **2020**, *9*, 601–604. [CrossRef]
45. Komolmit, P.; Charoensuk, K.; Thanapirom, K.; Suksawatamnuay, S.; Thaimai, P.; Chirathaworn, C.; Poovorawan, Y. Correction of vitamin D deficiency facilitated suppression of IP-10 and DPP IV levels in patients with chronic hepatitis C: A randomised double-blinded, placebo-control trial. *PLoS ONE* **2017**, *12*, e0174608. [CrossRef]
46. Fabri, M.; Realegeno, S.E.; Jo, E.-K.; Modlin, R.L. Role of autophagy in the host response to microbial infection and potential for therapy. *Curr. Opin. Immunol.* **2011**, *23*, 65–70. [CrossRef]
47. Campbell, G.R.; Spector, S.A. Toll-like receptor 8 ligands activate a vitamin D mediated autophagic response that inhibits human immunodeficiency virus type 1. *PLoS Pathog.* **2012**, *8*, e1003017. [CrossRef]
48. Jothimani, D.; Kailasam, E.; Danielraj, S.; Nallathambi, B.; Ramachandran, H.; Sekar, P.; Manoharan, S.; Ramani, V.; Narasimhan, G.; Kaliamoorthy, I.; et al. COVID-19: Poor outcomes in patients with zinc deficiency. *Int. J. Infect. Dis.* **2020**, *100*, 343–349. [CrossRef]
49. Anuk, A.T.; Polat, N.; Akdas, S.; Erol, S.A.; Tanacan, A.; Biriken, D.; Keskin, H.L.; Tekin, O.M.; Yazihan, N.; Sahin, D. The Relation Between Trace Element Status (Zinc, Copper, Magnesium) and Clinical Outcomes in COVID-19 Infection During Pregnancy. *Biol. Trace Element Res.* **2020**, *199*, 3608–3617. [CrossRef]
50. Shakeri, H.; Azimian, A.; Ghasemzadeh-Moghaddam, H.; Safdari, M.; Haresabadi, M.; Daneshmand, T.; Ahmadabad, H.N. Evaluation of the relationship between serum levels of zinc, vitamin B12, vitamin D, and clinical outcomes in patients with COVID-19. *J. Med. Virol.* **2021**. [CrossRef]
51. Joachimiak, M.P. Zinc against COVID-19? Symptom surveillance and deficiency risk groups. *PLoS Neglected Trop. Dis.* **2021**, *15*, e0008895. [CrossRef]
52. Read, S.; Obeid, S.; Ahlenstiel, C.; Ahlenstiel, G. The Role of Zinc in Antiviral Immunity. *Adv. Nutr.* **2019**, *10*, 696–710. [CrossRef] [PubMed]
53. Aggarwal, R.; Sentz, J.; Miller, M.A. Role of Zinc Administration in Prevention of Childhood Diarrhea and Respiratory Illnesses: A Meta-analysis. *Pediatrics* **2007**, *119*, 1120–1130. [CrossRef] [PubMed]
54. Finzi, E.; Harrington, A. Zinc treatment of outpatient COVID-19: A retrospective review of 28 consecutive patients. *J. Med. Virol.* **2021**, *93*, 2588–2590. [CrossRef] [PubMed]
55. Finzi, E. Treatment of SARS-CoV-2 with high dose oral zinc salts: A report on four patients. *Int. J. Infect. Dis.* **2020**, *99*, 307–309. [CrossRef] [PubMed]
56. Skalny, A.V.; Rink, L.; Ajsuvakova, O.P.; Aschner, M.; Gritsenko, V.A.; Alekseenko, S.I.; Svistunov, A.A.; Petrakis, D.; Spandidos, D.A.; Aaseth, J.; et al. Zinc and respiratory tract infections: Perspectives for COVID-19. *Int. J. Mol. Med.* **2020**, *46*, 17–26. [CrossRef] [PubMed]
57. Asl, S.H.; Nikfarjam, S.; Majidi Zolbanin, N.; Nassiri, R.; Jafari, R. Immunopharmacological perspective on zinc in SARS-CoV-2 infection. *Int. Immunopharmacol.* **2021**, *96*, 107630. [CrossRef] [PubMed]

58. Velthuis, A.T.; Worm, S.H.E.V.D.; Sims, A.C.; Baric, R.S.; Snijder, E.J.; Van Hemert, M.J. Zn2+ Inhibits Coronavirus and Arterivirus RNA Polymerase Activity In Vitro and Zinc Ionophores Block the Replication of These Viruses in Cell Culture. *PLoS Pathog.* **2010**, *6*, e1001176. [CrossRef] [PubMed]
59. Carlucci, P.M.; Ahuja, T.; Petrilli, C.; Rajagopalan, H.; Jones, S.; Rahimian, J. Zinc sulfate in combination with a zinc ionophore may improve outcomes in hospitalized COVID-19 patients. *J. Med. Microbiol.* **2020**, *69*, 1228–1234. [CrossRef] [PubMed]
60. Sandstead, H.H.; Prasad, A.S. Zinc intake and resistance to H1N1 influenza. *Am. J. Public Health* **2010**, *100*, 970–971. [CrossRef]
61. Xu, L.; Tong, J.; Wu, Y.; Zhao, S.; Lin, B.-L. Targeted oxidation strategy (TOS) for potential inhibition of coronaviruses by disulfiram—A 70-year old anti-alcoholism drug. *ChemRxiv* **2020**. [CrossRef]
62. Korant, B.D.; Kauer, J.C.; Butterworth, B.E. Zinc ions inhibit replication of rhinoviruses. *Nature* **1974**, *248*, 588–590. [CrossRef]
63. Cakman, I.; Kirchner, H.; Rink, L. Zinc Supplementation Reconstitutes the Production of Interferon-α by Leukocytes from Elderly Persons. *J. Interf. Cytokine Res.* **1997**, *17*, 469–472. [CrossRef]
64. Oyagbemi, A.A.; Ajibade, T.O.; Aboua, Y.G.; Gbadamosi, I.T.; Adedapo, A.D.A.; Aro, A.O.; Adejumobi, O.A.; Thamahane-Katengua, E.; Omobowale, T.O.; Falayi, O.O.; et al. Potential health benefits of zinc supplementation for the management of COVID-19 pandemic. *J. Food Biochem.* **2021**, *45*, e13604. [CrossRef]
65. Lang, C.J.; Hansen, M.; Roscioli, E.; Jones, J.; Murgia, C.; Ackland, M.L.; Zalewski, P.; Anderson, G.; Ruffin, R. Dietary zinc mediates inflammation and protects against wasting and metabolic derangement caused by sustained cigarette smoke exposure in mice. *BioMetals* **2010**, *24*, 23–39. [CrossRef]
66. Wessels, I.; Haase, H.; Engelhardt, G.; Rink, L.; Uciechowski, P. Zinc deficiency induces production of the proinflammatory cytokines IL-1β and TNFα in promyeloid cells via epigenetic and redox-dependent mechanisms. *J. Nutr. Biochem.* **2013**, *24*, 289–297. [CrossRef] [PubMed]
67. Khera, D.; Singh, S.; Purohit, P.; Sharma, P.; Singh, K. Prevalence of Zinc deficiency and effect of Zinc supplementation on prevention of acute respiratory infections: A non-randomized open label study. *Turk. Thorac. J.* **2020**, *21*, 371–376. [CrossRef] [PubMed]
68. Zabetakis, I.; Lordan, R.; Norton, C.; Tsoupras, A. COVID-19: The Inflammation Link and the Role of Nutrition in Potential Mitigation. *Nutrients* **2020**, *12*, 1466. [CrossRef] [PubMed]
69. Karakochuk, C.D.; Barr, S.; Boy, E.; Bahizire, E.; Tugirimana, P.L.; Akilimali, P.Z.; Houghton, L.; Green, T.J. The effect of inflammation on serum zinc concentrations and the prevalence estimates of population-level zinc status among Congolese children aged 6–59 months. *Eur. J. Clin. Nutr.* **2017**, *71*, 1467–1470. [CrossRef] [PubMed]
70. MacDonell, S.; Miller, J.C.; Harper, M.J.; Reid, M.R.; Haszard, J.J.; Gibson, R.S.; Houghton, L. A comparison of methods for adjusting biomarkers of iron, zinc, and selenium status for the effect of inflammation in an older population: A case for interleukin 6. *Am. J. Clin. Nutr.* **2018**, *107*, 932–940. [CrossRef] [PubMed]

Article

Consumption of Food Supplements during the Three COVID-19 Waves in Poland—Focus on Zinc and Vitamin D

Anna Puścion-Jakubik *,†, Joanna Bielecka †, Monika Grabia, Anita Mielech, Renata Markiewicz-Żukowska, Konrad Mielcarek , Justyna Moskwa, Sylwia K. Naliwajko , Jolanta Soroczyńska, Krystyna J. Gromkowska-Kępka, Patryk Nowakowski and Katarzyna Socha

Department of Bromatology, Faculty of Pharmacy with the Division of Laboratory Medicine, Medical University of Białystok, Mickiewicza 2D Street, 15-222 Białystok, Poland; joanna.bielecka@umb.edu.pl (J.B.); monika.grabia@umb.edu.pl (M.G.); anita.mielech@umb.edu.pl (A.M.); renmar@poczta.onet.pl (R.M.-Ż.); konrad.mielcarek@umb.edu.pl (K.M.); justyna.moskwa@umb.edu.pl (J.M.); sylwia.naliwajko@umb.edu.pl (S.K.N.); jolanta.soroczynska@umb.edu.pl (J.S.); krystyna.gromkowska.kepka@gmail.com (K.J.G.-K.); patryk.nowakowski@umb.edu.pl (P.N.); katarzyna.socha@umb.edu.pl (K.S.)

* Correspondence: anna.puscion-jakubik@umb.edu.pl; Tel.: +48-857-485-469

† These authors contributed equally to this work.

Abstract: Food supplements (FS) are a concentrated source of vitamins, minerals, or other ingredients with nutritional or other physiological effects. Due to their easy availability, widespread advertising, and sometimes low price, increased consumption of this group of preparations has been observed. Therefore, the aim of the study was to assess the knowledge and intake of FS during the COVID-19 pandemic in Poland, with particular reference to FS containing zinc and vitamin D. It was noted that both of the above ingredients were used significantly more often by people with higher education (59.0%), with a medical background or related working in the medical field (54.5%), and/or exercising at home (60.1%). Preparations containing vitamin D were used by 22.8% of the respondents in the first wave, 37.6% in the second wave, and 32.9% in the third wave. To sum up, we showed the highest consumption of vitamin and mineral supplements, and preparations containing zinc and vitamin D were taken significantly more often by people with higher medical and related education. This indicates a high awareness of health aspects and the need for preventive measures in these groups.

Keywords: food supplements; immunity; COVID-19; zinc; vitamin D; lifestyle; Poland

Citation: Puścion-Jakubik, A.; Bielecka, J.; Grabia, M.; Mielech, A.; Markiewicz-Żukowska, R.; Mielcarek, K.; Moskwa, J.; Naliwajko, S.K.; Soroczyńska, J.; Gromkowska-Kępka, K.J.; et al. Consumption of Food Supplements during the Three COVID-19 Waves in Poland—Focus on Zinc and Vitamin D. *Nutrients* **2021**, *13*, 3361. https://doi.org/10.3390/nu13103361

Academic Editors: William B. Grant and Ronan Lordan

Received: 2 September 2021
Accepted: 23 September 2021
Published: 25 September 2021

Publisher's Note: MDPI stays neutral with regard to jurisdictional claims in published maps and institutional affiliations.

1. Introduction

According to the definition of the European Food Safety Authority (EFSA), a food supplement (FS) is a foodstuff intended to be a complement to a normal diet and is a concentrated source of nutrients (vitamins, minerals) or other substances with a nutritional or physiological effect. FS could contain specific substances separately or in a complex combination. There are different forms of FS: pills, tablets, capsules, powder sachets, liquid ampoules, dropper bottles, and other. In the European Union, FS are regulated as foods and must be safe to consume. The regulations determine the maximum level for vitamins and minerals and some other substances. However, in Poland only maximum levels of vitamins and minerals are determined. FS could be used to correct nutritional deficiencies or to maintain an adequate intake of nutrients. Their intake is not a substitute for a varied and balanced diet [1]. Prevention and treatment are physiological activities that are not allowed to be attributed to FS. Moreover, the label must not state that they are recommended for the treatment of diseases.

In Poland, the first case of SARS-CoV-2 infection was diagnosed on 4 March 2020. One week later, the World Health Organization declared the COVID-19 pandemic (on 11 March 2020). To date, more than 209 million COVID-19 cases and over 4.4 million deaths have been reported worldwide, while in Poland over the total is 2.8 million cases and 75,000 deaths (as of 23 August 2021). Three waves of the pandemic can be distinguished in Poland so far. The first was between March and May 2020, the second between September and November 2020, and the third from February to April 2021. Most of the infected patients experienced mild to moderate symptoms such as tiredness, fever, cough, headache, and loss of smell or taste. However, the list of possible symptoms is getting longer due to the research being conducted on this topic. There is a higher risk of severe COVID-19 among the elderly and those suffering from chronic illnesses (e.g., diabetes, cancer, and chronic respiratory diseases) [2].

Adequate diet and nutritional status are key elements in the maintenance of the proper functioning of the immune system. SARS-CoV-2 infection is usually associated with a decreased immune response, leading to pneumonic inflammation. Among the notably important components improving immune functions, vitamins (A, B_6, B_9, B_{12}, C, D, and E), microelements (Fe, Cu, Se, and Zn), and n-3 long-chain polyunsaturated fatty acids (PUFAs) are indicated [3]. It was demonstrated that vitamin D supplementation was related to a lower risk of respiratory infections [4]. In a randomized trial, it was observed that humoral immunity was improved by the supplementation of vitamins A and D among pediatric patients who received influenza vaccination [5]. However, there is still insufficient evidence to provide exact recommendations on vitamin C, D, and E supplementation for the prevention and treatment of COVID-19 [6]. The positive influence of vitamin B_6 on immunity involves activation of innate or adaptive immunity and the influence on the proliferation of immune cells [7]. Zn is important for the development and functioning of neutrophils and natural killer cells [8]. Fe modulates the differentiation and proliferation of T-cells and the production of reactive oxygen species, which take part in removing infectious agents. The influence of PUFAs on viral infections is not well established and requires further research [9]. Sufficient Se intake supports the immune system, while Se deficiency impairs innate and acquired immunity by the negative influence on cellular as well as humoral immunity (i.e., the production of antibodies) [10].

During the pandemic, negative changes in eating habits and lifestyle such as increased consumption of alcohol, sweets, and fast food or reduced physical activity were reported [11,12]. Taking into account the positive aspects, an increased intake of fruits, vegetables, nuts, legumes, and fish was also observed [13]. On the other hand, greater interest in searching for information on improving the immune system by food products or FS was observed [14,15].

In Poland, before the pandemic (in 2017), the worth of the FS market was estimated at 4.4 billion PLN (approximately 113 million USD) and over 70% of Poles used FS. It is estimated that in 2025 the global supplements market will reach 300 billion USD [16].

Currently, several anti-COVID-19 vaccines have been approved for use in humans to protect against the disease. However, vaccination hesitancy or resistance is observed among different populations [17]. Therefore, supporting immunity through adequate nutrition rich in essential nutrients and developing effective therapy against COVID-19 still seem to be crucial.

This study aimed at an assessment of the changes in the FS intake patterns, with a special focus on the supplements influencing immunity during the three waves of the COVID-19 pandemic in Poland.

2. Materials and Methods

2.1. Participants

This study was carried out among 935 Polish residents during three pandemic waves: $n = 236$ people answered questions about the first waves of the pandemic; the second: $n = 364$, and the third: $n = 335$. Each survey was conducted for about one month after the end of the period it covered. The study was conducted from July 2020 to April 2021. Responses of people living abroad ($n = 9$) were rejected. The inclusion criteria were being a resident of Poland, an adult (over 18 years of age), and answering all the questions. Each participant was informed that their participation was completely voluntary, the questionnaire was anonymous, and they could resign from participation in the study at any time. The researchers did not collect any data that could be used to identify people, including personal data. Each participant was allowed to complete the questionnaire only once. Consent to participate in the study was expressed by writing down the responses and sending them to the researchers.

2.2. Questionnaire

The questionnaire (containing questions and answers) was included as an attachment to the publication. The three questionnaires contained the same questions, but the third contained one additional question concerning the respondents' knowledge about the possibility of preventing viral infections (number 35) (Appendix A).

2.3. Statistical Analysis

Statistical analysis of the results was performed using Statistica software (TIBCO Software Inc., Palo Alto, CA, USA) and calculator for the chi-square test [18]. The dependencies between the qualitative features were assessed using the Chi-square test of independence. The level of significance was $p < 0.05$.

3. Results

Most of the respondents were women (during the first wave: 80.0%, during the second wave: 81.9%, and during the third wave: 79.7%). Our survey was anonymous and voluntary, and we had no option to select a gender group. A larger percentage of women participating in the study may indicate, at the same time, greater interest in aspects of health and social life among people of this gender.

Residents of all 16 voivodeships participated in the study, but the vast majority were inhabitants of Podlaskie and Mazowieckie voivodeships; the remaining inhabitants accounted for less than 5%.

Adults with an average age of 31 ± 11, 28 ± 9, and 28 ± 10 years, mainly with higher education (66.9%, 59.3%, and 47.5%, respectively), participated in the survey. Most respondents lived in a large city of over 250,000 inhabitants (37.3%, 40.1%, and 36.1%) or a village (33.9%, 24.9%, and 30.7%). About half of the respondents from each group described their financial situation as rather good (56.8%, 53.3%, and 49.5%), and most households were comprised of 2–4 people. It is noteworthy that, during the first wave of the pandemic, as many as 50.0% of the respondents worked at their usual office or worksite, while during the second and third waves the percentage was lower (38.5% and 36.4%). About three-quarters of the respondents described their level of physical activity as low during each of the three periods (68.2%, 78.8%, and 78.2%) (Table 1).

It was shown that, during the first round of the survey, the highest percentage of respondents described their health as very good (39.8% vs. 27.0% and 21.5%). In the first half of 2020, the respondents significantly more often answered that they did not suffer from COVID-19. It is disturbing that, during the third wave, as many as 40.0% of respondents noticed an increase in their body weight, and only 42.7% undertook physical activity at home (Table 2).

Table 1. Characteristics of the study group.

Variable	First Wave n = 236 % (n)	Second Wave n = 364 % (n)	Third Wave n = 335 % (n)
Gender			
Female	80.0 (189)	81.9 (298)	79.7 (267)
Male	20.0 (47)	18.1 (66)	20.3 (68)
Anthropometric measurements			
Age (years)	31 ± 11	28 ± 9	28 ± 10
Mass (kg)	69 ± 15	69 ± 18	69 ± 15
Weight (m)	1.69 ± 0.08	1.70 ± 0.09	1.70 ± 10
BMI (kg/m^2)	25.06 ± 4.41	23.91 ± 7.45	23.72 ± 4.45
Education			
Primary school	2.1 (5)	1.1 (4)	2.4 (8)
Higher	66.9 (158)	59.3 (216)	47.5 (159)
Secondary	31.0 (73)	39.6 (144)	50.1 (168)
Type of education			
Medical and related	47.5 (112)	56.6 (206)	45.1 (151)
Nonmedical	36.4 (86)	22.3 (81)	25.4 (85)
Not applicable	16.1 (38)	21.1 (77)	29.5 (99)
Place of residence			
City with up to 150,000 inhabitants	21.6 (51)	26.7 (97)	26.3 (88)
City with 150,000–250,000 inhabitants	7.2 (17)	8.3 (30)	6.9 (23)
City with over 250,000 inhabitants	37.3 (88)	40.1 (146)	36.1 (121)
Village	33.9 (80)	24.9 (91)	30.7 (103)
Subjective assessment of the material situation			
Very good	20.8 (49)	25.8 (94)	22.1 (74)
Average	20.8 (49)	20.1 (73)	26.0 (87)
Rather good	56.8 (134)	53.3 (194)	49.5 (166)
Rather bad	1.6 (4)	0.5 (2)	2.1 (7)
Bad	0.0 (0)	0.3 (1)	0.3 (1)
Number of people in the household			
1	4.2 (10)	7.5 (28)	7.8 (26)
2	24.6 (58)	22.8 (83)	20.6 (69)
3	29.2 (69)	25.3 (92)	19.7 (66)
4	24.6 (58)	30.5 (111)	29.9 (100)
5	10.6 (25)	10.2 (37)	13.7 (46)
6	2.5 (9)	2.2 (8)	2.7 (9)
7	2.5 (6)	0.3 (1)	3.6 (12)
8	0.5 (1)	0.8 (3)	1.2 (4)
10	0.0 (0)	0.3 (1)	0.9 (3)
Professional activity			
Unemployed person	7.2 (17)	1.9 (7)	3.6 (12)
Person working in office	50.0 (118)	38.5 (140)	36.4 (122)
Person working remotely	8.5 (20)	8.2 (30)	6.0 (20)
Student	34.3 (81)	51.4 (187)	54.0 (181)
Physical activity			
Inactivity (sedentary)	0.0 (0)	0.0 (0)	0.0 (0)
Low (occasional exercise, 1–3 times a week)	68.2 (161)	78.8 (287)	78.2 (262)
Moderate (1 h of exercise per day)	25.4 (60)	18.9 (68)	17.3 (58)
High (hard physical work and daily workouts)	6.4 (15)	2.3 (9)	4.5 (15)

Table 2. Assessment of health and physical activity during COVID-19.

Variable	First Wave (n = 236)	Second Wave (n = 364)	Third Wave (n = 335)
How would you rate your health at the beginning of the pandemic in Poland?			
Very good	39.8 (94)	27.0 (98)	21.5 (72)
Good	45.3 (107)	58.2 (212)	61.8 (207)
Medium	11.4 (27)	12.9 (47)	14.6 (49)
Poor	3.5 (8)	1.9 (7)	2.1 (7)
Have you had COVID-19?			
Yes	0.0 (0)	11.8 (43)	17.6 (59)
No	86.0 (203) ***	59.9 (218)	48.7 (163)
It is difficult to say unequivocally	14.0 (33)	28.3 (103)	33.7 (113)
Has your body weight changed during the pandemic?			
No	48.3 (114)	49.5 (180)	43.9 (147)
Increased	37.3 (88)	31.9 (116)	40.0 (134)
Decreased	14.4 (34)	18.6 (68)	16.1 (54)
Did you exercise at home during the pandemic?			
Yes	42.4 (100)	48.4 (176)	42.7 (143)
No	57.6 (136)	51.6 (188)	57.3 (192)

Differences between the various pandemic waves: *** $p < 0.001$.

It is satisfactory that almost all respondents correctly answered what a dietary supplement is—that it only supplements nutritional deficiencies (99.2%, 100.0%, and 97.6%)—and know the difference between FS and medications; this answer was indicated by 91.5%, 97.5%, and 95.8%. It is surprising that the most frequently chosen category of FS during all three waves of the pandemic in Poland was vitamin and mineral preparations (40.3%, 60.2%, and 54.3%), and preparations affecting immunity came in second place. It was shown that preparations from this category were consumed by twice as many people during the second and third wave than during the first wave (18.2%, 37.4%, and 34.9%) (Table 3).

During the first wave, a significantly greater percentage of respondents declared not taking food supplements with zinc and vitamin D (63.6% vs. 30.0% and 39.4%), and the most important reason cited for using them was the desire to supplement deficiencies of vitamins and minerals (36.0%, 27.5%, and 54.3%) (Table 3).

The authority of pharmacists' recommendations was noticeable during the second wave—as many as 13.5% of respondents chose these preparations at the recommendation of a pharmacist. An important fact is that the vast majority (over 90%) drink FS and medications with water. As many as 62.3% of respondents declared that they had not noticed an increase in the number of advertisements for FS during the pandemic. The vast majority of respondents used supplements as recommended (58.1%, 70.5%, and 63.6%). More than 85% of respondents (86.9%, 94.2%, and 88.4%) were aware of the side effects, and over 90% of the risk of overdose (91.1%, 96.1%, and 94.0%). It should also be emphasized that over 70% of respondents indicated that FS should be used only in the case of diagnosed deficiencies (70.8%, 77.5%, and 73.4%). A significantly higher percentage (95.9%) indicated an awareness of interactions between FS and medications in the second wave. In the third round of the survey, a question was added regarding awareness of the beneficial effect of the use of preparations containing zinc and vitamin D in the prevention of viral infections—79.7% of respondents indicated that they had heard such reports (Table 3).

Table 3. Assessment of knowledge about food supplements and their consumption during the COVID-19 pandemic.

Variable	First Wave (n = 236)	Second Wave (n = 364)	Third Wave (n = 335)
What is a food supplement?			
A preparation that treats nutritional deficiencies	0.8 (2)	0.0 (0)	2.4 (8)
A preparation that only replenishes nutritional deficiencies	99.2 (234)	100.0 (364)	97.6 (327)
Do you think food supplements differ from medications?			
Yes	91.5 (216)	97.5 (355)	95.8 (321)
No	2.9 (7)	2.2 (8)	2.4 (8)
I do not know	5.6 (13)	0.3 (1)	1.8 (6)
What categories of food supplements did you use during the pandemic?#			
Vitamin–mineral supplements	40.3 (95)	60.2 (219)	54.3 (182)
Probiotics	13.1 (31)	18.1 (66)	15.5 (52)
Prebiotics	3.4 (8)	2.7 (10)	3.3 (11)
Supporting immunity	18.2 (43)	37.4 (136)	34.9 (117)
Supporting weight loss	3.0 (7)	1.9 (7)	3.0 (10)
Improving the condition of the hair, skin, and nails	14.4 (34)	19.5 (71)	22.4 (75)
Supporting the functioning of the urinary tract	2.1 (5)	0.8 (3)	3.3 (11)
Supporting the heart	1.7 (4)	1.4 (5)	4.2 (14)
Supporting memory	2.5 (6)	4.7 (17)	8.4 (28)
Supporting lowering cholesterol levels	1.3 (3)	0.8 (3)	1.2 (4)
Vision support	1.7 (4)	3.0 (11)	3.3 (11)
Supporting the functioning of the joints	5.1 (12)	4.7 (17)	1.8 (6)
Relieving the symptoms of menopause	0.0 (0)	0.0 (0)	0.6 (2)
Supporting the digestive tract	3.8 (9)	3.8 (14)	5.4 (18)
Improving well-being	3.4 (8)	4.4 (16)	4.8 (16)
Facilitating sedation and sleep	6.8 (16)	6.9 (25)	11.3 (38)
Supporting libido	1.3 (3)	0.8 (3)	0.3 (1)
Supporting alcohol metabolism	0.4 (1)	1.4 (5)	0.3 (1)
For athletes	3.8 (9)	2.7 (10)	5.1 (17)
Removing excess water	0.4 (1)	0.0 (0)	0.0 (0)
Other	0.0 (0)	1.6 (6)	0.0 (0)
I did not use dietary supplements	42.8 (101)	19.7 (72)	23.9 (80)
Have you used zinc and vitamin D food supplements since March 2020?			
No	63.6 (150) **	30.0 (109)	39.4 (132)
Only drugs	5.1 (12)	16.5 (60)	13.1 (44)
Yes both	7.2 (17)	14.0 (51)	12.5 (42)
Only zinc	1.3 (3)	1.9 (7)	2.1 (7)
Only vitamin D	22.8 (54)	37.6 (137)	32.9 (110)
Why did you use such food supplements?#			
Not applicable	47.9 (113)	24.7 (90)	26.9 (90)
To improve health	22.2 (52)	21.4 (78)	33.7 (113)
Due to a pharmacist's recommendation	1.7 (4)	13.5 (49)	3.0 (10)
Due to a doctor's recommendation	3.8 (9)	0.0 (0)	0.3 (1)
To supplement deficiencies of vitamins and minerals	36.0 (85)	27.5 (100)	54.3 (182)
To supplement the therapy prescribed by doctor	2.5 (6)	9.9 (36)	10.1 (34)
Due to a friend's recommendation	5.1 (12)	6.0 (22)	3.0 (10)
Because I was encouraged by TV/media/Internet advertising	0.0 (0)	0.3 (1)	1.2 (4)
Other	0.0 (0)	0.0 (0)	2.1 (7)

Table 3. *Cont.*

Variable	First Wave (n = 236)	Second Wave (n = 364)	Third Wave (n = 335)
What do you usually use to wash down food supplements and medications?#			
Tea	8.1 (19)	11.3 (41)	22.4 (75)
Cola	1.7 (4)	0.5 (2)	1.5 (5)
Not applicable	0.8 (2)	4.7 (17)	6.0 (20)
I do not drink	5.1 (12)	0.5 (2)	1.2 (4)
Juice	3.8 (9)	4.7 (17)	4.8 (16)
Water	93.2 (220)	93.4 (340)	90.1 (302)
Coffee	0.4 (1)	2.5 (9)	4.5 (15)
Milk	0.4 (1)	0.0 (0)	0.9 (3)
Other	0.4 (1)	0.5 (2)	0.6 (2)
Do you think there were more advertisements for food supplements during the pandemic?			
No	5.1 (12)	1.4 (5)	3.6 (12)
Yes	32.6 (77)	42.9 (156)	59.1 (198)
I did not notice a change	62.3 (147) **	55.7 (203)	37.3 (125)
Do you use food supplements in the amount recommended on the package?			
I do not use it	39.9 (80)	19.0 (69)	23.0 (77)
No, I use lower doses	3.8 (9)	4.7 (17)	6.0 (20)
No, I use higher doses	4.2 (10)	5.8 (21)	7.5 (25)
Yes	58.1 (137)	70.5 (257)	63.6 (213)
Do you think food supplements can have side effects?			
No, taking them is absolutely safe	13.1 (31)	5.8 (21)	11.6 (39)
Yes	86.9 (205)	94.2 (343)	88.4 (296)
How do you assess the advisability of using food supplements?			
They should be used only in the event of identified deficiencies	70.8 (167)	77.5 (282)	73.4 (246)
Their use is unnecessary	13.1 (31)	7.7 (28)	9.0 (30)
I have no opinion	16.1 (38)	14.8 (54)	17.6 (59)
Do you think food supplements can be overdosed on?			
No, they're safe	8.9 (21)	3.8 (14)	6.0 (20)
Yes	91.1 (215)	96.1 (350)	94.0 (315)
Do you think that food supplements can interact with medications prescribed by your doctor, and thus affect the effectiveness of therapy?			
No, they're safe	10.2 (24)	4.1 (15)	9.0 (30)
Yes	89.8 (212)	95.9 (349) ***	91.0 (305)
Has the pandemic affected your use of food supplements?			
No	90.7 (214) ***	61.8 (225)	81.8 (274)
Yes, I use fewer	0.4 (1)	1.4 (5)	1.2 (4)
Yes, I use more	8.9 (21)	23.1 (84)	17.0 (57)
I did not use food supplements	0.0 (0)	13.7 (50)	0.0 (0)
Have you heard that preparations containing zinc and vitamin D can support immunity and be helpful in the prevention of viral infections?#			
No	-	-	20.3 (68)
Yes	-	-	79.7 (267)

Differences between the various pandemic waves: ** $p < 0.01$, *** $p < 0.001$, # multiple choice question.

In the following part, the entire study group was divided in terms of the use of FS containing only zinc, only vitamin D, or both. FS with both ingredients were chosen significantly more often by people with higher education (59.0%) and with medical and related education (54.5%) (Table 3). Among the inhabitants of large cities—with a population of over 250,000—the highest percentage of respondents used preparations containing both vitamin D and zinc. Preparations containing only zinc were significantly more often used by people assessing their financial situation as rather good (64.7%) and by students (70.6%). Four-person families used both these components (73.9%) as prophylaxis. The highest percentage of people who suffered from COVID-19 consumed both zinc and vitamin D (no statistical significance); however, the degree of dependence, i.e., whether these ingredients were used before or after infection, was not found (Table 4).

Table 4. Consumption of food supplements with zinc, vitamin D and both ingredients depending on various factors.

Variable	Only Zinc $n = 17$ % (n)	Only Vitamin D $n = 301$ % (n)	Both Food Supplements $n = 110$
Gender			
Female	88.2 (15)	84.4 (254)	77.3 (85)
Male	11.8 (2)	15.6 (47)	22.7 (25)
Education			
Primary school	11.8 (2)	1.0 (3)	0.9 (1)
Secondary	58.8 (10)	41.7 (126)	40.1 (45)
Higher	29.4 (5)	57.3 (172)	59.0 (64) ***
Type of education			
Medical and related	35.3 (6)	50.1 (152)	54.5 (60) *
Nonmedical	29.4 (5)	27.9 (84)	27.3 (30)
Not applicable	35.3 (6)	22.0 (65)	18.2 (20)
Place of residence			
A city with up to 150,000 inhabitants	23.5 (4)	29.6 (89)	20.9 (23)
A city with 150,000–250,000 inhabitants	17.6 (3)	8.3 (25)	3.6 (4)
A city with over 250,000 inhabitants	35.4 (6)	39.2 (118)	47.3 (52)
Village	23.5 (4)	22.9 (69)	28.2 (31)
Subjective assessment of material situation			
Very good	23.5 (4)	23.6 (71)	30.0 (33)
Rather good	64.7 (11) ***	51.8 (156)	18.3 (55)
Average	11.8 (2)	23.6 (71)	15.5 (17)
Rather bad	0.0 (0)	1.0 (3)	4.5 (5)
Bad	0.0 (0)	0.0 (0)	0.0 (0)
Number of people in the household			
1	5.9 (1)	6.6 (20)	1.3 (4)
2	23.5 (4)	26.6 (80)	10.6 (32)
3	11.8 (2)	22.6 (68)	9.3 (28)
4	41.1 (7)	28.6 (86)	73.9 (31)
5	11.8 (2)	10.2 (31)	4.3 (13)
6	0.0 (0)	1.7 (5)	0.3 (1)
7	0.0 (0)	1.7 (5)	0.3 (1)
8	5.9 (1)	1.0 (3)	0.0 (0)
9	0.0 (0)	0.7 (2)	0.0 (0)
10	0.0 (0)	0.3 (1)	0.0 (0)
Professional activity			
Unemployed person	0.0 (0)	3.7 (11)	0.9 (5)
Person working in office	29.4 (5)	39.9 (120)	42.7 (47)
Person working remotely	0.0 (0)	8.3 (25)	8.2 (9)
Student	70.6 (12) *	48.1 (145)	48.2 (49)
Physical activity			
Inactivity (sedentary)	0.0 (0)	0.0 (0)	0.0 (0)
Low (occasional exercise, 1–3 times a week)	70.5 (12)	81.4 (245)	72.7 (80)
Moderate (1 h of training per day)	23.6 (4)	14.6 (44)	20.9 (23)
High (hard physical work and daily workouts)	5.9 (1)	3.9 (12)	6.4 (7)
How would you rate your health at the beginning of the pandemic in Poland?			
Very good	17.6 (3)	27.6 (83)	28.2 (31)
Good	64.7 (11)	57.8 (174)	56.4 (62)
Medium	11.8 (2)	11.9 (36)	11.8 (13)
Poor	5.9 (1)	2.7 (8)	3.6 (4)

Table 4. *Cont.*

Variable	Only Zinc $n = 17$ % (n)	Only Vitamin D $n = 301$ % (n)	Both Food Supplements $n = 110$
Have you had COVID-19?			
Yes	11.8 (2)	12.0 (36)	15.5 (17)
No	46.5 (8)	58.8 (177)	62.2 (69)
It is difficult to say unequivocally	41.7 (7)	29.2 (88)	21.8 (24)
Has your body weight changed during the pandemic?			
No	47.1 (8)	47.2 (142)	40.0 (44)
Increased	47.1 (8)	33.9 (102)	41.8 (46)
Decreased	5.8 (1)	18.9 (57)	18.2 (20)
Did you exercise at home during the pandemic?			
Yes	41.2 (7)	48.8 (147)	60.1 (67)*
No	58.8 (10)	51.2 (154)	39.9 (43)
What is a food supplement?			
A preparation that treats nutritional deficiencies	0.0 (0)	2.0 (6)	0.9 (1)
A preparation that supplements nutritional deficiencies	100 (17)	98.0 (295)	99.1 (109)
Do you think food supplements differ from medications?			
Yes	100 (17)	97.0 (292)	99.1 (109)
No	0.0 (0)	0.0 (0)	0.9 (1)
I do not know	0.0 (0)	3.0 (9)	0.0 (0)
What categories of food supplements did you use during the pandemic?#			
Vitamin–mineral supplements	88.2 (15)	79.4 (239)	85.5 (94)
Probiotics	11.8 (2)	22.6 (68)	28.2 (31)
Prebiotics	0.0 (0)	5.0 (15)	5.5 (6)
Supporting immunity	29.4 (5)	45.5 (137)	55.5 (61)
Supporting weight loss	0.0 (0)	2.7 (8)	2.7 (3)
Improving the condition of hair, skin, and nails	35.3 (6)	18.6 (56)	41.8 (46)
Supporting the functioning of the urinary tract	5.9 (1)	3.0 (9)	2.7 (3)
Supporting the heart	5.9 (1)	3.7 (11)	5.5 (6)
Supporting memory	5.9 (1)	6.3 (19)	9.1 (10)
Supporting lowering cholesterol levels	0.0 (0)	2.0 (6)	1.8 (2)
Vision support	5.9 (1)	2.0 (6)	10.0 (11)
Supporting the functioning of the joints	5.9 (1)	5.3 (16)	7.3 (8)
Relieving the symptoms of menopause	0.0 (0)	0.0 (0)	0.9 (1)
Supporting the digestive tract	5.9 (1)	4.3 (13)	6.4 (7)
Improving well-being	0.0 (0)	5.0 (15)	9.1 (10)
Facilitating sedation and sleep	5.9 (1)	11.3 (34)	12.7 (14)
Supporting libido	0.0 (0)	0.7 (2)	1.8 (2)
Supporting alcohol metabolism	0.0 (0)	1.7 (5)	0.9 (1)
For athletes	5.9 (1)	4.3 (13)	6.4 (7)
Removing excess water	0.0 (0)	0.0 (0)	0.0 (0)
Other	0.0 (0)	0.3 (1)	0.0 (0)
I did not use food supplements	0.0 (0)	0.0 (0)	0.0 (0)

Table 4. *Cont.*

Variable	Only Zinc $n = 17$ % (n)	Only Vitamin D $n = 301$ % (n)	Both Food Supplements $n = 110$
Why did you use such food supplements?#			
Not applicable	0.0 (0)	0.0 (0)	0.0 (0)
To improve health	35.3 (6)	54.8 (165)	54.5 (60)
Due to a pharmacist's recommendation	17.6 (3)	3.0 (9)	5.5 (6)
Due to a doctor's recommendation	17.6 (3)	0.3 (1)	0.0 (0)
To supplement deficiencies of vitamins and minerals	52.9 (9)	72.1 (217)	69.1 (76)
To supplement the therapy prescribed by doctor	0.0 (0)	13.3 (40)	14.5 (16)
Due to a friend's recommendation	5.9 (1)	6.6 (20)	0.9 (1)
Because I was encouraged by TV/media/Internet advertising	0.0 (0)	0.7 (2)	0.9 (1)
Other	0.0 (0)	1.0 (3)	1.8 (2)
What do you usually use to wash down food supplements and medications?#			
Tea	11.8 (2)	13.3 (40)	18.2 (20)
Cola	5.9 (1)	0.6 (2)	1.8 (2)
Not applicable	0.0 (0)	0.0 (0)	0.0 (0)
I do not drink	0.0 (0)	0.0 (0)	6.4 (7)
Juice	11.8 (2)	5.3 (16)	0.0 (0)
Water	94.1 (16)	98.0 (295) ***	97.3 (107)
Coffee	17.6 (3)	3.3 (10)	0.9 (1)
Milk	0.0 (0)	0.0 (0)	0.0 (0)
Other	0.0 (0)	0.0 (0)	0.0 (0)
Do you think there were more advertisements for food supplements during the pandemic?			
No	0.0 (0)	2.7 (8)	0.9 (1)
Yes	35.3 (6)	41.5 (125)	37.3 (41)
I did not notice a change	64.7 (11)	55.8 (168)	61.8 (68)
Do you use food supplements in the amount recommended on the package?			
I do not use it	0.0 (0)	0.0 (0)	0.0 (0)
No, I use lower doses	0.0 (0)	6.3 (19)	7.3 (8)
No, I use higher doses	0.0 (0)	8.3 (25)	13.6 (15)
Yes	100.0 (17) ***	85.4 (257)	79.1 (87)
Do you think food supplements can have side effects?			
No, taking them is absolutely safe	11.8 (2)	10.3 (31)	10.0 (11)
Yes	88.2 (15)	89.7 (270)	90.0 (99)
How do you assess the advisability of using food supplements?			
They should be used only in the event of identified deficiencies	94.1 (16) **	82.1 (247)	83.6 (92)
Their use is unnecessary	5.9 (1)	5.0 (15)	0.9 (1)
I have no opinion	0.0 (0)	12.9 (39)	15.5 (17)
Do you think food supplements can be overdosed on?			
No, they're safe	11.8 (2)	5.3 (16)	7.3 (8)
Yes	88.2 (15)	94.7 (285)	92.7 (102)
Do you think that food supplements can interact with medications prescribed by your doctor, and thus affect the effectiveness of therapy?			
No, they're safe	23.6 (4)	8.0 (24)	10.9 (12)
Yes	76.4 (13)	92.0 (277) **	89.1 (98)
Has the pandemic affected your use of food supplements?			
No	76.5 (13)	73.7 (222)	59.1 (65)
Yes, I use fewer	0.0 (0)	0.7 (2)	1.8 (2)
Yes, I use more	23.5 (4)	25.6 (77)	39.1 (43)
I did not use dietary supplements	0.0 (0)	0.0 (0)	0.0 (0)

Differences between the various pandemic waves: * $p < 0.05$, ** $p < 0.01$, *** $p < 0.001$, # multiple choice question.

High health awareness may be indicated by the fact that people choosing both ingredients in food supplements were also active and exercised at home (60.1%). People who took both zinc and vitamin D also used other vitamin and mineral ingredients (85.5% of the respondents) and other FS affecting immunity (55.5%). It is surprising that a large percentage of the respondents also took preparations supporting the appearance of hair, skin, and nails (41.8%). It should be emphasized that 100% of respondents using zinc took the preparations in accordance with the recommendations, and 94.1% indicated that food supplements should be used only in the case of proven deficiencies (Table 4).

In our research, we found that the highest percentage of people during all three waves used food supplements containing only vitamin D, while searches on Google Trends indicate that, during the first wave, information about zinc was more popular—the importance of vitamin D increased during the second wave (Figure 1).

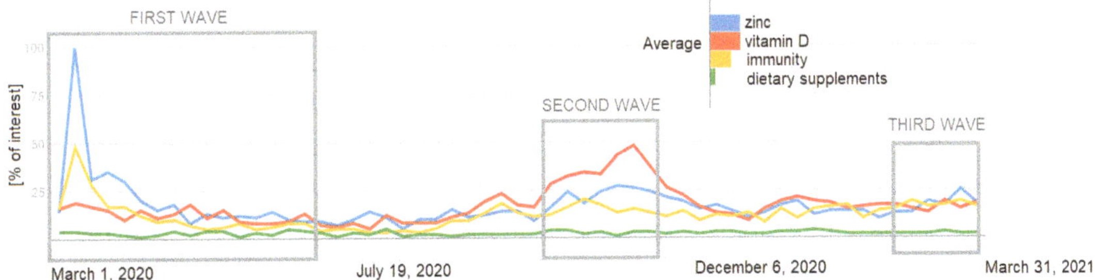

Figure 1. The popularity of searching for selected terms (own design based on data from Google Trends).

4. Discussion

There are several dietary and lifestyle factors that could influence immunity in positive as well as negative ways. The COVID-19 pandemic has prompted people to search for natural methods to boost immunity, including FS usage. Several vaccines are currently approved for human use, but there is still a need to develop effective therapies to treat COVID-19 and alleviate the negative health consequences of the disease.

Currently, there are quite a few studies available on the impact of the pandemic on lifestyles and nutrition. However, there are not many studies dealing with COVID-19 and dietary supplement intake. To the best of our knowledge, our study is the first to assess the consumption of dietary supplements in Poland during the three COVID-19 waves.

A Google Trends analysis showed that, in Poland, in relation to the coronavirus, the following terms were searched for: vitamin C, vitamin D, and *Glycyrrhiza glabra*; globally, there were also search terms such as vitamin K, selenium, zinc, garlic, onion, elderberry, lactoferrin, echinacea, and *Nigella sativa* L. Polish residents were trying to find antiviral properties for turmeric, garlic, and iodine as well as immune-boosting properties for fish oil [14].

Kamarli Altun et al. conducted a cross-sectional study among Turkish dietitians concerning the supplements, functional foods, and herbal medicines they used to protect themselves against SARS-CoV-2 infection. Nearly 90% of the study participants found that proper nutrition could affect the clinical course of the disease and almost all respondents (94.5%) declared FS intake. Less than half of dietitians (46.1%) started using herbal medicine, while nearly one-third included functional foods into the diet (34.9%). Fish oil was the most commonly chosen FS (81.9%). Women were twice as likely to use FS as men [19]. In this study we reported a lower prevalence of FS intake: most of the participants declared usage of FS 57.2%, 80.3%, and 76.1% in the first, second, and third waves, respectively, of the pandemic in Poland. The most often chosen type of FS was preparations with vitamins and minerals. More people reported that pandemic affect on using these supplements in the second wave compared to the first and third waves (23.1% vs. 8.9% and 17.0%).

Another cross-sectional study was carried out by Alfawaz et al. among Saudi Arabian residents, focused on changes in FS usage before the pandemic and during lockdown. Males tended to use FS (multivitamin, selenium, zinc, and vitamin D) more frequently than females. Among the subgroup of COVID-19 patients, men used more multivitamin and zinc supplements than women, while women had a higher intake of supplements with vitamins D and C. The male study participants 26–35 years of age declared a significantly higher use of multivitamin supplements than females (30.1 vs. 22.6%; $p < 0.054$) of the same age group. As determinants of FS usage, researchers distinguished the influence of age, level of education, and income [20].

The supplementation pattern among COVID-19 patients in Teheran was analyzed by Bagheri et al. Significantly higher vitamin D intake was reported in outpatients (30%) compared to hospitalized patients (16.5%). It was observed that vitamin D intake was related to a reduced risk of exacerbation of the disease. Moreover, a relevant difference was found considering zinc intake—9% vs. 2% in outpatients and inpatients, respectively. However, none of the patients declared the usage of multivitamins, vitamin C, vitamin E, folic acid, iron, omega-3, and omega-6 fatty acids [21]. In this research, none of the participants had COVID-19 in the first wave, 11.8% in the second, and 17.6% in the third waves; 14%, 28.3%, and 33.7% of respondents could not equivocally say whether they had SARS-CoV-2 infection. Considering the supplementation of vitamin D and zinc at the beginning of the pandemic, most of the participants (63.6%) did not take them during the first wave, which is contrary to the responses about the second (30.1%) and third (39.4%) waves of the pandemic. This difference was statistically significant. Vitamin D intake was declared by 22.8%, 37.6%, and 32.9%, while zinc was taken by 1.3%, 1.9%, and 2.1%; both compounds were used by 7.2%, 14.0%, and 12.5% of participants during the three waves of COVID-19 in Poland. Our results indicate a higher prevalence of intake of FS with vitamin D and zinc among the Polish population than among the Iranian population.

While in many countries increased interest in diet supplementation was observed, the findings of the cross-sectional study among the Lebanese population showed a decreased supplement intake. Before the pandemic, over 73% of the respondents used FS, while after the COVID-19 outbreak it was 69.9%. However, for specific subgroups of FS, increased intake was reported. Noticeably higher usage of antioxidants (14% vs. 15.6%), vitamin C (35.3% vs. 42.1%), vitamin D (35.5% vs. 41%), vitamin E (15.2% vs. 17.5%), and zinc (18.8% vs. 29.3%) was reported [22]. Our results indicate that the COVID-19 pandemic breakout did not generally influence the pattern of supplementation among Polish residents. The vast majority of the respondents (90.7%, 61.8%, and 81.8% in the three waves, respectively) did not change their FS usage. On the other hand, we found that if the study participants decided to modify something in their diet supplementation, they tended to use more preparations, especially during the second wave (23.1%).

Another analysis considering supplementation patterns during the pandemic was carried out on the basis of the results of the application-based community survey. This study involved 175,652 supplement users and 197,068 nonusers. The risk of COVID-19 infection among women who declared intake of probiotics, omega-3 fatty acids, multivitamins, and vitamin D was lowered by 14%, 12%, 13%, and 9%, respectively. No protective association was observed among men. Moreover, no positive effect was found for respondents taking vitamin C, zinc, or garlic FS [23].

Vitamin D supplementation had a positive influence on recovery from symptoms in patients with mild to moderate COVID-19. Sabico et al., in a randomized control trial, administrated two weeks of oral supplementation of vitamin D (1000 UI vs. 5000 UI) to patients with suboptimal vitamin D status. In the group that received a higher dose, a reduced time to recovery from cough and sensory loss was found. Based on these findings, it seems reasonable to recommend vitamin D as an adjuvant to COVID-19 therapy for patients with mild to moderate symptoms [24].

Vitamin D supplementation and the risk of COVID-19 were assessed in a prospective study by Hao et al. based on data from the UK Biobank cohort study. Habitual use of vitamin D supplements was related to a 34% lower risk of infection [25].

Szarpak et al. carried out a meta-analysis of four studies, comprising 1474 patients, focusing on the influence of zinc on COVID-19 patient outcomes. In the group of patients who received zinc supplementation, survival to hospital discharge was 56.8%, while in the group to which supplementation was not administered it was 75.9%. Moreover, patients who were given supplementation had a higher percentage of in-hospital mortality (22.3% vs. 13.6%) and longer hospital stay (7.7 days vs. 7.2 days). Based on these findings, zinc supplementation does not have a beneficial impact on the abovementioned outcomes [26]. Dubourg et al. observed that median blood Zn levels were significantly lower in COVID-19 patients with poor clinical outcomes in comparison to patients with good clinical outcomes (840 µg/L vs. 970 µg/L). Those results may indicate the importance of Zn supplementation during SARS-CoV-2 infection [27].

A positive correlation was shown between Zn deficiency and COVID-19 cases per million among Asian countries in a retrospective study by Ali et al. The prevalence of Zn deficiency was nearly twice as high among Asians compared to the European population (17.5% vs. 8.9%). On the other hand, a significantly negative correlation between serum Zn levels and COVID-19 deaths per million was recognized among the European population [28]. Hoverer, cohort studies are needed to confirm these observations.

In research conducted by Adbelmaksoud et al., Zn supplementation (220 mg of zinc sulfate twice a day) was related a shortened time of smell recovery after SARS-CoV-2 infection, without an influence on the total recovery of the disease. Moreover, serum Zn levels were similar considering subgroups in the case of disease severity or the presence or absence of olfactory and/or gustatory dysfunction [29].

Thomas et al. carried out a randomized control trial among ambulatory patients (n = 214) suffering from COVID-19. Patients were randomized in a 1:1:1:1 allocation ratio every 10 days. In the first intervention group, patients were given 50 mg of zinc gluconate, the second 8000 mg of ascorbic acid, the third both, and in the last group a standard treatment regimen was observed. Researchers did not observe a significant difference in secondary outcomes among the studied groups. The results of the study by Thomas et al. do not confirm the assumption of the study by Dubourg et al. [30].

It should be emphasized that the vast majority of respondents took dietary supplements, and not drugs containing zinc and vitamin D. Drugs were used by 5.1% of the respondents during the first wave; during the second wave it was 15.1%, and during the third 13.1%. The respondents themselves noticed the need for supplementation, and the advice of specialists (doctors, pharmacists) was much less frequently cited. Therefore, education on the proper selection of a preparation (the right chemical form, with good digestibility) and the right dose (in accordance with the recommendations corresponding to the daily requirement for supplemented dietary components) seems to be important. It was estimated that many respondents did not provide the names of the zinc and vitamin D preparations used—this may indicate that they do not pay attention to it, while one of the important criteria is the price of FS.

The limitations of this study include the unequal gender proportions (the predominance of women)—if the majority of participants in the study were men, the results could be different. However, greater female participation is a fairly common problem in volunteer-based surveys. Moreover, our survey was a retrospective study, so incorrect recall of information by survey participants may be an important problem.

5. Conclusions

The popularity of dietary supplements, especially vitamin and mineral supplements, is gradually increasing in Poland. During the COVID-19 pandemic, the consumption of dietary supplements containing zinc and vitamin D increased, especially among people with higher education, or medical and paramedical education, which indicates the increased awareness of this social group regarding pro-health prophylaxis. Due to the nonrestrictive registration procedures of dietary supplements, it seems necessary to educate consumers in terms of the selection of appropriate preparations, proper nutrition, and balanced supplementation.

Author Contributions: Conceptualization, A.P.-J.; methodology, A.P.-J. and J.B.; software, A.P.-J., J.B., M.G., K.M., A.M., J.M., S.K.N., R.M.-Ż. and P.N.; validation, K.S., R.M.-Ż. and A.P.-J.; formal analysis, K.S.; investigation, A.P.-J., J.B., M.G. and A.M.; resources, K.S. and A.P.-J.; data curation, A.P.-J., J.B., M.G., R.M.-Ż., K.M., J.M., S.K.N., A.M., K.J.G.-K. and J.S.; writing—original draft preparation, A.P.-J. and J.B.; writing—review and editing, K.S., R.M.-Ż. and S.K.N.; visualization, A.P.-J. and J.B.; supervision, K.S.; project administration, A.P.-J., J.B. and K.S. All authors have read and agreed to the published version of the manuscript.

Funding: This research received no external funding.

Institutional Review Board Statement: Not applicable.

Informed Consent Statement: The participants of the study were informed that completing the anonymous questionnaire, which did not contain personally identifiable information, was tantamount to consenting to participation in the study.

Data Availability Statement: Excel spreadsheets with data are available from the authors.

Conflicts of Interest: The authors declare no conflict of interest.

Appendix A

Table A1. Questionnaire.

Question	Answer
1. Gender	Female/male
2. Age (year)	
3. Height (cm)	
4. Body weight (kg)	
5. Education level	Primary school/higher/secondary
6. Type of education	Medical and related/nonmedical/not applicable
7. Place of living	Village/City < 150,000 inhabitants/City 150,000–250,000 inhabitants/City > 250,000 inhabitants
8. Subjective assessment of the financial situation of the household	Very good/rather good/average/rather bad/bad
9. Number of people in the household	
10. Professional activity	Unemployed person/person working in office /person working online/student
11. Physical activity	Inactivity (sedentary)/low (occasional exercise, 1–3 times a week)/moderate (1 h of training per day)/high (hard physical work and daily workouts)
12. Voivodeship	
13. How would you rate your health at the beginning of the pandemic in Poland?	Very good/good/medium/poor
14. Have you had COVID-19?	Yes/no
15. Chronic diseases	Hypertension/type 1 diabetes/type 2 diabetes/atherosclerosis/gout/hypothyroidism/overactive thyroid gland/allergy/food intolerances/obesity/insulin resistance/cancer/I am not sick
16. Has your body weight changed during the pandemic?	No/increased/decreased
17. Did you exercise at home during the pandemic?	Yes/no
18. What is a food supplement?	A preparation that treats nutritional deficiencies/preparation, which supplements nutritional deficiencies
19. Do you think food supplements differ from medications?	Yes/no
20. Do you think food supplements differ from medications? If so, please specify how.	
21. What categories of food supplements did you use during the pandemic?	Vitamins and minerals/probiotics/prebiotics/supporting immunity/supporting weight loss/improving the condition of hair, skin, and nails/supporting the functioning of the urinary tract/supporting the heart/supporting memory/supporting lowering cholesterol levels/vision support/supporting the functioning of the joints/relieving the symptoms of menopause/supporting the digestive tract/improving well-being/facilitating sedation and sleep/supporting alcohol metabolism/for athletes/removing excess water/I did not use dietary supplements

Table A1. *Cont.*

Question	Answer
22. Have you used zinc and vitamin D food supplements since March 2020/September 2020/February 2021?	No/only drugs/yes both/only zinc/only vitamin D
23. Why did you use such food supplements?	To improve health/to supplement deficiencies of vitamins and minerals/to supplement the therapy prescribed by my doctor/because a pharmacist recommended them/because I was encouraged by TV, media, and/or Internet advertising /someone I knew recommended them to me/not applicable
24. The names of all food supplements currently used, together with the name of the manufacturer	
25. The names of chronic medicines used (name and dose)	
26. Why did you use such food supplements?	Not applicable/someone I knew recommended them to me/to improve health
27. What do you usually use to wash down food supplements and medications?	Tea/cola/not applicable/I don't drink/juice/water
28. Do you think there were more advertisements for food supplements during the pandemic?	Yes/no/I didn't notice a change
29. Do you use food supplements in the amount recommended on the package?	I do not use them at all/no, I use lower doses/no, I use higher doses/yes
30. Do you think food supplements can have side effects?	No, taking them is absolutely safe/yes
31. How do you assess the advisability of using food supplements?	They should be used in the event of identified deficiencies/their use is unnecessary/I have no opinion
32. Do you think food supplements can be overdosed on?	No, they're safe/yes
33. Do you think that food supplements can interact with medications prescribed by your doctor, and thus affect the effectiveness of therapy?	No, they're safe/yes
34. Has the pandemic affected your use of food supplements?	No/yes, I use fewer/yes, I use more
35. Have you heard that preparations containing zinc and vitamin D can support immunity and be helpful in the prevention of viral infections?36. Any additional comments	Yes/no

References

1. European Food Safety Authority. Available online: https://www.efsa.europa.eu/en/topics/topic/food-supplements (accessed on 20 August 2021).
2. World Health Organization, Coronavirus Disease (COVID-19) Pandemic. Available online: https://www.who.int/emergencies/diseases/novel-coronavirus-2019 (accessed on 23 August 2021).
3. Berger, M.M.; Herter-Aeberli, I.; Zimmermannm, M.B.; Spieldenner, J.; Eggersdorfer, M. Strengthening the immunity of the Swiss population with micronutrients: A narrative review and call for action. *Clin. Nutr. ESPEN* **2021**, *43*, 39–48. [CrossRef] [PubMed]
4. Martineau, A.R.; Jolliffe, D.A.; Hooper, R.L.; Greenberg, L.; Aloia, J.F.; Bergman, P.; Dubnov-Raz, G.; Esposito, S.; Ganmaa, D.; Ginde, A.A.; et al. Vitamin D supplementation to prevent acute respiratory tract infections: Systematic review and meta-analysis of individual participant data. *BMJ* **2017**, *356*, i6583. [CrossRef] [PubMed]
5. Patel, N.; Penkert, R.R.; Jones, B.G.; Sealy, R.E.; Surman, S.L.; Sun, Y.; Tang, L.; DeBeauchamp, J.; Webb, A.; Richardson, J.; et al. Baseline serum Vitamin A and D levels determine benefit of oral vitamin A&D supplements to humoral immune responses following pediatric influenza vaccination. *Viruses* **2019**, *11*, 907. [CrossRef]
6. Subedi, L.; Tchen, S.; Gaire, B.P.; Hu, B.; Hu, K. Adjunctive nutraceutical therapies for COVID-19. *Int. J. Mol. Sci.* **2021**, *22*, 1963. [CrossRef]
7. Ueland, P.M.; McCann, A.; Midttun, Ø.; Ulvik, A. Inflammation, vitamin B6 and related pathways. *Mol. Aspects Med.* **2017**, *53*, 10–27. [CrossRef]
8. Shankar, A.H.; Prasad, A.S. Zinc and immune function: The biological basis of altered resistance to infection. *Am. J. Clin. Nutr.* **1998**, *68*, 447S–463S. [CrossRef] [PubMed]
9. Ebrahimzadeh-Attari, V.; Panahi, G.; Hebert, J.R.; Ostadrahimi, A.; Saghafi-Asl, M.; Lotfi-Yaghin, N.; Baradaran, B. Nutritional approach for increasing public health during pandemic of COVID-19: A comprehensive review of antiviral nutrients and nutraceuticals. *Health Promot. Perspect.* **2021**, *11*, 119–136. [CrossRef]
10. Sheridan, P.A.; Zhong, N.; Carlson, B.A.; Perella, C.M.; Hatfield, D.L.; Beck, M.A. Decreased selenoprotein expression alters the immune response during influenza virus infection in mice. *J. Nutr.* **2007**, *137*, 1466–1471. [CrossRef]
11. Skotnicka, M.; Karwowska, K.; Kłobukowski, F.; Wasilewska, E.; Małgorzewicz, S. Dietary habits before and during the COVID-19 epidemic in selected European Countries. *Nutrients* **2021**, *13*, 1690. [CrossRef]
12. Prete, M.; Luzzetti, A.; Augustin, L.S.A.; Porciello, G.; Montagnese, C.; Calabrese, I.; Ballarin, G.; Coluccia, S.; Patel, L.; Vitale, S.; et al. Changes in lifestyle and dietary habits during COVID-19 lockdown in Italy: Results of an online survey. *Nutrients* **2021**, *13*, 1923. [CrossRef]
13. Grant, F.; Scalvedi, M.L.; Scognamiglio, U.; Turrini, A.; Rossi, L. Eating habits during the COVID-19 lockdown in Italy: The nutritional and lifestyle side effects of the pandemic. *Nutrients* **2021**, *13*, 2279. [CrossRef]

14. Hamulka, J.; Jeruszka-Bielak, M.; Górnicka, M.; Drywień, M.E.; Zielińska-Pukos, M.A. Dietary supplements during COVID-19 outbreak. Results of Google Trends analysis supported by PLifeCOVID-19 online studies. *Nutrients* **2021**, *13*, 54. [CrossRef] [PubMed]
15. Lee, J.; Kwan, Y.; Lee, J.Y.; Shin, J.I.; Lee, K.H.; Hong, S.H.; Han, Y.J.; Kronbichler, A.; Smith, L.; Koyanagi, A.; et al. Public interest in immunity and the justification for intervention in the early stages of the COVID-19 pandemic: Analysis of Google Trends data. *J. Med. Internet Res.* **2021**, *23*, e26368-10. [CrossRef] [PubMed]
16. Czerwiński, A.; Liebers, D. Regulation of the market regulation of the dietary supplements market. In *Does Poland Have a Chance to Become a European Leader?* Polish Economic Institute: Warsaw, Poland, 2020.
17. Tavolacci, M.P.; Dechelotte, P.; Ladner, J. COVID-19 vaccine acceptance, hesitancy, and resistancy among University students in France. *Vaccines* **2021**, *9*, 654. [CrossRef] [PubMed]
18. Preacher, K.J. Calculation for the Chi-Square Test: An Interactive Calculation Tool for Chi-Square Tests of Goodness of Fit and Independence (Computer Software). Available online: http://quantpsy.org (accessed on 29 August 2021).
19. Kamarli Altun, H.; Karacil Ermumcu, M.; Seremet Kurklu, N. Evaluation of dietary supplement, functional food and herbal medicine use by dietitians during the COVID-19 pandemic. *Public Health Nutr.* **2021**, *24*, 861–869. [CrossRef] [PubMed]
20. Alfawaz, H.A.; Khan, N.; Aljumah, G.A.; Hussain, S.D.; Al-Daghri, N.M. Dietary intake and supplement use among Saudi Residents during COVID-19 lockdown. *Int. J. Environ. Res. Public Health* **2021**, *18*, 6435. [CrossRef]
21. Bagheri, M.; Haghollahi, F.; Shariat, M.; Jafarabadi, M.; Aryamloo, P.; Rezayof, E. Supplement usage pattern in a group of COVID-19 patients in Tehran. *J. Family Reprod. Health* **2020**, *14*, 158–165. [CrossRef]
22. Mohsen, H.; Yazbeck, N.; Al-Jawaldeh, A.; Bou Chahine, N.; Hamieh, H.; Mourad, Y.; Skaiki, F.; Salame, H.; Salameh, P.; Hoteit, M. Knowledge, attitudes, and practices related to dietary supplementation, before and during the COVID-19 pandemic: Findings from a cross-sectional survey in the Lebanese population. *Int. J. Environ. Res. Public Health.* **2021**, *18*, 8856. [CrossRef]
23. Louca, P.; Murray, B.; Klaser, K.; Graham, M.S.; Mazidi, M.; Leeming, E.R.; Thompson, E.; Bowyer, R.; Drew, D.A.; Nguyen, L.H.; et al. Modest effects of dietary supplements during the COVID-19 pandemic: Insights from 445 850 users of the COVID-19 symptom study app. *BMJ Nutr. Prev. Health.* **2021**, *4*, 149–157. [CrossRef]
24. Sabico, S.; Enani, M.A.; Sheshah, E.; Aljohani, N.J.; Aldisi, D.A.; Alotaibi, N.H.; Alshingetti, N.; Alomar, S.Y.; Alnaami, A.M.; Amer, O.E.; et al. Effects of a 2-Week 5000 IU versus 1000 IU Vitamin D3 supplementation on recovery of symptoms in patients with mild to moderate Covid-19: A randomized clinical trial. *Nutrients* **2021**, *13*, 2170. [CrossRef]
25. Hao, M.; Zhou, T.; Heianza, Y.; Qi, L. Habitual use of vitamin D supplements and risk of coronavirus disease 2019 (COVID-19) infection: A prospective study in UK Biobank. *Am. J. Clin. Nutr.* **2021**, *113*, 1275–1281. [CrossRef]
26. Szarpak, L.; Pruc, M.; Gasecka, A.; Jaguszewski, M.J.; Michalski, T.; Peacock, F.W.; Smereka, J.; Pytkowska, K.; Filipiak, K.J. Should we supplement zinc in COVID-19 patients? Evidence from meta-analysis. *Pol. Arch. Intern. Med.* **2021**. [CrossRef]
27. Dubourg, G.; Lagier, J.C.; Brouqui, P.; Casalta, J.P.; Jacomo, V.; La Scola, B.; Rolain, J.M.; Raoult, D. Low blood zinc concentrations in patients with poor clinical outcome during SARS-CoV-2 infection: Is there a need to supplement with zinc COVID-19 patients? *J. Microbiol. Immunol. Infect.* **2021**. [CrossRef] [PubMed]
28. Ali, N.; Fariha, K.A.; Islam, F.; Mohanto, N.C.; Ahmad, I.; Hosen, M.J.; Ahmed, S. Assessment of the role of zinc in the prevention of COVID-19 infections and mortality: A retrospective study in the Asian and European population. *J. Med. Virol.* **2021**, *93*, 4326–4333. [CrossRef] [PubMed]
29. Abdelmaksoud, A.A.; Ghweil, A.A.; Hassan, M.H.; Rashad, A.; Khodeary, A.; Aref, Z.F.; Sayed, M.A.A.; Elsamman, M.K.; Bazeed, S.E.S. Olfactory disturbances as presenting manifestation among Egyptian patients with COVID-19: Possible role of zinc. *Biol. Trace Elem. Res.* **2021**, 1–8. [CrossRef]
30. Thomas, S.; Patel, D.; Bittel, B.; Wolski, K.; Wang, Q.; Kumar, A.; Il'Giovine, Z.J.; Mehra, R.; McWilliams, C.; Nissen, S.E.; et al. Effect of high-dose zinc and ascorbic acid supplementation vs usual care on symptom length and reduction among ambulatory patients with SARS-CoV-2 infection: The COVID A to Z randomized clinical trial. *JAMA Network Open.* **2021**, *14*, e210369. [CrossRef]

MDPI

Review

Does Oxidative Stress Management Help Alleviation of COVID-19 Symptoms in Patients Experiencing Diabetes?

Alok K. Paul [1,2], Md K. Hossain [3], Tooba Mahboob [4], Veeranoot Nissapatorn [4], Polrat Wilairatana [5,*], Rownak Jahan [2], Khoshnur Jannat [2], Tohmina A. Bondhon [2], Anamul Hasan [2], Maria de Lourdes Pereira [6] and Mohammed Rahmatullah [2,*]

1 School of Pharmacy and Pharmacology, University of Tasmania, Private Bag 26, Hobart, TAS 7001, Australia; alok.paul@utas.edu.au
2 Department of Biotechnology & Genetic Engineering, University of Development Alternative, Lalmatia, Dhaka 1207, Bangladesh; rownak86@hotmail.com (R.J.); jannat.koli.22@gmail.com (K.J.); afrozebondhon@gmail.com (T.A.B.); anamulhasanoris@gmail.com (A.H.)
3 Institute for Health and Sports, Victoria University, Melbourne, VIC 3011, Australia; md.hossain18@live.vu.edu.au
4 World Union for Herbal Drug Discovery (WUHeDD) and Research Excellence Center for Innovation and Health Products (RECIHP), School of Allied Health Sciences, Walailak University, Nakhon Si Thammarat 80160, Thailand; tooba666@hotmail.com (T.M.); nissapat@gmail.com (V.N.)
5 Department of Clinical Tropical Medicine, Faculty of Tropical Medicine, Mahidol University, Bangkok 10400, Thailand
6 CICECO-Aveiro Institute of Materials & Department of Medical Sciences, University of Aveiro, 3810-193 Aveiro, Portugal; mlourdespereira@ua.pt
* Correspondence: polrat.wil@mahidol.ac.th (P.W.); rahamatm@hotmail.com (M.R.); Tel.: +66-023549168 (P.W.); +880-1715032621 (M.R.)

Citation: Paul, A.K.; Hossain, M.K.; Mahboob, T.; Nissapatorn, V.; Wilairatana, P.; Jahan, R.; Jannat, K.; Bondhon, T.A.; Hasan, A.; de Lourdes Pereira, M.; et al. Does Oxidative Stress Management Help Alleviation of COVID-19 Symptoms in Patients Experiencing Diabetes? *Nutrients* **2022**, *14*, 321. https://doi.org/10.3390/nu14020321

Academic Editors: William B. Grant and Ronan Lordan

Received: 30 November 2021
Accepted: 11 January 2022
Published: 13 January 2022

Abstract: Severe acute respiratory syndrome (SARS)-CoV-2 virus causes novel coronavirus disease 2019 (COVID-19) with other comorbidities such as diabetes. Diabetes is the most common cause of diabetic nephropathy, which is attributed to hyperglycemia. COVID-19 produces severe complications in people with diabetes mellitus. This article explains how SARS-CoV-2 causes more significant kidney damage in diabetic patients. Importantly, COVID-19 and diabetes share inflammatory pathways of disease progression. SARS-CoV-2 binding with ACE-2 causes depletion of ACE-2 (angiotensin-converting enzyme 2) from blood vessels, and subsequently, angiotensin-II interacts with angiotensin receptor-1 from vascular membranes that produce NADPH (nicotinamide adenine dinucleotide hydrogen phosphate) oxidase, oxidative stress, and constriction of blood vessels. Since diabetes and COVID-19 can create oxidative stress, we hypothesize that COVID-19 with comorbidities such as diabetes can synergistically increase oxidative stress leading to end-stage renal failure and death. Antioxidants may therefore prevent renal damage-induced death by inhibiting oxidative damage and thus can help protect people from COVID-19 related comorbidities. A few clinical trials indicated how effective the antioxidant therapy is against improving COVID-19 symptoms, based on a limited number of patients who experienced COVID-19. In this review, we tried to understand how effective antioxidants (such as vitamin D and flavonoids) can act as food supplements or therapeutics against COVID-19 with diabetes as comorbidity based on recently available clinical, preclinical, or in silico studies.

Keywords: COVID-19; diabetes mellitus; oxidative stress; kidney damage; antioxidant

1. Introduction

The novel coronavirus disease 2019 (COVID-19) is caused by severe acute respiratory syndrome CoV2 (SARS-CoV-2) virus and can be associated with infected patients with various comorbidities such as diabetes, hypertension, and cardiovascular disorders. Studies show that the viral infection triggers severe clinical symptoms and mortality with people experiencing comorbidities such as diabetes, cancer, and heart and lung disorders. Importantly among these people, diabetic patients experience the most severe clinical symptoms that cause the highest proportional death than non-diabetic patients after SARS-CoV-2 infection [1,2]. Along with diabetes, old age, congestive heart failure, smoking, β-blocker use, presence of bilateral lung infiltrates, elevated creatinine and severe vitamin D deficiency" are significant cause of mortality in COVID-19 patients [3]. In addition, high plasma lactate dehydrogenase level, a marker of oxidative stress, and advanced age 70 years or above) showed increased mortality, anxiety, and severity of COVID-19 symptoms in the clinic [4–6]. Several questions need to be answered to understand the pathophysiological connections between COVID-19 and diabetes mellitus, which leads to an increase in fatalities. Approximately four different pathogenesis are involved in SARS-CoV-2 infection, such as activation of the renin-angiotensin (RAS) pathway, oxidative stress, excess cytokines release, and dysfunction of endothelium. COVID-19 develops after SARS-CoV-2 entry in host's cells and RAS activation with oxidative bursts [7,8]. In this article, we give some insights on common features between diabetes and COVID-19-induced kidney damage and discuss the implications of increased oxidative stress in the process, which may help improve patient prognosis.

2. Is Oxidative Stress a Major Cause of Diabetes-Induced Kidney Damage?

Diabetes is one of the most common metabolic disorders influenced by several factors such as age, sex, ethnicity, genetic factors, and pregnancy and appears as a comorbidity with obesity, cardiovascular diseases, atherosclerosis, renal failure, cancer, and many other chronic diseases [9]. People with diabetes show an impaired function of insulin (insulin resistance) and therefore need an increased amount of insulin than β cells (in the pancreas of a person) can produce. As a result, the presence of higher blood glucose in the bloodstream is observed. It has been postulated that diabetic nephropathy develops due to localized oxidative stress, where the key initiator may be increased mitochondrial production of reactive oxygen species (ROS) arising from hyperglycemia and leading to various renal disorders [10]. Diabetic nephropathy is present in almost one-third of Type 1 and Type 2 diabetic patients [11]. Diabetic neuropathy, nephropathy, and retinopathy can arise from oxidative stress-induced complications in diabetes mellitus along with a host of other disorders like coronary artery disease [12].

Diabetes is considered to be one of the major indicators for severe COVID-19 prognosis, as more diabetic patients (diabetes type-2 is mainly evident, with limited evidence from diabetes Type-1) showed severe COVID-19 symptoms and deaths after exposure to SARS-CoV-2 virus [1,13–15]. A meta-analysis concluded that the diabetic patients showed a 200% increased probability of death with severe COVID-19 symptoms than non-diabetic patients [16]. Importantly, Toll-like receptor 4 (TLR4) is responsible for initiating diabetes by expressing the transcriptional factor nuclear factor-kappaB (NF-κB) and the enzyme nicotinamide adenine dinucleotide phosphate (NADPH) oxidase to produce ROS, which also induce activation of endothelial nitric oxide synthase (eNOS) and xanthine oxidase enzymes [17]. Together these enzymes produce excess ROS and can be the causative agent(s) for diabetes-like diseases [18]. Another recent study reported that presence of diabetes mellitus type 1 results in increased morbidity and mortality rates during coronavirus (COVID-19) disease [13]. Diabetic patients displayed higher cell counts of leukocytes and neutrophils in their blood during admission with comparatively severe COVID-19 symptoms than non-diabetic patients. The diabetic patients also required more antibiotic therapy and artificial ventilation, but still resulted in more deaths during their stay in the healthcare facilities in China [1]. Oxidative stress also causes decreased use of glucose by muscles and adipose tissues. An increase of 8-epi-prostaglandin F2α, an oxidative stress indicator, is positively correlated with insulin resistance [19] (Figure 1). Insulin resistance is also thoroughly interrelated with inflammation as a preclinical study showed increased tumor necrosis factor α (TNF-α) from adipose tissues of obese and diabetic animals), a proinflammatory cytokine that can cause insulin resistance; suppression of TNF-α helps recovery of insulin resistance [20] (Figure 1). NLRP3 (nucleotide-binding oligomerization domain-like receptor family pyrin domain containing 3), a polyprotein complex inflammasome found in macrophages, is also responsible for causing diabetes and the release of inflammatory cytokines. NLRP3 is stimulated by the activation of NF-κB (nuclear factor-kappa B, which is triggered by TNF-α) and causes the secretion of proinflammatory cytokines pro-IL-1β and pro-IL-18 (Figure 1). NLRP3 matures by PAMPs (pathogen-associated molecular patterns) and DAMPs (damage-associated molecular patterns) or lipopolysaccharides. The maturation of NLRP3 causes the release of cytokines such as IL (interleukin)-1β and IL-18 and inflammation in the body [21]. Adipose tissues mainly produce inflammatory biomarkers such as TNF- α, and macrophages and other immune cells are partially responsible for insulin resistance. Type-2 diabetic patients show increased inflammatory cytokines and autoimmune responses in the pancreatic islet cells and can cause insulin resistance and decreased insulin secretion, although the whole mechanism is not yet clearly understood [22]. Oxidative stress, insulin resistance, inflammation, and kidney cell damage are interrelated and part of a chronic pathophysiological mechanism.

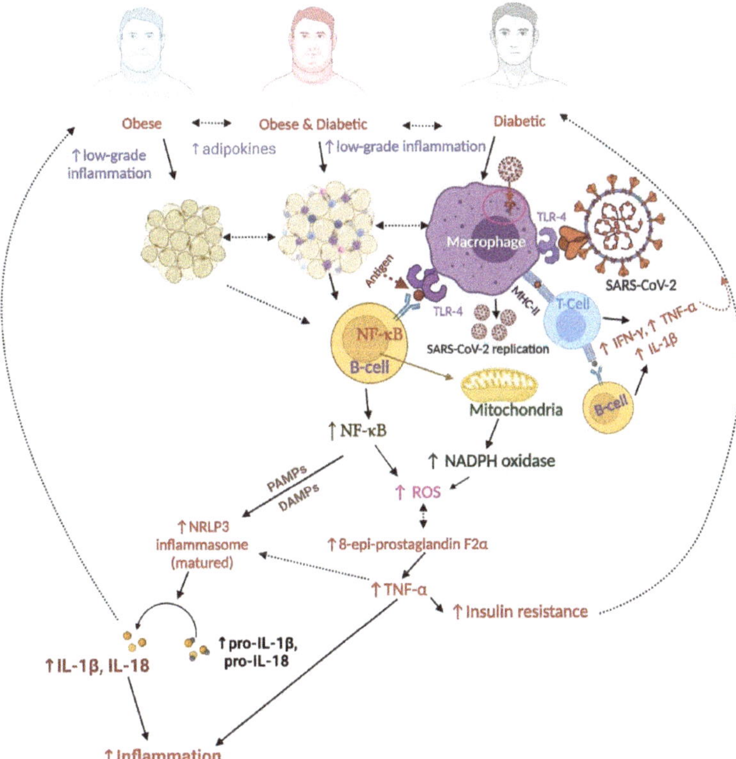

Figure 1. Correlation between SARS-CoV-2, oxidative stress, diabetes, and obesity. Abbreviations: ↑: increase; ROS, reactive oxygenated species; TNF-α, tumor necrosis factor α, TLR, Toll-like receptor; IL, interleukin; NADPH, nicotinamide adenine dinucleotide phosphate oxidase; IFNγ, gamma interferon; NF-κB, nuclear factor-kappa B; NLRP3, nucleotide-binding oligomerization domain-like receptor family pyrin domain containing 3 inflammasome; PAMPs, pathogen-associated molecular patterns; DAMPs, damage-associated molecular patterns; MHC-II, major histocompatibility complex class II. (The figure was made with www.biorender.com, accessed on 13 December 2021).

3. What Is the Clinical Evidence on the Relationships between COVID-19 and Diabetes?

Some studies suggest that the COVID-19 vaccination should be prioritized in diabetic patients (both type 1 and 2) as they have a poorer prognosis with COVID-19 compared to COVID-19 patients without diabetes [23,24]. A recent randomized clinical trial on the Scottish population (a population cohort study) in the first wave found increased severity in COVID-19 symptoms and admitted for fatal and critical care units for treatment with diabetes compared with those without diabetes [25]. The overall odds ratio for diabetes was 1·395, calculated against patients without diabetes, which means diabetes was strongly positively correlated with the severity of COVID-19 patients. Noticeably, the odds ratio for the severity of Type-1 diabetic patients were much higher than Type-2 diabetic patients [25]. Another RCT with children with Type-1 diabetes in the US suggested that preintervention and social support improved the children to manage COVID-19 pandemic-related stress and depressive symptoms for the children and their parents [26].

Another ongoing RCT with COVID-19 patients introduced "telemetric continuous glucose monitoring" for patients with positive diabetes suggested remote glucose monitoring may provide similar results to conventional finger-prick test (n = ~36 each group) but better outcomes as it needs less exposure of healthcare workers and fewer risk of cross-contaminations or reinfections [27].

A further RCT in Taiwan tried to educate and guide patients with diabetes Type-2 to maintain their health during the COVID-19 pandemic and found that the health-related coaching helped keep patient's glycosylated hemoglobin (HbA1 c) levels under control; they maintained physical exercises, and reduced eating out [28].

A systematic review investigated the relationships among periodontal diseases, diabetes, and COVID-19 and indicated that hyperglycemia (e.g., diabetes) might increase the possibilities of periodontitis development and influence excessive expression of angiotensin-converting enzyme 2 (ACE-2) in periodontal tissue of diabetes Type-2 patients [29]. In addition, the excessive ACE-2 can favor the SARS-CoV-2 virus to develop COVID-19 [29]. Therefore, periodontal diseases or diabetes type 2 can potentially influence the development of COVID-19 symptoms and go for mild to severe form depending on the physiological and pathological conditions of the patients. However, no proper randomized clinical trials are evident to date proving this relationship.

4. How Can SARS-CoV-2 Damage the Kidneys?

SARS-CoV-2 enters the host body interacting with the angiotensin-converting enzyme-2 (ACE-2), which is present in multiple organs, mainly kidneys, lungs, testis, breast, heart, and gastrointestinal systems [30]. SARS-CoV-2 interacts with angiotensin-converting enzyme 2 (ACE2) and causes an increase of angiotensin-2 in tissues that activates CD8+ and CD4+ T-lymphocytes macrophages and NK cells and releases pro-inflammatory and inflammatory cytokines such as IL-1β, IL-2, IL-4, IL-17, IL-21, and IFNγ (gamma interferon) [31]. SARS-CoV-2 interaction with TLR4 in macrophages can also activate major histocompatibility complex (MHC) class II molecules and thus in-crease T-cells- and B-cells-mediated secretions of proinflammatory cytokines (IL-1β, IFNγ, and TNF-α) (Figure 1). The released inflammatory cytokines from the lungs, kidneys, or elsewhere in the body because of SARS-CoV-2 infection, are transported through the bloodstream that causes quick acute inflammation in the capillaries of kidneys, lungs, heart, and all major organs.

ACE-2 receptor is expressed mainly in proximal tubular epithelial cells in both diabetic and healthy kidneys, but diabetic patients express higher ACE-2 receptors in their pancreatic islets than normoglycemic patients [32]. In COVID-19 patients, ACE-2 receptor containing proximal tubular epithelial cells has been detected in urine samples, suggesting a common infection pattern of SARS-CoV-2 in patients with diabetes [33]. Importantly, overexpression of ACE-2 receptors in the proximal tubular epithelium of diabetic patients may cause severe SARS-CoV-2 associated clinical symptoms and damage to kidneys as microscopic examination of COVID-19 infected kidneys showed proximal tubular injury and acute tubular necrosis [34]. Another study indicated that acute injury in the kidney is responsible for the increased morbidity and mortality of SARS-CoV-2 infected patients [35].

ACE-2 binding of SARS-CoV-2 causes depletion of ACE-2 receptors that may facilitate the binding of angiotensin-II with angiotensin receptor-1 from blood vascular membranes that produce NADPH (nicotinamide adenine dinucleotide hydrogen phosphate) oxidase, oxidative stress and cause constriction of blood vessels, platelet aggression, the release of proinflammatory cytokines (i.e., inflammation), and increase the severity of the infection [36,37]. SARS-CoV-2 induced severe infection also causes a high neutrophil/lymphocyte ratio that generates increased reactive oxygen species levels. The oxidative stress further induces platelet dysfunction and tissue damage in the lung, kidney, and other major organs [38].

In a cross-sectional study conducted with 50 COVID-19 patients in Nigeria, oxidative stress marker, 8-isoprostaglandin F2α, was found to be significantly higher ($p = 0.049$); on the other hand, malondialdehyde (MDA) was lower ($p < 0.001$) in COVID-19 patients than controls. The authors further concluded that COVID-19 infections and other comorbidities such as diabetes, malaria, and hypertension increased the risks of developing oxidative stress [39]. Furthermore, increased oxidative stress could be responsible for "amplifying and perpetuating the cytokine storm, coagulopathy, and cell hypoxia" in COVID-19 patients [40]. Oxidative stress has also been described as a 'key player' in the pathogenesis, severity, and mortality risk in SARS-CoV-2 infections [41]. A systematic review and me-ta-analysis showed that acute respiratory distress syndrome development in COVID-19 patients accelerated the development of acute kidney injury (AKI) and higher mortality rate [42].

5. Synergistic Kidney Damage and Morbidity Due to COVID-19 and Diabetes

Both diabetes and COVID-19 cause oxidative damage and inflammation in tissues and share common molecular pathways to generate clinical symptoms. As discussed, the presence of both diseases, COVID-19 and diabetes can cause synergistic oxidative stress, severe inflammation, vasoconstriction, and thrombosis in capillary blood vessels, mainly in the kidney and lungs, and therefore cause synergistic damage in these organs that leads to death. A study conducted on 174 COVID-19 patients (24 patients among them diabetic) found that diabetic patients with COVID-19 were at an increased risk of poor prognosis due to higher risks of severe pneumonia and out-of-control inflammatory responses [43]. Another study reported that the chance of developing COVID-19 pneumonia is 87.9% higher in patients with diabetic nephropathy, and the probability of ventilation is 101.7% higher, probability of a fatal outcome is 20.8% more compared to chronic kidney disease alone [44]. Noticeably in this regard, a recent study found significantly lower mortality in metformin-administered COVID-19 diabetic patients (3/104, 2.9%) than in the non-metformin-administered COVID-19 diabetic group (22/179, 12.3%, $p = 0.01$), suggesting that blood sugar control is a significant factor in reducing mortality rates when diabetes is a comorbid factor with COVID-19 [45]. However, metformin can act through a secondary mechanism. Since the drug acts through AMPK (AMP-activated protein kinase) activation, such activation can lead to phosphorylation of ACE2, the receptor for SARS-CoV-2 [46]. This in turn can lead to conformational and functional changes in ACE2 leading to decreased binding ability of the SARS-CoV-2 spike protein receptor binding domain (S-RBD), leading to decreased entry of the virus into human host cells. The presence of a large phosphate moiety on ACE2 due to phosphorylation by AMPK can further be a factor in decreased binding ability of S-RBD to ACE2 because of steric hindrance [46]. Furthermore, COVID-19 can by itself cause AKI, a fact recognized earlier on following the outbreak of the pandemic [47]. We hypothesize that COVID-19 and diabetes increase oxidative stress that can play a synergistic role in damage to the kidneys, when present as comorbidities (Figure 2) [48,49].

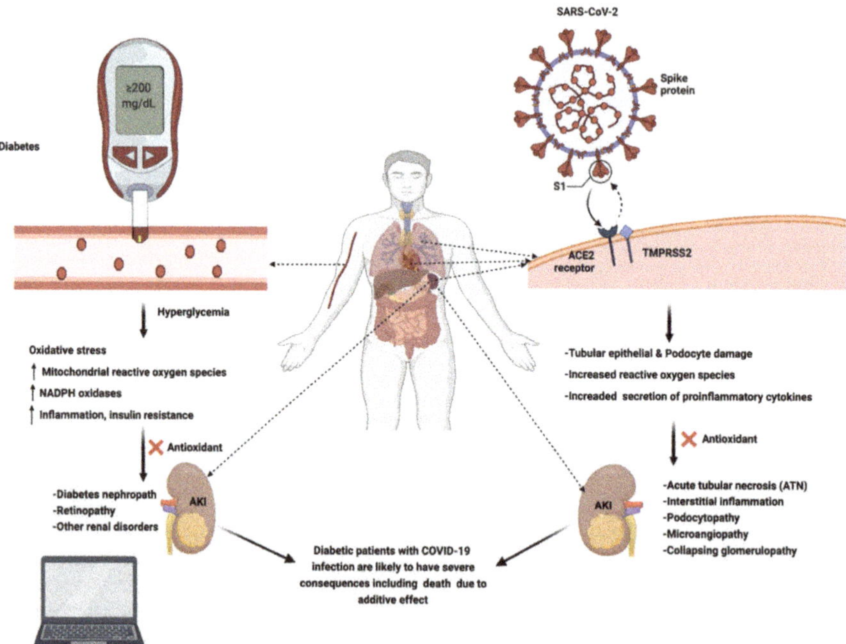

Figure 2. SARS-CoV-2 and diabetes induce kidney damage via oxidative stress: the role of antioxidants. Abbreviations: ↑: increase; CKD, chronic kidney disease AKI, acute kidney injury; SARS-CoV-2, severe acute respiratory syndrome coronavirus-2; COVID-19, coronavirus disease 2019; TMPRSS2, Transmembrane protease serine 2. (The figure was made with www.biorender.com, accessed on 13 December 2021).

Interestingly, some antioxidants like flavonoids have been suggested as a complementary therapy for COVID-19 [50] and diabetes [51], which could be beneficial in ameliorating kidney damage during COVID-19 infection with diabetes as a comorbidity [52]. For example, the flavonoid apigenin reportedly attenuated renal dysfunction, oxidative stress and fibrosis in streptozotocin-induced diabetic rats [53]. Apigenin has also been shown in in silico studies to be an inhibitor of Mpro, the main protease of SARS-CoV-2 and which plays a vital role in viral replication [54]. Apigenin is not the only example of its type. The flavonoid quercetin reportedly acts as a prophylactic to COVID-19 [55,56], as well as an antidiabetic and antioxidant compound. Moreover, recently, a preclinical study showed quercetin's renal protective effects [57]. Intragastric administration of quercetin (1.5 and 3 g per kg body weight daily for eight weeks) effectively reduced apoptosis of renal cells and plasma levels of blood urea nitrogen, creatinine, and uric acid in male Sprague Dawley rat model of chronic renal failure [57]. The study also reported that quercetin treated rats showed reduced inflammation by preventing phosphoinositide 3-kinase (PI3 k)/Akt (protein kinase B) signaling pathway by targeting phosphoinositide 3-kinase regulatory subunit 1 (PIK3 R1) and reduced expression NLRP3, p-PI3 k, Phospho-Akt (p-Akt), and caspase1 in kidney tissues [57]. Another study reported that in a mouse model of renal dysplasia, quercetin treatment increased the epithelial organization of developing nephrons, inhibited nuclear beta-catenin, and thus improved renal dysplasia [58]. A report showed that combined pretreatment of 30 mg/kg resveratrol and 50 mg/kg quercetin over a period of seven days prevented paracetamol-induced (2 g/kg body weight) acute renal failure via reducing plasma creatinine, urea, and inflammatory markers (e.g., MDA, IL-6, and TNF-α) [59].

Modlinger and colleagues show that oxidative stress can cause salt retention in kidneys by promoting the expression of vasoconstrictor molecules and NADPH oxidase, and thus it can cause acute to chronic renal failure [60]. Another report mentioned that COVID-19 causes activation of the innate immune response and secretion of inflammatory cytokines due to the development of oxidative stress [61]. The cytokine storm seen repeatedly in COVID-19 patients has been hypothesized to be a consequence of oxidative stress [50]; as such, it can be expected that antioxidants such as flavonoid compounds would relieve COVID-19 severity, similar to antioxidant flavonoid effects on ameliorating diabetic cardiac myopathy through alleviation of oxidative stress [62] and diabetic nephropathy through a similar mechanism. Quercetin, apigenin, baicalin, luteolin, hesperidin, genistein, proanthocyanidin and eriodictyol have been found to be capable of alleviating oxidative stress in diabetic nephropathy [63]. Incidentally, all the above flavonoid compounds have been reported to bind to SARS-CoV-2 protein components or the receptor hACE2 [54–56,64–69]. These flavonoids are also antioxidants suggesting a common mode of action in both COVID-19 and diabetes, which in all probability is through reducing oxidative stress.

There are also recommendations on using Chinese herbal medicines and polyphenolic compounds containing antioxidants as an adjuvant to reduce the severity and mortality of COVID-19 patients with diabetes [70,71]. Besides flavonoids, phenolic compounds, which have antioxidant capacity and are present in essential oils of plants, may play a similar beneficial role in reducing oxidative stress during diabetes and COVID-19. Eugenol, a phenolic compound present in clove (*Syzygium aromaticum* (L.) Merr. & L. M. Perry, family: Myrtaceae), has been shown to ameliorate insulin resistance, oxidative stress, and inflammation in high fat diet/streptozotocin-induced diabetic rat [72], inhibit pancreatic α-amylase [73], and inhibited α-glucosidase activity and formation of advanced glycation end-products [74]. Antioxidant therapy prevented the cardiovascular disorders of patients who require dialysis, but the effect was not seen in patients with chronic kidney disease (CKD). Importantly, Jun and colleagues reported that antioxidants could reduce the development of kidney disease (late-stage) and serum creatinine levels by improving serum clearance of creatinine. The study reported that antioxidant therapy did not increase life-threatening adverse events, indicating its possible safety, although it needs validation from a larger population cohort and more comprehensive observational studies [75].

A recent RCT investigated the effect of 1 g of quercetin (along with standard care) over a period of four weeks in COVID-19 patients (n = 76, per group) and observed reduced severity of COVID-19 symptoms, duration of hospitalization, artificial ventilation, and fewer deaths in comparison with patients with standard care (without quercetin supplementation) [76]. Another pilot RCT from the same group of authors found that 600 mg of quercetin supplement over a period of 2 weeks improved COVID-19 related clinical symptoms and relevant plasma parameters on a small number of patients and compared against standard care group (n = 21) [77]. On the other hand, another RCT did not observe any effect of the antioxidant, ascorbic acid on a small number of COVID-19 patients (n = ~53 each group) treated over a period of 10 days with ascorbic acid (8 g), zinc gluconate (50 mg), or both agents, and none (standard of care) [78]. Similarly, a second RCT with 6 g/day (1.5 g, four times daily) intravenous ascorbic acid supplement with standard care for 5 days produced no improvement against patients with standard care (n = 30 per group) [79]. Another RCT planned to administer 24 g/day vitamin C for 7 days intravenously on COVID-19 patients but finished the study without reporting any results [80]. From these limited numbers of available clinical trials, the reports were based on small numbers of patients. More extensive studies are required over an extended period to make any fruitful comment on the effectiveness of these antioxidant compounds against SARS-CoV-2.

Molecular docking studies showed that the compound (quercetin) has high binding affinities to various targets in SARS-CoV-2 [81], and can be a potential nutraceutical against COVID-19 [82–84]. It is evident that both diabetes and COVID-19 induce the over-production of reactive oxygen species, which ultimately may cause damage to many vital organs, including the kidney, heart, and lungs [17]. It is also evident from some studies that antioxidants can reduce kidney disease. There are increased hospitalization and mortality rate with COVID-19 patients with diabetes. It is hypothesized that antioxidant therapy may reduce the fatality of COVID-19 patients with diabetes by reducing the over-production of the reactive oxygen species. However, this concept is in an early stage and needs many studies to validate this concept. The case can then be made for antioxidants (flavonoids and phenolic compounds) for use as therapeutic or nutraceutical in the case of COVID-19 patients and who have diabetes as a comorbidity for these compounds antioxidative capacities (Table 1).

Table 1. Several dietary flavonoids with anti-COVID-19, antioxidant and antidiabetic properties.

Flavonoid	Anti-COVID-19	Antioxidant	Antidiabetic
Quercetin	In silico and in vitro studies demonstrated that quercetin can interfere with various stages of the coronavirus entry and replication cycle such as PLpro, 3CLpro, and NTPase/helicase [85,86].	Significantly increased antioxidant enzyme activities in streptozotocin (STZ)-induced diabetic rats [87]. DPPH and ABTS radical scavenging activities reported [88].	Pre-treatment prevented STZ-induced diabetes in rats [87].
Kaempferol	In silico studies showed that kaempferol can inhibit Spike glycoprotein of SARS-CoV-2 [89].	DPPH and ABTS radical scavenging activities reported [88]. Antioxidant effect observed in DPPH (2,2-diphenyl-1-picrylhydrazyl), $ABTS^+$ radical scavenging and xanthine oxidase inhibition assays [90].	Dipeptidyl peptidase IV (DPP-4) and α-glucosidase inhibitory effect was observed [90].
Myricetin	Inhibition of SARS-CoV-2 replication by targeting Mpro (in silico) and ameliorating pulmonary inflammation (reducing bleomycin-induced pulmonary inflammation in mice) [91].	Antioxidant effect observed in DPPH (2,2-diphenyl-1-picrylhydrazyl), $ABTS^+$ radical scavenging and xanthine oxidase inhibition assays [89].	Dipeptidyl peptidase IV (DPP-4) and α-glucosidase inhibitory effect was reported [89].
Luteolin	In silico studies show luteolin to bind strongly to Mpro, PLpro, and ACE-2 [65]. In silico studies indicated that luteolin can bind to S2 unit of spike protein (S) of SARS-CoV-2 [92].	DPPH and ABTS radical scavenging activities reported [88].	Luteolin ameliorated diabetes in mice. Luteolin improved blood glucose, HbA1c (hemoglobin A1c), and insulin levels. Anti-inflammatory and anti-oxidative effects of luteolin were also observed [93].
Apigenin	In silico studies indicated that apigenin can bind to S2 unit of spike protein (S) of SARS-CoV-2 [92].	DPPH and ABTS radical scavenging activities reported [88].	The beneficial roles played by apigenin in diabetes mellitus have been reviewed. The compound is an antioxidant; metabolism of glucose and transfer to peripheral tissues are enhanced; pancreatic secretion of insulin is increased; activities of gluconeogenic enzymes and aldose reductase enzyme are suppressed leading to prevention of diabetic complications like cataract, retinopathy, and neuropathy [94].
Naringenin	In silico evidence of Mpro inhibition and reduction of angiotensin-converting enzyme receptors activity, reviewed by Tutunchi et al. [95].	Antioxidant and anti-diabetic effects observed in STZ-nicotinamide-induced diabetic rats as shown by significantly lower mean levels of fasting blood glucose and glycosylated hemoglobin, significantly elevated serum insulin levels, significantly higher mean activities of pancreatic enzymatic antioxidants, significantly higher mean levels of plasma non-enzymatic antioxidants, lower mean pancreatic tissue levels of MDA and lower mean activities of alanine aminotransferase (ALT), aspartate aminotransferase (AST), alkaline phosphatase (ALP) and lactate dehydrogenase (LDH) in serum [96].	Antioxidant and anti-diabetic effects observed in STZ-nicotinamide-induced diabetic rats as shown by significantly lower mean levels of fasting blood glucose and glycosylated hemoglobin, significantly elevated serum insulin levels, significantly higher mean activities of pancreatic enzymatic antioxidants, significantly higher mean levels of plasma non-enzymatic antioxidants, lower mean pancreatic tissue levels of MDA and lower mean activities of alanine aminotransferase (ALT), aspartate aminotransferase (AST), alkaline phosphatase (ALP) and lactate dehydrogenase (LDH) in serum [96].

Table 1. *Cont.*

Flavonoid	Anti-COVID-19	Antioxidant	Antidiabetic
Hesperidin	In silico studies indicate that hesperidin may bind to multiple components of SARS-CoV-2 (like Mpro, PLpro, Spike protein) and its human receptor ACE2, reviewed by Agrawal et al. [97].	Antioxidant and anti-diabetic effects observed in nicotinamide-STZ-induced diabetic rata [98].	Antioxidant and anti-diabetic effects observed in nicotinamide-STZ-induced diabetic rata [98].
Catechin	As shown in in silico studies, catechin can bind to S protein of SARS-CoV-2 and hACE2, thus inhibiting viral entry [99].	Catechin showed antioxidant activity such as free radical scavenging activity against DPPH and ABTS free radicals [100].	Catechin inhibited activity of α-amylase and α-glucosidase; catechin also significantly decreased the different lipid parameters, hepatic, and renal function enzyme levels along with Hb1c level in diabetic rats [100].

Abbreviations: ACE-2, angiotensin-converting enzyme 2; DPPH, 2,2-di(4-tert-octylphenyl)-1-picrylhydrazyl; PLpro, papain-like protease; 3CLpro, 3-chymotrypsin-like protease; NTPase, nucleoside-triphosphatase; Mpro, main protease; HbA1c, hemoglobin A1c; DPP-4, Dipeptidyl peptidase IV; STZ, streptozotocin; ABTS, 2,2′-azinobis-(3-ethylbenzthiazolin-6-sulfonic acid); SARS-CoV-2, severe acute respiratory syndrome coronavirus-2; COVID-19, coronavirus disease 2019.

Various in silico studies demonstrated that quercetin, luteolin, myricetin, naringenin, and hesperidin could interfere with various stages enzymes of SARS-CoV-2 (viral papain such as protease (PLpro) [85], and main protease (Mpro; 3 CLpro, also named 3-chymotrypsin-like protease) [92,97], NTPase/helicase) [99] of the coronavirus entry and replication cycle [85,92,95,97]. On the other hand, kaempferol [89], luteolin [92], apigenin [97], and catechin-like flavonoids [99] interact and inhibit (in silico) SARS-CoV-2 spike proteins (especially S2) and hACE-2 receptors, and thus can prevent viral entry inside the host cells [89,92,97,99] (Table 1).

It is noticeable that most of the antioxidant activities of flavonoids were measured (Table 1) using chemical reactions and assessing the kinetics or reaching the equilibrium state such as free radical scavenging activity against 2,2′-Azinobis-(3-ethylbenzothiazoline−6-sulfonic) acid (ABTS) and [2,2-di(4-tert-octylphenyl)-1-picrylhydrazyl] (DPPH) free radicals, as these reagents cause oxidative stress (overproduction of reactive oxygenated species, ROS) [101]. The main issue is that normal cells produce small amounts of ROS, which cannot be measured correctly using current colorimetric methods. Noticeably, some of these flavonoids were tested for antioxidant enzyme activities such as superoxide dismutase, glutathione peroxidase, and catalase enzymes in pancreatic cells using a STZ-induced diabetic rat model [87,102]. It needs to be further pointed out that flavonoids do not just reduce oxidative stress through scavenging of free radical species but also through inhibition of ROS producing enzymes such as xanthine oxidase [85] or through chelation of metal ions [87].

Flavonoids showed antidiabetic effects, such as quercetin inhibited glucose absorption from intestine, improved glucose use from peripheral tissues, as well as it simulated insulin secretion. Studies also suggest that consumption of quercetin displayed a long plasma half-life in humans [103]. Furthermore, a meta-analysis on the effects of quercetin showed that the flavonoid reduced blood glucose levels in a dose-dependent manner in experimentally induced (e.g., STZ-induced) diabetic animals, and it is effective at higher doses (10, 25 or 50 mg/kg body weight) [104]. Quercetin inhibits the enzymes dipeptidyl peptidase IV (DPP-IV) and thus shows antioxidant and antihyperglycemic properties [105]. Importantly, it is a generally recognized as a safe compound according to FDA [106]. Quercetin also inhibited TNF-α-mediated inflammation and insulin resistance in human adipose cells in an in vitro study [107]. Another flavonoid, kaempferol increases glucose uptake and glucose transporter 4 translocation via a Janus kinase 2-dependent pathway in skeletal (L6) myoblast cell line, which indicates kaempferol's hyperglycemic effect in vitro [108]. A clinical study showed that consumption of a formulation that contained myricetin, quercetin, chlorogenic acid (another group of polyphenol compounds) reduced plasma glucose levels in confirmed diabetes-2 patients, and cotreatment with metformin showed potentiation of metformin's antidiabetic activities [109]. Three times daily application of a topical formulation contained quercetin for four weeks improved numbness, jolting pain, and irritation, and quality of life of patients who experience symptomatic diabetic peripheral neuropathy in a small number of patients (total $n = 34$) [110]. Another clinical trial showed no effect of a flavonoid against placebo over a 12-week combined treatment of isoquercetin (225 mg once daily) and sodium nitrite (40 mg twice daily) in CKD patients ($n = 35$ per group) [111]. Noticeably, an antioxidant such as resveratrol (a stilbenoid compound) caused suppression of angiotensin-2 that may be used as an adjunct therapy to COVID-19 [112,113]. It seems that not all antioxidants are effective in preventing oxidative stress. The capability of preventing oxidative damage varies between compounds, which needs further extensive clinical trials to elucidate the efficacies of these compounds.

Vitamin D (a natural antioxidant) and magnesium deficiencies also exacerbate the underlying pathogenetic mechanisms in COVID-19 [114]. Vitamin D is essential to maintaining a healthy immune system [115]. Vitamin D levels were shown to be associated with blood glucose and body mass index of COVID-19 patients. As suggested by di Filippo and colleagues, a common pathophysiological mechanism might be involved with hyperglycemia, adiposity, and COVID-19 severity [116]. Magnesium activates vitamin D and protects cells from oxidative stress [114]. Severe COVID-19 patients showed lower vitamin D levels and higher oxidative stress parameters (like plasma LDH, peroxides, and oxidative stress index) than less severe COVID-19 patients [117]. A randomized clinical trial in Spain, oral supplement of calcifediol (25-hydroxyvitamin D3: 0.532 mg on day 1, 0.266 mg on days 3 and 7, and weekly afterwards) in COVID-19 patients ($n = 50$) along with standard treatment for COVID-19 in hospital reduced the severity of symptoms and admission to Intensive Care Units (ICU) than standard care group ($n = 26$) [118]. As the study was based on a small number of COVID-19 patients (total $n = 76$) and there was in-equality of sample sizes between control and treatment groups, the study requires further validations to comment on the efficacy of vitamin D against COVID-19. However, it is re-ally a promising study that antioxidants such as calcifediol helped reduction of COVID-19 severity and ICU admission [118]. Noticeably, people with inherited glucose-6-phosphate dehydrogenase (G6 PD) deficiency can cause of reduced circulatory 25-hydroxyvitamin D in blood, and can be vulnerable to excess oxidative stress, cytokine release, and pulmonary dysfunction due to COVID-19 infection [119]. It is important to note that there is no strong clinical evidence for flavonoids or vitamin C against protection from oxidative damage caused by COVID-19. Vitamin D can prevent oxidative damage produced by SARS-CoV-2 in people suffering from COVID-19. However, further evidence is required in larger population cohorts based on various geographical locations, age groups, food habits, and ethnicity.

Various fruits and vegetables are sources of flavonoids. Common vegetables such as tomatoes are natural sources of quercetin, kaempferol, and naringenin [120] (Figure 3). Broccoli, celery, cabbages, peppers, and parsley are sources for luteolin [121,122]. Noticeably onions and tea are main dietary sources of flavonols (e.g., quercetin and kaempferol) and flavones (apigenin and luteolin) [123]. Onions, parsley, sage, tea, citrus fruit (like oranges, lemons, and limes), apples, grapes, cherries, and berries are potential sources of quercetin and other flavonoids [123–126] (Figure 3). Noticeably parsley, onion, zinger (source of hesperidin), citrus fruit-peels, sage are sources of essential oils, which can improve the bioavailability of flavonoids (like quercetin microemulsion of peppermint, clove and rosemary oils) [127]. Essential oils are sources of antioxidants, improve the quality of life of diabetic patients, analgesics, and may have the capability to improve COVID-19 and related comorbidities [128–132]. Iddir and associates reported that poor nutrition stimulates increased oxidative stress and inflammation, which render poor immunity against pathogens. However, dietary protein intake can help antibody production, and micronutrients such as vitamins D, A, C, and E, flavonoids, carotenoids, and minerals such as zinc can prevent the expression of transcription factors (NF-kB and Nrf-2) related to inflammation [133]. This information is also supported by a clinical study with COVID-19 patients that showed reduced plasma antioxidant levels than people without SARS-CoV-2 infection [39].

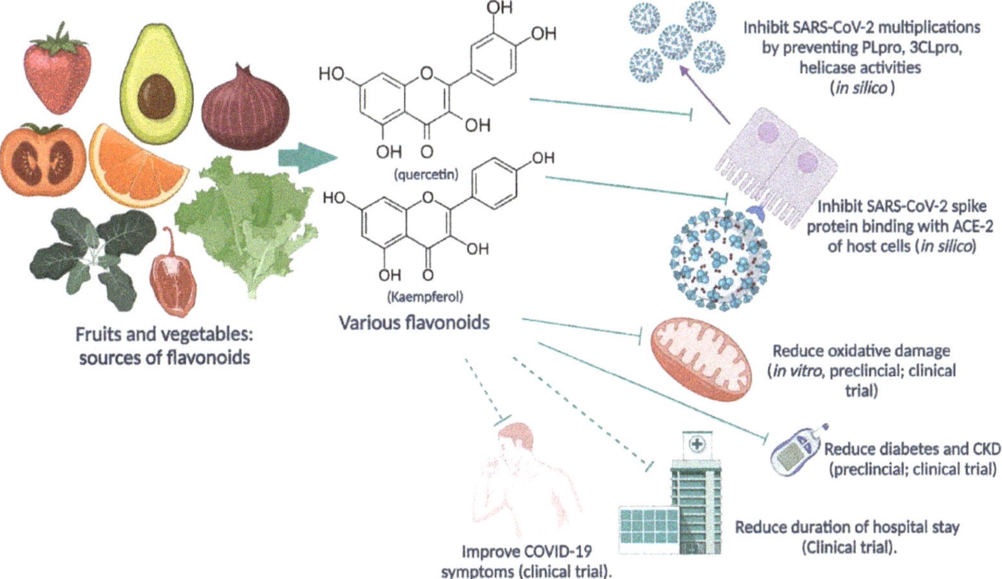

Figure 3. Roles of flavonoids against SARS-CoV-2 based on recent in silico, pre-clinical, and clinical studies. Abbreviations: CKD, chronic kidney disease; SARS-CoV-2, severe acute respiratory syndrome coronavirus-2; COVID-19, coronavirus disease 2019. (The figure was made with www.biorender.com, accessed on 13 December 2021).

6. Conclusions

COVID-19 and diabetic patients have a common feature of increased oxidative stress. Patients with both disorders generally end up with poor prognosis and death. A large part of this poor prognosis and death is caused by kidney failure. COVID-19 and diabetes may both be responsible by increasing oxidative stress in a synergistic manner. Flavonoids and polyphenols, because of the nature of their chemical structure are good antioxidants. These phytochemicals can scavenge reactive oxygen species (ROS) and inhibit enzymes responsible for making ROS. They also inhibit production of ROS through chelation of metal ions. We suggest that this oxidative stress factor of COVID-19 with diabetes as a comorbidity and vice versa has been overlooked largely. We further recommend that judicious use of vitamin D, flavonoids, and other antioxidants as possible therapeutics, may mitigate this oxidative stress effect and improve the prognosis of patients suffering from both COVID-19 and diabetes.

Author Contributions: Conceptualization, A.K.P., R.J., V.N., P.W. and M.R.; writing—original draft preparation, A.K.P., M.K.H., T.A.B. and M.R.; writing—review and editing, A.K.P., T.M., K.J., A.H., R.J., V.N., P.W., M.d.L.P. and M.R.; visualization, A.K.P. and M.K.H.; supervision, M.R. All authors have read and agreed to the published version of the manuscript.

Funding: This research received no external funding.

Institutional Review Board Statement: Not applicable.

Informed Consent Statement: Not applicable.

Data Availability Statement: Not applicable.

Acknowledgments: M.d.L.P. thanks project CICECO-Aveiro Institute of Materials, UIDB//2020 and UIDP//2020, national funds by FCT/MCTES.

Conflicts of Interest: The authors declare no conflict of interest.

References

1. Zhang, Y.; Cui, Y.; Shen, M.; Zhang, J.; Liu, B.; Dai, M.; Chen, L.; Han, D.; Fan, Y.; Zeng, Y.; et al. Association of diabetes mellitus with disease severity and prognosis in COVID-19: A retrospective cohort study. *Diabetes Res. Clin. Pract.* **2020**, *165*, 108227. [CrossRef]
2. Elias Ferreira, P.; Vinicius Carlos, I.; Antonio Adolfo Mattos de, C.; Anselmo Cordeiro de, S.; José Renato de Oliveira, L.; Eduardo, F.; Adriano Conrado, R.; Luiz Fernando de Oliveira, M. Mortality in patients with diabetes by COVID 19 a systematic review. *Res. Sq.* **2021**. preprint. [CrossRef]
3. Alguwaihes, A.M.; Al-Sofiani, M.E.; Megdad, M.; Albader, S.S.; Alsari, M.H.; Alelayan, A.; Alzahrani, S.H.; Sabico, S.; Al-Daghri, N.M.; Jammah, A.A. Diabetes and COVID-19 among hospitalized patients in Saudi Arabia: A single-centre retrospective study. *Cardiovasc. Diabetol.* **2020**, *19*, 205. [CrossRef]
4. Jovanovic, P.; Zoric, L.; Stefanovic, I.; Dzunic, B.; Djordjevic-Jocic, J.; Radenkovic, M.; Jovanovic, M. Lactate dehydrogenase and oxidative stress activity in primary open-angle glaucoma aqueous humour. *Bosn. J. Basic Med. Sci.* **2010**, *10*, 83–88. [CrossRef]
5. Acharya, D.; Lee, K.; Lee, D.S.; Lee, Y.S.; Moon, S.S. Mortality rate and predictors of mortality in hospitalized COVID-19 patients with diabetes. *Healthcare* **2020**, *8*, 338. [CrossRef]
6. Musche, V.; Kohler, H.; Bäuerle, A.; Schweda, A.; Weismüller, B.; Fink, M.; Schadendorf, T.; Robitzsch, A.; Dörrie, N.; Tan, S.; et al. COVID-19-related fear, risk perception, and safety behavior in individuals with diabetes. *Healthcare* **2021**, *9*, 480. [CrossRef]
7. Kouhpayeh, S.; Shariati, L.; Boshtam, M.; Rahimmanesh, I.; Mirian, M.; Esmaeili, Y.; Najaflu, M.; Khanahmad, N.; Zeinalian, M.; Trovato, M.; et al. The molecular basis of COVID-19 pathogenesis, conventional and nanomedicine therapy. *Int. J. Mol. Sci.* **2021**, *22*, 5438. [CrossRef]
8. Dalamaga, M.; Christodoulatos, G.S.; Karampela, I.; Vallianou, N.; Apovian, C.M. Understanding the Co-Epidemic of Obesity and COVID-19: Current evidence, comparison with previous epidemics, mechanisms, and preventive and therapeutic perspectives. *Curr. Obes. Rep.* **2021**, *10*, 214–243. [CrossRef]
9. Apicella, M.; Campopiano, M.C.; Mantuano, M.; Mazoni, L.; Coppelli, A.; Del Prato, S. COVID-19 in people with diabetes: Understanding the reasons for worse outcomes. *Lancet Diabetes Endocrinol.* **2020**, *8*, 782–792. [CrossRef]
10. Forbes, J.M.; Coughlan, M.T.; Cooper, M.E. Oxidative stress as a major culprit in kidney disease in diabetes. *Diabetes* **2008**, *57*, 1446–1454. [CrossRef]
11. Shahbazian, H.; Rezaii, I. Diabetic kidney disease: Review of the current knowledge. *J. Renal. Inj. Prev.* **2013**, *2*, 73–80. [CrossRef]
12. Phillips, M.; Cataneo, R.N.; Cheema, T.; Greenberg, J. Increased breath biomarkers of oxidative stress in diabetes mellitus. *Clin. Chim. Acta* **2004**, *344*, 189–194. [CrossRef]
13. Kountouri, A.; Korakas, E.; Ikonomidis, I.; Raptis, A.; Tentolouris, N.; Dimitriadis, G.; Lambadiari, V. Type 1 Diabetes Mellitus in the SARS-CoV-2 Pandemic: Oxidative stress as a major pathophysiological mechanism linked to adverse clinical outcomes. *Antioxidants* **2021**, *10*, 752. [CrossRef]
14. Lim, S.; Bae, J.H.; Kwon, H.S.; Nauck, M.A. COVID-19 and diabetes mellitus: From pathophysiology to clinical management. *Nat. Rev. Endocrinol.* **2021**, *17*, 11–30. [CrossRef]
15. Holman, N.; Knighton, P.; Kar, P.; O'Keefe, J.; Curley, M.; Weaver, A.; Barron, E.; Bakhai, C.; Khunti, K.; Wareham, N.J.; et al. Risk factors for COVID-19-related mortality in people with type 1 and type 2 diabetes in England: A population-based cohort study. *Lancet Diabetes Endocrinol.* **2020**, *8*, 823–833. [CrossRef]
16. Kumar, A.; Arora, A.; Sharma, P.; Anikhindi, S.A.; Bansal, N.; Singla, V.; Khare, S.; Srivastava, A. Is diabetes mellitus associated with mortality and severity of COVID-19? A meta-analysis. *Diabetes Metab. Syndr.* **2020**, *14*, 535–545. [CrossRef]
17. De Oliveira, A.A.; Nunes, K.P. Crosstalk of TLR4, vascular NADPH oxidase, and COVID-19 in diabetes: What are the potential implications? *Vascul. Pharmacol.* **2021**, *139*, 106879. [CrossRef]
18. Konior, A.; Schramm, A.; Czesnikiewicz-Guzik, M.; Guzik, T.J. NADPH oxidases in vascular pathology. *Antioxid. Redox Signal.* **2014**, *20*, 2794–2814. [CrossRef]
19. Luc, K.; Schramm-Luc, A.; Guzik, T.J.; Mikolajczyk, T.P. Oxidative stress and inflammatory markers in prediabetes and diabetes. *J. Physiol. Pharmacol.* **2019**, *70*, 809–824. [CrossRef]
20. Hotamisligil, G.S.; Shargill, N.S.; Spiegelman, B.M. Adipose expression of tumor necrosis factor-alpha: Direct role in obesity-linked insulin resistance. *Science* **1993**, *259*, 87–91. [CrossRef]
21. Ding, S.; Xu, S.; Ma, Y.; Liu, G.; Jang, H.; Fang, J. Modulatory mechanisms of the NLRP3 inflammasomes in diabetes. *Biomolecules* **2019**, *9*, 850. [CrossRef]
22. Lontchi-Yimagou, E.; Sobngwi, E.; Matsha, T.E.; Kengne, A.P. Diabetes mellitus and inflammation. *Curr. Diab. Rep.* **2013**, *13*, 435–444. [CrossRef]
23. Chee, Y.J.; Tan, S.K.; Yeoh, E. Dissecting the interaction between COVID-19 and diabetes mellitus. *J. Diabetes Investig.* **2020**, *11*, 1104–1114. [CrossRef]
24. Pal, R.; Bhadada, S.K.; Misra, A. COVID-19 vaccination in patients with diabetes mellitus: Current concepts, uncertainties and challenges. *Diabetes Metab. Syndr.* **2021**, *15*, 505–508. [CrossRef]
25. McGurnaghan, S.J.; Weir, A.; Bishop, J.; Kennedy, S.; Blackbourn, L.A.K.; McAllister, D.A.; Hutchinson, S.; Caparrotta, T.M.; Mellor, J.; Jeyam, A.; et al. Risks of and risk factors for COVID-19 disease in people with diabetes: A cohort study of the total population of Scotland. *Lancet Diabetes Endocrinol.* **2021**, *9*, 82–93. [CrossRef]

26. Wang, C.H.; Hilliard, M.E.; Carreon, S.A.; Jones, J.; Rooney, K.; Barber, J.R.; Tully, C.; Monaghan, M.; Streisand, R. Predictors of mood, diabetes-specific and COVID-19-specific experiences among parents of early school-age children with type 1 diabetes during initial months of the COVID-19 pandemic. *Pediatr. Diabetes* **2021**, *22*, 1071–1080. [CrossRef]
27. Klarskov, C.K.; Lindegaard, B.; Pedersen-Bjergaard, U.; Kristensen, P.L. Remote continuous glucose monitoring during the COVID-19 pandemic in quarantined hospitalized patients in Denmark: A structured summary of a study protocol for a randomized controlled trial. *Trials* **2020**, *21*, 968. [CrossRef]
28. Lin, C.L.; Huang, L.C.; Chang, Y.T.; Chen, R.Y.; Yang, S.H. Under COVID-19 pandemic: A quasi-experimental trial of observation on diabetes patients' health behavior Aafected by the pandemic from a coaching intervention program. *Front. Public Health* **2021**, *9*, 580032. [CrossRef]
29. Casillas Santana, M.A.; Arreguín Cano, J.A.; Dib Kanán, A.; Dipp Velázquez, F.A.; Munguía, P.; Martínez Castañón, G.A.; Castillo Silva, B.E.; Sámano Valencia, C.; Salas Orozco, M.F. Should we be concerned about the association of diabetes mellitus and periodontal disease in the risk of infection by SARS-CoV-2? A systematic review and hypothesis. *Medicina* **2021**, *57*, 493. [CrossRef]
30. Fu, J.; Zhou, B.; Zhang, L.; Balaji, K.S.; Wei, C.; Liu, X.; Chen, H.; Peng, J.; Fu, J. Expressions and significances of the angiotensin-converting enzyme 2 gene, the receptor of SARS-CoV-2 for COVID-19. *Mol. Biol. Rep.* **2020**, *47*, 4383–4392. [CrossRef]
31. Gavriatopoulou, M.; Korompoki, E.; Fotiou, D.; Ntanasis-Stathopoulos, I.; Psaltopoulou, T.; Kastritis, E.; Terpos, E.; Dimopoulos, M.A. Organ-specific manifestations of COVID-19 infection. *Clin. Exp. Med.* **2020**, *20*, 493–506. [CrossRef]
32. Taneera, J.; El-Huneidi, W.; Hamad, M.; Mohammed, A.K.; Elaraby, E.; Hachim, M.Y. Expression profile of SARS-CoV-2 host receptors in human pancreatic islets revealed upregulation of ACE2 in diabetic donors. *Biology* **2020**, *9*, 215. [CrossRef] [PubMed]
33. Menon, R.; Otto, E.A.; Sealfon, R.; Nair, V.; Wong, A.K.; Theesfeld, C.L.; Chen, X.; Wang, Y.; Boppana, A.S.; Luo, J.; et al. SARS-CoV-2 receptor networks in diabetic and COVID-19-associated kidney disease. *Kidney Int.* **2020**, *98*, 1502–1518. [CrossRef]
34. Werion, A.; Belkhir, L.; Perrot, M.; Schmit, G.; Aydin, S.; Chen, Z.; Penaloza, A.; De Greef, J.; Yildiz, H.; Pothen, L.; et al. SARS-CoV-2 causes a specific dysfunction of the kidney proximal tubule. *Kidney Int.* **2020**, *98*, 1296–1307. [CrossRef]
35. Moitinho, M.S.; Belasco, A.; Barbosa, D.A.; Fonseca, C.D.D. Acute kidney injury by SARS-CoV-2 virus in patients with COVID-19: An integrative review. *Rev. Bras. Enferm.* **2020**, *73* (Suppl. S2), e20200354. [CrossRef]
36. Gan, R.; Rosoman, N.P.; Henshaw, D.J.E.; Noble, E.P.; Georgius, P.; Sommerfeld, N. COVID-19 as a viral functional ACE2 deficiency disorder with ACE2 related multi-organ disease. *Med. Hypotheses* **2020**, *144*, 110024. [CrossRef]
37. Beltrán-García, J.; Osca-Verdegal, R.; Pallardó, F.V.; Ferreres, J.; Rodríguez, M.; Mulet, S.; Sanchis-Gomar, F.; Carbonell, N.; García-Giménez, J.L. Oxidative stress and inflammation in COVID-19-associated sepsis: The potential role ofaAnti-oxidant therapy in avoiding disease progression. *Antioxidants* **2020**, *9*, 936. [CrossRef]
38. Laforge, M.; Elbim, C.; Frère, C.; Hémadi, M.; Massaad, C.; Nuss, P.; Benoliel, J.J.; Becker, C. Tissue damage from neutrophil-induced oxidative stress in COVID-19. *Nat. Rev. Immunol.* **2020**, *20*, 515–516. [CrossRef]
39. Muhammad, Y.; Kani, Y.A.; Iliya, S.; Muhammad, J.B.; Binji, A.; El-Fulaty Ahmad, A.; Kabir, M.B.; Umar Bindawa, K.; Ahmed, A. Deficiency of antioxidants and increased oxidative stress in COVID-19 patients: A cross-sectional comparative study in Jigawa, Northwestern Nigeria. *SAGE Open Med.* **2021**, *9*, 2050312121991246. [CrossRef]
40. Cecchini, R.; Cecchini, A.L. SARS-CoV-2 infection pathogenesis is related to oxidative stress as a response to aggression. *Med. Hypotheses* **2020**, *143*, 110102. [CrossRef]
41. Delgado-Roche, L.; Mesta, F. Oxidative stress as key player in severe acute respiratory syndrome Coronavirus (SARS-CoV) infection. *Arch. Med. Res.* **2020**, *51*, 384–387. [CrossRef]
42. Alenezi, F.K.; Almeshari, M.A.; Mahida, R.; Bangash, M.N.; Thickett, D.R.; Patel, J.M. Incidence and risk factors of acute kidney injury in COVID-19 patients with and without acute respiratory distress syndrome (ARDS) during the first wave of COVID-19: A systematic review and meta-analysis. *Ren. Fail.* **2021**, *43*, 1621–1633. [CrossRef]
43. Guo, W.; Li, M.; Dong, Y.; Zhou, H.; Zhang, Z.; Tian, C.; Qin, R.; Wang, H.; Shen, Y.; Du, K.; et al. Diabetes is a risk factor for the progression and prognosis of COVID-19. *Diabetes Metab. Res. Rev.* **2020**, *36*, e3319. [CrossRef] [PubMed]
44. Leon-Abarca, J.A.; Memon, R.S.; Rehan, B.; Iftikhar, M.; Chatterjee, A. The impact of COVID-19 in diabetic kidney disease and chronic kidney disease: A population-based study. *Acta Biomed.* **2020**, *91*, e2020161. [CrossRef]
45. Luo, P.; Qiu, L.; Liu, Y.; Liu, X.L.; Zheng, J.L.; Xue, H.Y.; Liu, W.H.; Liu, D.; Li, J. Metformin treatment was associated with decreased mortality in COVID-19 patients with diabetes in a retrospective analysis. *Am. J. Trop. Med. Hyg.* **2020**, *103*, 69–72. [CrossRef]
46. Sharma, S.; Ray, A.; Sadasivam, B. Metformin in COVID-19: A possible role beyond diabetes. *Diabetes Res. Clin. Pract.* **2020**, *164*, 108183. [CrossRef] [PubMed]
47. Rudnick, M.R.; Hilburg, R. Acute kidney injury in COVID-19: Another challenge for nephrology. *Am. J. Nephrol.* **2020**, *51*, 761–763. [CrossRef] [PubMed]
48. Jha, J.C.; Banal, C.; Chow, B.S.; Cooper, M.E.; Jandeleit-Dahm, K. Diabetes and kidney disease: Role of oxidative stress. *Antioxid. Redox Signal.* **2016**, *25*, 657–684. [CrossRef] [PubMed]
49. De Las Heras, N.; Martín Giménez, V.M.; Ferder, L.; Manucha, W.; Lahera, V. Implications of oxidative stress and potential role of mitochondrial dysfunction in COVID-19: Therapeutic effects of vitamin D. *Antioxidants* **2020**, *9*, 897. [CrossRef] [PubMed]
50. Soto, M.E.; Guarner-Lans, V.; Soria-Castro, E.; Manzano Pech, L.; Pérez-Torres, I. Is antioxidant therapy a useful complementary measure for COVID-19 treatment? An Algorithm for Its Application. *Medicina* **2020**, *56*, 386. [CrossRef] [PubMed]

51. Golbidi, S.; Ebadi, S.A.; Laher, I. Antioxidants in the treatment of diabetes. *Curr. Diabetes Rev.* **2011**, *7*, 106–125. [CrossRef]
52. Miller, A.L. Antioxidant flavonoids: Structure, function and clinical usage. *Alt. Med. Rev.* **1996**, *1*, 103–111.
53. Vargas, F.; Romecín, P.; García-Guillén, A.I.; Wangesteen, R.; Vargas-Tendero, P.; Paredes, M.D.; Atucha, N.M.; García-Estañ, J. Flavonoids in kidney health and disease. *Front. Physiol.* **2018**, *9*, 394. [CrossRef] [PubMed]
54. Alzaabi, M.M.; Hamdy, R.; Ashmawy, N.S.; Hamoda, A.M.; Alkhayat, F.; Khademi, N.N.; Al Joud, S.M.A.; El-Keblawy, A.A.; Soliman, S.S.M. Flavonoids are promising safe therapy against COVID-19. *Phytochem. Rev.* **2021**, 1–22, online ahead of print. [CrossRef]
55. Brito, J.C.M.; Lima, W.G.; da Cruz Nizer, W.S. Quercetin as a potential nutraceutic against coronavirus disease 2019 (COVID-19). *Ars. Pharm. (Internet)* **2021**, *62*, 85–89.
56. Xu, D.; Hu, M.J.; Wang, Y.Q.; Cui, Y.L. Antioxidant activities of quercetin and its complexes for medicinal application. *Molecules* **2019**, *24*, 1123. [CrossRef]
57. Tu, H.; Ma, D.; Luo, Y.; Tang, S.; Li, Y.; Chen, G.; Wang, L.; Hou, Z.; Shen, C.; Lu, H.; et al. Quercetin alleviates chronic renal failure by targeting the PI3k/Akt pathway. *Bioengineered* **2021**, *12*, 6538–6558. [CrossRef] [PubMed]
58. Cunanan, J.; Deacon, E.; Cunanan, K.; Yang, Z.; Ask, A.; Morikawa, L.; Todorova, E.; Bridgewater, D. Quercetin treatment reduces the severity of renal dysplasia in a beta-catenin dependent manner. *PLoS ONE* **2020**, *15*, e0234375. [CrossRef]
59. Dallak, M.; Dawood, A.F.; Haidara, M.A.; Abdel Kader, D.H.; Eid, R.A.; Kamar, S.S.; Shams Eldeen, A.M.; Al-Ani, B. Suppression of glomerular damage and apoptosis and biomarkers of acute kidney injury induced by acetaminophen toxicity using a combination of resveratrol and quercetin. *Drug Chem. Toxicol.* **2020**, 1–7, online ahead of print. [CrossRef] [PubMed]
60. Modlinger, P.S.; Wilcox, C.S.; Aslam, S. Nitric oxide, oxidative stress, and progression of chronic renal failure. *Semin. Nephrol.* **2004**, *24*, 354–365. [CrossRef]
61. Chernyak, B.V.; Popova, E.N.; Prikhodko, A.S.; Grebenchikov, O.A.; Zinovkina, L.A.; Zinovkin, R.A. COVID-19 and oxidative stress. *Biochemistry (Moscow)* **2020**, *85*, 1543–1553. [CrossRef]
62. Jubaidi, F.F.; Zainalabidin, S.; Taib, I.S.; Hamid, Z.A.; Budin, S.B. The potential role of Ffavonoids in ameliorating diabetic cardiomyopathy via alleviation of cardiac oxidative stress, inflammation and apoptosis. *Int. J. Mol. Sci.* **2021**, *22*, 5094. [CrossRef] [PubMed]
63. Hu, Q.; Qu, C.; Xiao, X.; Zhang, W.; Jiang, Y.; Wu, Z.; Song, D.; Peng, X.; Ma, X.; Zhao, Y. Flavonoids on diabetic nephropathy: Advances and therapeutic opportunities. *Chin. Med.* **2021**, *16*, 74. [CrossRef]
64. Lin, C.; Tsai, F.J.; Hsu, Y.M.; Ho, T.J.; Wang, G.K.; Chiu, Y.J.; Ha, H.A.; Yang, J.S. Study of baicalin toward COVID-19 Treatment: In silico target analysis and in vitro inhibitory effects on SARS-CoV-2 proteases. *Biomed. Hub.* **2021**, *6*, 122–137. [CrossRef] [PubMed]
65. Shawan, M.; Halder, S.K.; Hasan, M.A. Luteolin and abyssinone II as potential inhibitors of SARS-CoV-2: An in silico molecular modeling approach in battling the COVID-19 outbreak. *Bull. Natl. Res. Cent.* **2021**, *45*, 27. [CrossRef] [PubMed]
66. Cheng, F.J.; Huynh, T.K.; Yang, C.S.; Hu, D.W.; Shen, Y.C.; Tu, C.Y.; Wu, Y.C.; Tang, C.H.; Huang, W.C.; Chen, Y.; et al. Hesperidin is a potential inhibitor against SARS-CoV-2 infection. *Nutrients* **2021**, *13*, 2800. [CrossRef]
67. Marmitt, D.J.; Goettert, M.I.; Rempel, C. Compounds of plants with activity against SARS-CoV-2 targets. *Expert. Rev. Clin. Pharmacol.* **2021**, *14*, 623–633. [CrossRef]
68. Wang, Y.; Fang, S.; Wu, Y.; Cheng, X.; Zhang, L.K.; Shen, X.R.; Li, S.Q.; Xu, J.R.; Shang, W.J.; Gao, Z.B.; et al. Discovery of SARS-CoV-2-E channel inhibitors as antiviral candidates. *Acta Pharmacol. Sin.* **2021**, 1–7. [CrossRef]
69. Deshpande, R.R.; Tiwari, A.P.; Nyayanit, N.; Modak, M. In silico molecular docking analysis for repurposing therapeutics against multiple proteins from SARS-CoV-2. *Eur. J. Pharmacol.* **2020**, *886*, 173430. [CrossRef]
70. Yang, Y.; Islam, M.S.; Wang, J.; Li, Y.; Chen, X. Traditional Chinese medicine in the treatment of patients infected with 2019-new Coronavirus (SARS-CoV-2): A review and perspective. *Int. J. Biol. Sci.* **2020**, *16*, 1708–1717. [CrossRef]
71. Tabatabaei-Malazy, O.; Abdollahi, M.; Larijani, B. Beneficial effects of anti-oxidative herbal medicines in diabetic patients infected with COVID-19: A hypothesis. *Diabetes Metab. Syndr. Obes.* **2020**, *13*, 3113–3116. [CrossRef]
72. Al-Trad, B.; Alkhateeb, H.; Alsmadi, W.; Al-Zoubi, M. Eugenol ameliorates insulin resistance, oxidative stress and inflammation in high fat-diet/streptozotocin-induced diabetic rat. *Life Sci.* **2019**, *216*, 183–188. [CrossRef]
73. Mnafgui, K.; Kaanich, F.; Derbali, A.; Hamden, K.; Derbali, F.; Slama, S.; Allouche, N.; Elfeki, A. Inhibition of key enzymes related to diabetes and hypertension by eugenol in vitro and in alloxan-induced diabetic rats. *Arch. Physiol. Biochem.* **2013**, *119*, 225–233. [CrossRef]
74. Singh, P.; Jayaramaiah, R.H.; Agawane, S.B.; Vannuruswamy, G.; Korwar, A.M.; Anand, A.; Dhaygude, V.S.; Shaikh, M.L.; Joshi, R.S.; Boppana, R.; et al. Potential dual role of eugenol in inhibiting advanced glycation end products in diabetes: Proteomic and mechanistic insights. *Sci. Rep.* **2016**, *6*, 18798. [CrossRef]
75. Jun, M.; Venkataraman, V.; Razavian, M.; Cooper, B.; Zoungas, S.; Ninomiya, T.; Webster, A.C.; Perkovic, V. Antioxidants for chronic kidney disease. *Cochrane Database Syst. Rev.* **2012**, *10*, Cd008176. [CrossRef] [PubMed]
76. Di Pierro, F.; Derosa, G.; Maffioli, P.; Bertuccioli, A.; Togni, S.; Riva, A.; Allegrini, P.; Khan, A.; Khan, S.; Khan, B.A.; et al. Possible therapeutic effects of adjuvant quercetin supplementation against early-stage COVID-19 infection: A prospective, randomized, controlled, and open-label study. *Int. J. Gen. Med.* **2021**, *14*, 2359–2366. [CrossRef]
77. Di Pierro, F.; Iqtadar, S.; Khan, A.; Ullah Mumtaz, S.; Masud Chaudhry, M.; Bertuccioli, A.; Derosa, G.; Maffioli, P.; Togni, S.; Riva, A.; et al. Potential clinical benefits of quercetin in the early stage of COVID-19: Results of a second, pilot, randomized, controlled and open-label clinical trial. *Int. J. Gen. Med.* **2021**, *14*, 2807–2816. [CrossRef]

78. Thomas, S.; Patel, D.; Bittel, B.; Wolski, K.; Wang, Q.; Kumar, A.; Il'Giovine, Z.J.; Mehra, R.; McWilliams, C.; Nissen, S.E.; et al. Effect of high-dose zinc and ascorbic acid supplementation vs. usual care on symptom length and reduction among ambulatory patients with SARS-CoV-2 infection: The COVID A to Z randomized clinical trial. *JAMA Netw. Open* **2021**, *4*, e210369. [CrossRef] [PubMed]
79. JamaliMoghadamSiahkali, S.; Zarezade, B.; Koolaji, S.; SeyedAlinaghi, S.; Zendehdel, A.; Tabarestani, M.; Sekhavati Moghadam, E.; Abbasian, L.; Dehghan Manshadi, S.A.; Salehi, M.; et al. Safety and effectiveness of high-dose vitamin C in patients with COVID-19: A randomized open-label clinical trial. *Eur. J. Med. Res.* **2021**, *26*, 20. [CrossRef]
80. Liu, F.; Zhu, Y.; Zhang, J.; Li, Y.; Peng, Z. Intravenous high-dose vitamin C for the treatment of severe COVID-19: Study protocol for a multicentre randomised controlled trial. *BMJ Open* **2020**, *10*, e039519. [CrossRef] [PubMed]
81. Asif, M.; Saleem, M.; Saadullah, M.; Yaseen, H.S.; Al Zarzour, R. COVID-19 and therapy with essential oils having antiviral, anti-inflammatory, and immunomodulatory properties. *Inflammopharmacology* **2020**, *28*, 1153–1161. [CrossRef]
82. Savant, S.; Srinivasan, S.; Kruthiventi, A.K. Potential nutraceuticals for COVID-19. *Nutr. Diet. Suppl.* **2021**, *13*, 25. [CrossRef]
83. Paul, A.K.; Jahan, R.; Bondhon, T.A.; Jannat, K.; Hasan, A.; Rahmatullah, M.; Nissapatorn, V.; Pereira, M.L.; Wiart, C. Potential role of flavonoids against SARS-CoV-2 induced diarrhea. *Trop. Biomed.* **2021**, *38*, 360–365. [CrossRef] [PubMed]
84. Jannat, K.; Paul, A.K.; Bondhon, T.A.; Hasan, A.; Nawaz, M.; Jahan, R.; Mahboob, T.; Nissapatorn, V.; Wilairatana, P.; de Lourdes Pereira, M.; et al. Nanotechnology applications of flavonoids for viral diseases. *Pharmaceutics* **2021**, *13*, 1895. [CrossRef] [PubMed]
85. Agrawal, P.K.; Agrawal, C.; Blunden, G. Quercetin: Antiviral significance and possible COVID-19 integrative considerations. *Nat. Prod. Commun.* **2020**, *15*. [CrossRef]
86. Diniz, L.R.L.; de Santana Souza, M.T.; Duarte, A.B.S.; de Sousa, D.P. Mechanistic aspects and therapeutic potential of quercetin against COVID-19-associated acute kidney injury. *Molecules* **2020**, *25*, 5772. [CrossRef] [PubMed]
87. Abdelmoaty, M.A.; Ibrahim, M.A.; Ahmed, N.S.; Abdelaziz, M.A. Confirmatory studies on the antioxidant and antidiabetic effect of quercetin in rats. *Indian J. Clin. Biochem.* **2010**, *25*, 188–192. [CrossRef]
88. Tian, C.; Liu, X.; Chang, Y.; Wang, R.; Lv, T.; Cui, C.; Liu, M. Investigation of the anti-inflammatory and antioxidant activities of luteolin, kaempferol, apigenin and quercetin. *South Afr. J. Bot.* **2021**, *137*, 257–264. [CrossRef]
89. Tallei, T.E.; Tumilaar, S.G.; Niode, N.J.; Fatimawali; Kepel, B.J.; Idroes, R.; Effendi, Y.; Sakib, S.A.; Emran, T.B. Potential of Plant Bioactive Compounds as SARS-CoV-2 Main protease (M(pro)) and spike (S) glycoprotein Iihibitors: A molecular docking study. *Scientifica* **2020**, *2020*, 6307457. [CrossRef]
90. Sarian, M.N.; Ahmed, Q.U.; Mat So'ad, S.Z.; Alhassan, A.M.; Murugesu, S.; Perumal, V.; Syed Mohamad, S.N.A.; Khatib, A.; Latip, J. Antioxidant and antidiabetic effects of flavonoids: A structure-activity relationship based study. *Biomed. Res. Int.* **2017**, *2017*, 8386065. [CrossRef]
91. Xiao, T.; Cui, M.; Zheng, C.; Wang, M.; Sun, R.; Gao, D.; Bao, J.; Ren, S.; Yang, B.; Lin, J.; et al. Myricetin inhibits SARS-CoV-2 viral replication by targeting M(pro) and ameliorates pulmonary inflammation. *Front. Pharmacol.* **2021**, *12*, 669642. [CrossRef] [PubMed]
92. Pandey, P.; Rane, J.S.; Chatterjee, A.; Kumar, A.; Khan, R.; Prakash, A.; Ray, S. Targeting SARS-CoV-2 spike protein of COVID-19 with naturally occurring phytochemicals: An in silico study for drug development. *J. Biomol. Struct. Dyn.* **2021**, *39*, 6306–6316. [CrossRef] [PubMed]
93. Zang, Y.; Igarashi, K.; Li, Y. Anti-diabetic effects of luteolin and luteolin-7-O-glucoside on KK-A(y) mice. *Biosci. Biotechnol. Biochem.* **2016**, *80*, 1580–1586. [CrossRef]
94. Barky, A.; Ezz, A.; Mohammed, T. The Potential role of apigenin in diabetes mellitus. *Int. J. Clin. Case Rep. Rev.* **2020**, *3*, 032. [CrossRef]
95. Tutunchi, H.; Naeini, F.; Ostadrahimi, A.; Hosseinzadeh-Attar, M.J. Naringenin, a flavanone with antiviral and anti-inflammatory effects: A promising treatment strategy against COVID-19. *Phytother. Res.* **2020**, *34*, 3137–3147. [CrossRef]
96. Annadurai, T.; Muralidharan, A.R.; Joseph, T.; Hsu, M.J.; Thomas, P.A.; Geraldine, P. Antihyperglycemic and antioxidant effects of a flavanone, naringenin, in streptozotocin-nicotinamide-induced experimental diabetic rats. *J. Physiol. Biochem.* **2012**, *68*, 307–318. [CrossRef]
97. Agrawal, P.K.; Agrawal, C.; Blunden, G. Pharmacological significance of hesperidin and hesperetin, two citrus flavonoids, as promising antiviral compounds for prophylaxis against and combating COVID-19. *Nat. Prod. Commun.* **2021**, *16*. [CrossRef]
98. Ali, A.M.; Gabbar, M.A.; Abdel-Twab, S.M.; Fahmy, E.M.; Ebaid, H.; Alhazza, I.M.; Ahmed, O.M. Antidiabetic potency, antioxidant effects, and mode of actions of *Citrus reticulata* fruit peel hydroethanolic extract, hesperidin, and quercetin in nicotinamide/streptozotocin-induced Wistar diabetic rats. *Oxid. Med. Cell. Longev.* **2020**, *2020*, 1730492. [CrossRef]
99. Jena, A.B.; Kanungo, N.; Nayak, V.; Chainy, G.B.N.; Dandapat, J. Catechin and curcumin interact with S protein of SARS-CoV2 and ACE2 of human cell membrane: Insights from computational studies. *Sci. Rep.* **2021**, *11*, 2043. [CrossRef]
100. Nazir, N.; Zahoor, M.; Ullah, R.; Ezzeldin, E.; Mostafa, G.A.E. Curative effect of catechin isolated from *Elaeagnus umbellata* Thunb. berries for diabetes and related complications in streptozotocin-induced diabetic rats model. *Molecules* **2020**, *26*, 137. [CrossRef] [PubMed]
101. Munteanu, I.G.; Apetrei, C. Analytical methods used in determining antioxidant activity: A review. *Int. J. Mol. Sci.* **2021**, *22*, 3380. [CrossRef]
102. Kakkar, R.; Kalra, J.; Mantha, S.V.; Prasad, K. Lipid peroxidation and activity of antioxidant enzymes in diabetic rats. *Mol. Cell. Biochem.* **1995**, *151*, 113–119. [CrossRef] [PubMed]

103. Eid, H.M.; Haddad, P.S. The antidiabetic potential of quercetin: Underlying mechanisms. *Curr. Med. Chem.* **2017**, *24*, 355–364. [CrossRef]
104. Bule, M.; Abdurahman, A.; Nikfar, S.; Abdollahi, M.; Amini, M. Antidiabetic effect of quercetin: A systematic review and meta-analysis of animal studies. *Food Chem. Toxicol.* **2019**, *125*, 494–502. [CrossRef] [PubMed]
105. Singh, A.K.; Patel, P.K.; Choudhary, K.; Joshi, J.; Yadav, D.; Jin, J.O. Quercetin and coumarin inhibit dipeptidyl peptidase-IV and exhibits antioxidant properties: In silico, in vitro, ex vivo. *Biomolecules* **2020**, *10*, 207. [CrossRef]
106. Serban, M.C.; Sahebkar, A.; Zanchetti, A.; Mikhailidis, D.P.; Howard, G.; Antal, D.; Andrica, F.; Ahmed, A.; Aronow, W.S.; Muntner, P.; et al. Effects of quercetin on blood pressure: A systematic review and meta-analysis of randomized controlled trials. *J. Am. Heart Assoc.* **2016**, *5*, e002713. [CrossRef] [PubMed]
107. Chuang, C.C.; Martinez, K.; Xie, G.; Kennedy, A.; Bumrungpert, A.; Overman, A.; Jia, W.; McIntosh, M.K. Quercetin is equally or more effective than resveratrol in attenuating tumor necrosis factor-{alpha}-mediated inflammation and insulin resistance in primary human adipocytes. *Am. J. Clin. Nutr.* **2010**, *92*, 1511–1521. [CrossRef]
108. Kitakaze, T.; Jiang, H.; Nomura, T.; Hironao, K.Y.; Yamashita, Y.; Ashida, H. Kaempferol promotes glucose uptake in myotubes through a JAK2-dependent pathway. *J. Agric. Food Chem.* **2020**, *68*, 13720–13729. [CrossRef]
109. Ahrens, M.J.; Thompson, D.L. Effect of emulin on blood glucose in type 2 diabetics. *J. Med. Food* **2013**, *16*, 211–215. [CrossRef]
110. Valensi, P.; Le Devehat, C.; Richard, J.L.; Farez, C.; Khodabandehlou, T.; Rosenbloom, R.A.; LeFante, C. A multicenter, double-blind, safety study of QR-333 for the treatment of symptomatic diabetic peripheral neuropathy. A preliminary report. *J. Diabetes Complicat.* **2005**, *19*, 247–253. [CrossRef]
111. Chen, J.; Hamm, L.L.; Bundy, J.D.; Kumbala, D.R.; Bodana, S.; Chandra, S.; Chen, C.S.; Starcke, C.C.; Guo, Y.; Schaefer, C.M.; et al. Combination treatment with sodium nitrite and isoquercetin on endothelial dysfunction among patients with CKD: A randomized phase 2 pilot trial. *Clin. J. Am. Soc. Nephrol.* **2020**, *15*, 1566–1575. [CrossRef]
112. Babajani, F.; Kakavand, A.; Mohammadi, H.; Sharifi, A.; Zakeri, S.; Asadi, S.; Afshar, Z.M.; Rahimi, Z.; Sayad, B. COVID-19 and renin angiotensin aldosterone system: Pathogenesis and therapy. *Health Sci. Rep.* **2021**, *4*, e440. [CrossRef]
113. Marinella, M.A. Indomethacin and resveratrol as potential treatment adjuncts for SARS-CoV-2/COVID-19. *Int. J. Clin. Pract.* **2020**, *74*, e13535. [CrossRef] [PubMed]
114. DiNicolantonio, J.J.; O'Keefe, J.H. Magnesium and vitamin D deficiency as a potential cause of immune dysfunction, cytokine storm and disseminated intravascular coagulation in COVID-19 patients. *Mo. Med.* **2021**, *118*, 68–73.
115. Slominski, R.M.; Stefan, J.; Athar, M.; Holick, M.F.; Jetten, A.M.; Raman, C.; Slominski, A.T. COVID-19 and vitamin D: A lesson from the skin. *Exp. Dermatol.* **2020**, *29*, 885–890. [CrossRef] [PubMed]
116. Di Filippo, L.; Allora, A.; Doga, M.; Formenti, A.M.; Locatelli, M.; Rovere Querini, P.; Frara, S.; Giustina, A. Vitamin D levels are associated with blood glucose and BMI in COVID-19 patients, predicting disease severity. *J. Clin. Endocrinol. Metab.* **2022**, *107*, e348–e360. [CrossRef]
117. Atanasovska, E.; Petrusevska, M.; Zendelovska, D.; Spasovska, K.; Stevanovikj, M.; Kasapinova, K.; Gjorgjievska, K.; Labachevski, N. Vitamin D levels and oxidative stress markers in patients hospitalized with COVID-19. *Redox Rep.* **2021**, *26*, 184–189. [CrossRef]
118. Entrenas Castillo, M.; Entrenas Costa, L.M.; Vaquero Barrios, J.M.; Alcalá Díaz, J.F.; López Miranda, J.; Bouillon, R.; Quesada Gomez, J.M. Effect of calcifediol treatment and best available therapy versus best available therapy on intensive care unit admission and mortality among patients hospitalized for COVID-19: A pilot randomized clinical study. *J. Steroid. Biochem. Mol. Biol.* **2020**, *203*, 105751. [CrossRef]
119. Jain, S.K.; Parsanathan, R.; Levine, S.N.; Bocchini, J.A.; Holick, M.F.; Vanchiere, J.A. The potential link between inherited G6PD deficiency, oxidative stress, and vitamin D deficiency and the racial inequities in mortality associated with COVID-19. *Free Radic. Biol. Med.* **2020**, *161*, 84–91. [CrossRef] [PubMed]
120. Ali, M.Y.; Sina, A.A.; Khandker, S.S.; Neesa, L.; Tanvir, E.M.; Kabir, A.; Khalil, M.I.; Gan, S.H. Nutritional composition and bioactive compounds in tomatoes and their impact on human health and disease: A review. *Foods* **2020**, *10*, 45. [CrossRef]
121. Miean, K.H.; Mohamed, S. Flavonoid (myricetin, quercetin, kaempferol, luteolin, and apigenin) content of edible tropical plants. *J. Agric. Food Chem.* **2001**, *49*, 3106–3112. [CrossRef]
122. Lin, Y.; Shi, R.; Wang, X.; Shen, H.M. Luteolin, a flavonoid with potential for cancer prevention and therapy. *Curr. Cancer Drug Targets* **2008**, *8*, 634–646. [CrossRef]
123. Panche, A.N.; Diwan, A.D.; Chandra, S.R. Flavonoids: An overview. *J. Nutr. Sci.* **2016**, *5*, e47. [CrossRef]
124. Yanishlieva-Maslarova, N.; Heinonen, I. Sources of natural antioxidants: Vegetables, fruits, herbs, spices and teas. *Antioxid. Food Pract. Appl.* **2001**, *10*, 210–266.
125. González-Sarrías, A.; Tomás-Barberán, F.A.; García-Villalba, R. Structural diversity of polyphenols and distribution in foods. *Diet. Polyphen. Metab. Health Eff.* **2020**, *1*, 1–29.
126. Tanwar, B.; Modgil, R. Flavonoids: Dietary occurrence and health benefits. *Spatula Dd* **2012**, *2*, 59–68. [CrossRef]
127. Lv, X.; Liu, T.; Ma, H.; Tian, Y.; Li, L.; Li, Z.; Gao, M.; Zhang, J.; Tang, Z. Preparation of essential oil-based microemulsions for improving the solubility, pH stability, photostability, and skin permeation of quercetin. *AAPS PharmSciTech* **2017**, *18*, 3097–3104. [CrossRef]
128. Jahan, R.; Paul, A.K.; Bondhon, T.A.; Hasan, A.; Jannat, K.; Mahboob, T.; Nissapatorn, V.; de L. Pereira, M.; Wiart, C.; Wilairatana, P.; et al. *Zingiber officinale*: Ayurvedic uses of the plant and in silico binding studies of selected phytochemicals with Mpro of SARS-CoV-2. *Nat. Prod. Commun.* **2021**, *16*. [CrossRef]

129. Jahan, R.; Paul, A.K.; Jannat, K.; Rahmatullah, M. Plant essential oils: Possible COVID-19 therapeutics. *Nat. Prod. Commun.* **2021**, *16*. [CrossRef]
130. Wilkin, P.J.; Al-Yozbaki, M.; George, A.; Gupta, G.K.; Wilson, C.M. The undiscovered potential of essential oils for treating SARS-CoV-2 (COVID-19). *Curr. Pharm. Des.* **2020**, *26*, 5261–5277. [CrossRef]
131. Nasiri Lari, Z.; Hajimonfarednejad, M.; Riasatian, M.; Abolhassanzadeh, Z.; Iraji, A.; Vojoud, M.; Heydari, M.; Shams, M. Efficacy of inhaled *Lavandula angustifolia* Mill. essential oil on sleep quality, quality of life and metabolic control in patients with diabetes mellitus type II and insomnia. *J. Ethnopharmacol.* **2020**, *251*, 112560. [CrossRef] [PubMed]
132. Gok Metin, Z.; Arikan Donmez, A.; Izgu, N.; Ozdemir, L.; Arslan, I.E. Aromatherapy massage for neuropathic pain and quality of life in diabetic patients. *J. Nurs. Sch.* **2017**, *49*, 379–388. [CrossRef] [PubMed]
133. Iddir, M.; Brito, A.; Dingeo, G.; Fernandez Del Campo, S.S.; Samouda, H.; La Frano, M.R.; Bohn, T. Strengthening the immune system and reducing inflammation and oxidative stress through diet and nutrition: Considerations during the COVID-19 Crisis. *Nutrients* **2020**, *12*, 1562. [CrossRef] [PubMed]

 nutrients

Article

Hesperidin Is a Potential Inhibitor against SARS-CoV-2 Infection

Fang-Ju Cheng [1], Thanh-Kieu Huynh [2], Chia-Shin Yang [3], Dai-Wei Hu [2], Yi-Cheng Shen [2,4], Chih-Yen Tu [4,5], Yang-Chang Wu [6,7,8], Chih-Hsin Tang [2,5], Wei-Chien Huang [1,2,8,9,*], Yeh Chen [1,3,*] and Chien-Yi Ho [10,11,*]

1 Drug Development Center, China Medical University, Taichung 404, Taiwan; fangju27@gmail.com
2 Graduate Institute of Biomedical Science, China Medical University, Taichung 404, Taiwan; huynhkieuthanh@gmail.com (T.-K.H.); hudebby1024@gmail.com (D.-W.H.); greywolf0127@gmail.com (Y.-C.S.); chtang@mail.cmu.edu.tw (C.-H.T.)
3 Institute of New Drug Development, China Medical University, Taichung 404, Taiwan; xin740103@hotmail.com
4 Department of Internal Medicine, Division of Pulmonary and Critical Care Medicine, China Medical University Hospital, Taichung 404, Taiwan; chesttu@gmail.com
5 School of Medicine, China Medical University, Taichung 404, Taiwan
6 Chinese Medicine Research and Development Center, China Medical University Hospital, Taichung 404, Taiwan; yachwu@mail.cmu.edu.tw
7 Graduate Institute of Integrated Medicine, College of Chinese Medicine, China Medical University, Taichung 404, Taiwan
8 Department of Medical Laboratory Science and Biotechnology, College of Medical and Health Science, Asia University, Taichung 404, Taiwan
9 Research Center for Cancer Biology and Center for Molecular Medicine, China Medical University, Taichung 404, Taiwan
10 Department of Biomedical Imaging and Radiological Science, China Medical University, Taichung 404, Taiwan
11 Division of Family Medicine, Physical Examination Center, Department of Medical Research, China Medical University Hsinchu Hospital, Hsinchu 302, Taiwan
* Correspondence: whuang@mail.cmu.edu.tw (W.-C.H.); chyeah6599@mail.cmu.edu.tw (Y.C.); samsam172@yahoo.com.tw (C.-Y.H.); Tel.: +886-4-22052121 (ext. 7931) (W.-C.H.); +886-4-22023366 (ext. 6513) (Y.C.); +886-3-5580558 (C.-Y.H.)

Citation: Cheng, F.-J.; Huynh, T.-K.; Yang, C.-S.; Hu, D.-W.; Shen, Y.-C.; Tu, C.-Y.; Wu, Y.-C.; Tang, C.-H.; Huang, W.-C.; Chen, Y.; et al. Hesperidin Is a Potential Inhibitor against SARS-CoV-2 Infection. *Nutrients* **2021**, *13*, 2800. https://doi.org/10.3390/nu13082800

Academic Editors: William B. Grant and Ronan Lordan

Received: 5 June 2021
Accepted: 14 August 2021
Published: 16 August 2021

Abstract: Hesperidin (HD) is a common flavanone glycoside isolated from citrus fruits and possesses great potential for cardiovascular protection. Hesperetin (HT) is an aglycone metabolite of HD with high bioavailability. Through the docking simulation, HD and HT have shown their potential to bind to two cellular proteins: transmembrane serine protease 2 (TMPRSS2) and angiotensin-converting enzyme 2 (ACE2), which are required for the cellular entry of severe acute respiratory syndrome coronavirus 2 (SARS-CoV-2). Our results further found that HT and HD suppressed the infection of VeroE6 cells using lentiviral-based pseudo-particles with wild types and variants of SARS-CoV-2 with spike (S) proteins, by blocking the interaction between the S protein and cellular receptor ACE2 and reducing ACE2 and TMPRSS2 expression. In summary, hesperidin is a potential TMPRSS2 inhibitor for the reduction of the SARS-CoV-2 infection.

Keywords: hesperidin; TMPRSS2; ACE2; SARS-CoV-2; D614G; 501Y.v2

1. Introduction

The severe acute respiratory syndrome coronavirus 2 (SARS-CoV-2), a single-stranded and positive-sense RNA virus in the betacoronavirus family, caused the global pandemic of coronavirus disease in 2019 (COVID-19) [1,2]. Compared to SARS-CoV and the Middle East respiratory syndrome coronavirus (MERS-CoV), SARS-CoV-2 exhibited a higher rate of human-to-human transmission and has threatened global health with a high mortality rate [3]. The development of an effective strategy in controlling the rapid spread of COVID-19 is an urgent issue.

The cell entry of SARS-CoV-2 depends on the binding of the viral spike (S) protein to the cellular receptor angiotensin-converting enzyme2 (ACE2). Through the binding between the S1 subunit and ACE2, the S protein is cleaved into the domains of S1 and S2 by host transmembrane serine protease 2 (TMPRSS2) to expose the fusion peptide on S2 for the subsequent membrane fusion and cellular entry of viral RNA [4,5]. After the cellular entry, both the viral 3-chymotrypsin-like protease (3CLpro; also called Mpro) and papain-like protease (PLpro) further cleave and process the viral polyproteins for the production of four essential structural proteins: the S protein, nucleocapsid (N) protein, membrane (M) protein, and envelope (E) protein, which are needed to compose a complete viral particle [6,7]. To date, COVID-19 has infected over 160 million people, resulting in 3 million deaths. It is imperative to search for potential and protective treatments against SARS-CoV-2 infection. For the development of preventive or therapeutic approaches against SARS-CoV-2 infection, the critical proteins involved in the cellular attachment and replication of SARS-CoV-2 are considered as effective targets [5,8–10].

Food has potential benefits to block COVID-19 by modulating the immune system or defending oxidative stress in response to virus infection [11]. Hesperidin (HD; 3,5,7-trihydroflavanone 7-rhamnoglucoside), a major functional flavanone in flavonoids, can be isolated from lemons and other citrus fruits [12], from which rhamnose sugar can also be removed by glycosyl hydrolases to form hesperetin (HT; 3′, 5,7,-trihydroxy-4′-methoxyflavanone) [13]. Due to the removal of rhamnose sugar, HT has a higher bioavailability compared to HD and can be absorbed directly into the small intestine [13,14]. It has been documented that HD possesses several pharmacological effects, primarily the promotion of anti-oxidation, suppression of pro-inflammatory cytokine production, and repression of cancer cell growth [15]. In addition, HD attenuated the influenza A virus (H1N1)-induced secretion of pro-inflammatory cytokines, contributing to the improvement of pulmonary function [16]. More recently, HD has been shown to possess the anti-SARS-CoV-2 infection activity in an in vitro cell line model [17], and is predicted to bind to viral S and 3CLpro proteins of SARS-CoV-2 and cellular ACE2 of host cells in the molecular simulation analysis [18,19], implying the anti-SARS CoV-2 potential of HD. However, there is no substantial evidence to address the activity of HD/HT against SARS-CoV-2 infection. In this study, we addressed the molecular mechanisms underlying the anti-SARS-CoV-2 infection of HD and HT.

2. Materials and Methods

2.1. Cell Lines and Cell Culture

The human Beas 2B lung cell line was grown in Dulbecco's Modified Eagle Medium (DMEM) with low glucose (Gibco, Waltham, MA, USA, Cat. No. 31600083); the human NCI-H460 lung cancer cell line was grown in RPMI 1640 medium (Gibco, Waltham, MA, USA, Cat. No. 31800089), and monkey VeroE6 kidney cells were cultured in DMEM medium with GlutaMax supplement (Gibco, Waltham, MA, USA, Cat. No. 31966021). All cell lines were cultured in the presence of 10% FBS, 1% streptomycin (100 μg/mL), 1.5 g/L sodium bicarbonate, 1 mM sodium pyruvate, and HEPES (25 mM) at 37 °C in a humidified 5% CO_2/95% incubator.

2.2. Measurement of Luciferase Intensity of SARS-CoV-2 Spike Protein-Pseudotyped Lentiviral Particles

VSV-G pseudotyped lentivirus and the virus particle pseudotyped (Vpp) of SARS-CoV-2 Spike protein SARS-CoV-2-S) with luciferase were obtained from RNA Technology Platform and Gene Manipulation Core, Academia Sinica in Taiwan. The veroE6 cell line was treated with 100 μM of hesperetin and hesperidin for 2 days and then was infected with nCoV-2 pseudovirus. After 24 h, the infected cells were lysed with One-Glo™ Luciferase assay buffer (Promega, Madison, WI, USA, Cat. No. E6120), and luciferase activity was measured using a luminometer.

2.3. Measurement of 3-(4,5-Dimethylthiazol-2-yl)-2,5-diphenyltetrazolium Bromide (MTT)

The cell viability was measured according to the manufacturer's protocol. VeroE6 cell line (8000 cells) in 96-well plates were treated with HT and HD in a dose-dependent manner for 2 days and then subjected to MTT assay (Sigma-Aldrich, Cat. No. M2003).

2.4. Docking-Pose Prediction

For the first step of docking simulation, the crystal structures of TMPRSS2 (PDB code: 1Z8A, accessed date: 14 March 2006), ACE2 (PDB code: 3D0G, accessed date: 8 July 2008), PLpro (PDB code: 3E9S, 7 October 2008), and 3CLpro (PDB code: 3AW0, 14 December 2011) were received from Protein Data Bank (PDB, https://www.rcsb.org/). The structures of HT (PDB code: 5JDC) and HD (PDB code: 6CCF) are also from PDB. Discovery Studio, a docking software using a genetic algorithm, was employed to simulate the automated docking of HT and HD with the catalytic sites of proteins as mentioned above.

2.5. FRET-Based Enzyme Activity Assay

The effects of HT and HD on the interaction between SARS-CoV-2 Spike S1 and human ACE2 were measured by TR-FRET assay according to the manufacturer's protocol (BPS Bioscience, Inc., San Diego, CA, USA, Catalog #79949-1). Briefly, ACE2 and Spike S1 proteins, with or without 60 μM tested compounds, were incubated at room temperature for 1 h. TR-FRET signals were recorded by detecting the emission at a wavelength 620 or 665 nm with excitation at a wavelength 340 nm. HT and HD almost had no effect on the fluorescence of individual proteins. The data were normalized with the effect of tested inhibitors on the fluorescence of ACE2. In the examination of the inhibitory effects of HT and HD on protease activity of human TMPRSS2, the reaction mixture containing 15 μg/mL recombinant protein (Creative Biomart Inc., Shirley, NY, USA, Cat. No. TMPRSS2-1856H) and 60 μM HT or HD in assay buffer (25 mM Tris 8.0, 150 mM NaCl) was pre-incubated at room temperature for 30 min. The reaction was initiated by the addition of 20 μM fluorescent protein substrate. Substrate cleavage was monitored continuously for 6 h by detecting mNeonGreen fluorescence (excitation: 506 nm/emission: 536 nm) using Synergy™ H1 hybrid multi-mode microplate reader (BioTek Instruments, Inc., Winooski, VT, USA). The first 1 h of the reaction was used to calculate initial velocity (V_0). The initial velocity with each compound was calculated and normalized to DMSO control. The effects of HT and HD on SARS-CoV-2 Papain-like Protease (PLpro) were examined by using Papain-like Protease Assay Kit (BPS Bioscience, San Diego, CA, USA, Catalog #79995-2) according to the manufacturer's protocol. Briefly, PLpro was incubated with 60 μM HT or HD for 1 h at 37 °C. The peptide substrate was then added to start the reaction. The fluorescence signal was monitored continuously for 1 h by detecting emission at a wavelength of 460 nm with excitation at a wavelength of 360 nm. The preparation of recombinant SARS-CoV-2 3CLpro and the measurement of its enzyme activity assay were described previously [20]. Briefly, 60 μM HT or HD was pre-incubated with SARS-CoV-2 3CLpro for 30 min at room temperature. The reaction was started by the addition of 20 μM fluorescent protein substrate. The fluorescent signal (Ex/Em: 434 nm/474 nm) was continuously monitored for 1 h. The first 15 min of the reaction was used to calculate initial velocity (V_0) and was normalized to DMSO control. All the data were shown as mean ± SEM from three independent experiments performed in at least three replicates.

2.6. Western Blotting

Cell lines treated with HT and HD in a dose-dependent manner for 2 days were lysed in RIPA buffer containing phosphatase and protease inhibitors. The protein lysates were separated by SDS-PAGE and were then transferred to PVDF membranes. The membranes were blocked in 5% milk in TBST buffer (TBS with 0.1% Tween 20), and were incubated with primary antibodies, including ACE2 (Genetex, Hsinchu, Taiwan, Cat. No. GTX101395), TMPRSS2 (Santa Cruz, CA, USA, Cat. No. sc-515727), and β-actin (Sigma-Aldrich, Darmstadt, Germany, Cat. No. A2228), for overnight at 4 °C followed by incubation with

HRP-conjugated second antibody for 1 h at room temperature. After washing with TBST buffer, the immunoreactive signals were visualized by using enhanced chemiluminescence with ECL reagent.

2.7. Statistical Analysis

Data were shown as the mean ± standard error of the mean (SEM). A one-way ANOVA was used for most comparisons. A p-value < 0.05 was considered statistically significant.

3. Results

3.1. Molecular Docking Reveals HT and HD as Potential Multiple-Target Inhibitors against COVID-19

To test whether HT and HD possess potential activity against SARS-CoV-2 infection, we performed a molecular docking simulation to predict the binding affinity of these two compounds to cellular proteins involved in the cellular entry of SARS-CoV-2 and to viral proteases of SARS-CoV2. The results showed that ACE2 (Figure 1A–F) and TMPRSS2 (Figure 1G–L), the cellular proteins involved in the entry of SARS-CoV2, could interact with HT and HD, and that the energy values of ACE2-HT, ACE2-HD, TMPRSS2-HT, and TMPRSS2-HD were −34.81, −1.65, −30.56, and −7.2 kCal/mol, respectively (Table 1). In addition, these two compounds also yielded an interaction with PLpro (Figure 2A–F) and 3CLpro (Figure 2G–L), the viral proteins involved in the replication of SARS-CoV2. The predicted energy values of PLpro-HT, PLpro-HD, 3CLpro-HT, and 3CLpro-HD were −13.69, 17, −17.94, and 6.21 kCal/mol, respectively (Table 1). Additionally, the binding intensity of HT to PLpro and 3CLpro was higher than that of HD (Table 1). Taken together, HT and HD were potentially natural agents against COVID-19 infection by interfering with the cellular entry and virus replication.

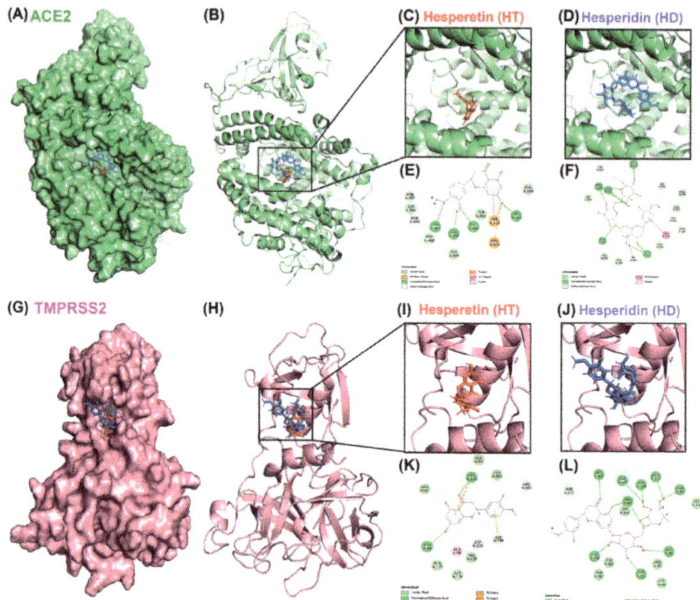

Figure 1. Molecular docking pose visualization for the interaction of ACE2/TMPRSS2 and HT/HD. Compound–protein interaction between HT (PDB code: 5JDC; red structure)/HD (PDB code: 6CCF; blue structure) and ACE2 protein (PDB code: 3D0G) in 3D (**A–D**) and 2D (**E**,**F**). Compound–protein interaction between HT/HD and TMPRSS2 protein (PDB code: 1Z8A) in 3D (**G–J**) and 2D (**K**,**L**).

Table 1. The best predicted energy values of hesperetin and hesperidin with proteins related to SARS-CoV2.

		c-Docker Energy Value (kcal/mol)	Residues	Distance (Å)	Types
ACE2	HT	−34.81	Ala348	1.9	H-bond (H34, O)
			Asp382	2.76	CH-bond (H31, OD1)
			Glu398	2.42	H-bond (H36, O)
				2.44	CH-bond (HA, O21)
			Glu402	2.23	H-bond (HN, O20)
	HD	−1.65	Asp206	2.97/2.97	CH-bond (H52, OD2)/(H54, OD2)
			Thr347	2.74	CH-bond (HA, O14)
			Ala348	2.66	H-bond (H73, O)
			Glu375	2.02/2.17	H-bond (H74, OE1)/(H75, OE1)
			Asp382	2.09	H-bond (HD2, O9)
				2.59	CH-bond (H57, OD1)
			Glu398	2.57	CH-bond (H52, OE1)
			His401	2.76	p-s
			Glu402	2.69	CH-bond (H65, OE1)
			Arg514	4.78	p-cation
TMPRSS2	HT	−30.56	Lys254	3.92	p-cation
			Gly378	2.15	H-bond (H67, O)
	HD	−7.2	His203	4.81	p-alkyl
			Lys254	1.94	H-bond (HZ1, O15)
				1.75	H-bond (HZ3, O14)
				4.08	p-cation
				2.65	CH-bond (HE1, O6)
			Glu301	2.68/2.74	H-bond (H72, O)/(H71, OE2)
				2.47	CH-bond (H56, OE2)
PLpro	HT	−13.69	Asp165	2.98	CH-bond (H31, OD2)
			Pro248	4.97	p-alkyl
	HD	17	Leu163	2.28	H-bond (H77, O)
			Gly164	5.21	Amide-π stracked
			Asp165	2.96	p-anion
			Glu168	2.94	H-bond (H71, OE2)
			Pro249	2.64	CH-bond (HD2, O6)
			Tyr265	5.47	p-p T shaped
			Gln270	2.44	CH-bond (H60, OE1)
				2.31	Unfavorable donor-donor (HE22, H75)
3CLpro	HT	−17.94	Phe140	1.94	H-bond (H34, O)
			Asn142	2.96	H-bond (HD21, O18)
			Glu166	2.75	H-bond (H34, OE1)
				3.07	p-cation
			Gln189	2.19	H-bond (HE22, O21)
	HD	6.21	His41	2.34	H-bond (HE2, O15)
				4.6	p-alkyl
			Asn142	2.63	CH-bond (H46, OD1)
				2.81/2.85	H-bond (HN, O2)/(HD21, O3)
			Gly143	2.8	H-bond (HN, O14)
			Cys145	2.27	H-bond (HG, O15)
			Met165	5.17	Alkyl
			Glu166	2.47/2.69	CH-bond (H61, O)/(H59, O)
				1.95	H-bond (H73, O)

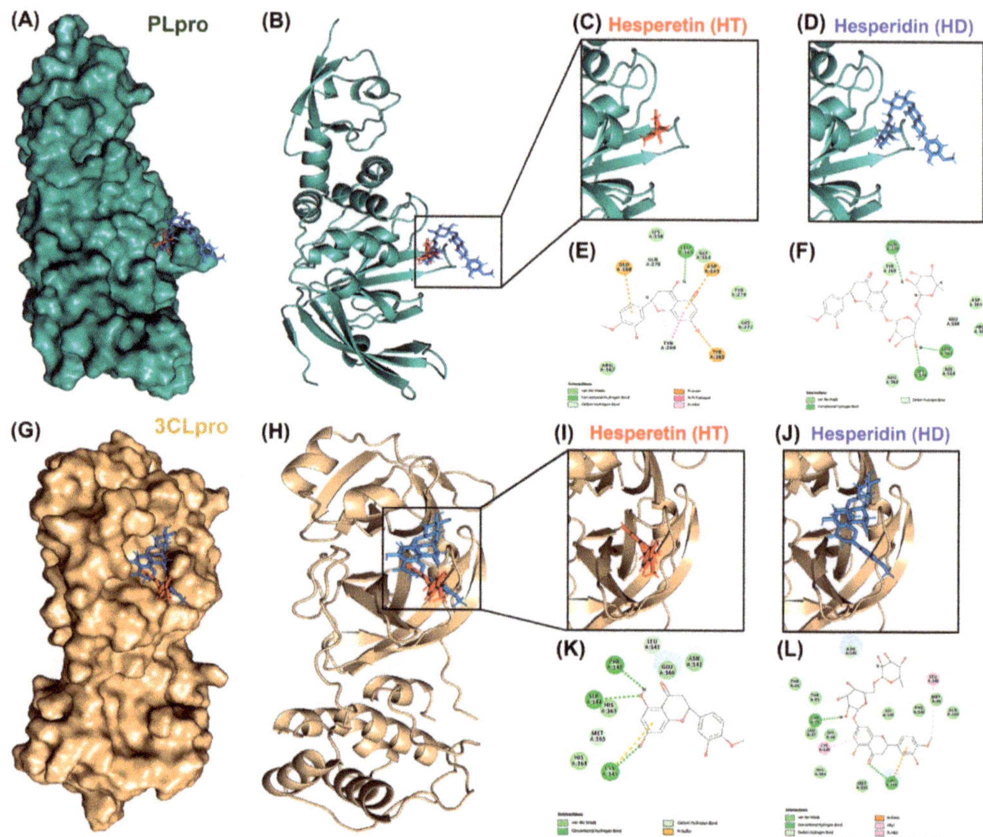

Figure 2. Molecular docking pose visualization for the interaction of PLpro/3CLpro and HT/HD. Compound–protein interaction between HT (PDB code: 5JDC; red)/HD (PDB code: 6CCF; blue) and PLpro protein (PDB code: 3E9S) in 3D (**A–D**) and 2D (**E**,**F**). Compound–protein interaction between HT/HD and 3CLpro protein (PDB code: 3AW0) in 3D (**G–J**) and 2D (**K**,**L**).

3.2. *HT Suppresses the Interaction between ACE2 and the Spike Protein of SARS-CoV-2 In Vitro*

To further confirm the inhibitory effects of HT and HD on SARS-CoV-2 infection, we performed a FRET assay to examine the binding affinity between human receptor ACE2 and the S protein, as well as the enzymatic activities of TMPRSS2, PLpro, and 3CLpro in the presence of HT and HD. The result displayed that the treatment with HT but not HD reduced the binding activity between ACE2 and the S protein (Figure 3A). The result of the docking simulation also showed that HT entered the pocket of human ACE2 bound to the S protein (Figure S1). In addition, HT and HD only slightly decreased the enzyme activity of TMPRSS2 (Figure 3B) but did not influence the enzyme activity of PLpro (Figure 3C) and 3CLpro (Figure 3D), suggesting that HT had the potential role of blocking the cellular entry of SARS-CoV-2 via impeding the binding of human receptor ACE2 with the S protein.

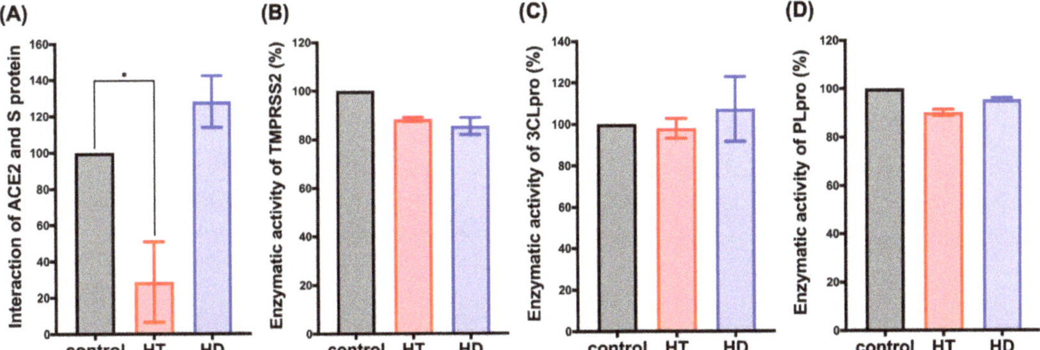

Figure 3. HT (Hesperetin) decreased the interaction of ACE2 and the spike protein. FRET assay was performed to determine the interaction between human receptor ACE2 and S protein (**A**). The in vitro enzymatic activity of TMPRSS2 (**B**), PLpro (**C**), and 3CLpro (**D**) was determined after 1 hr incubation with HT and HD (Hesperidin). Data are shown as mean ± SEM from 3 independent experiments with triplicates. * $p < 0.05$.

3.3. HT and HD Downregulated the Protein Expression of ACE2 and TMPRSS2 in Normal and Malignant Lung Cells

In order to examine the cytotoxicity of HT and HD, we performed MTT assays to test their effect on the cell viability of VeroE6 cells in a dose-dependent manner. The IC50 values of HT and HD in VeroE6 cell were 1491 and 1435 μM, respectively (Figure S2), indicating their low toxicity to cells. We next examined whether HT and HD influence the cellular protein expressions of ACE2 and TMPRSS2 in Western blot analysis. In Figure 4, treatments with HT and HD repressed the protein expressions of ACE2 and TMPRSS2 in normal lung epithelial Beas 2B cell (Figure 4A) and in H460 lung cancer cells (Figure 4B), suggesting that HT and HD not only disrupted the interaction of ACE2 and SARS-CoV-2 S protein but also inhibited the protein expressions of ACE2 and TMPRSS2 for the reduction of SARS-CoV-2 infection. As shown in Supplemental Figure S3, we unexpectedly found that both HT and HD increased but did not decrease the mRNA levels of ACE2 and TMPRSS2, suggesting a post-transcriptional downregulation of the two proteins by HT and HD. Interestingly, heat shock protein 70/90 (HSP70/90) was found to mediate the protein stabilization of ACE2 and thereby maintained the cell entry mechanism of SARS-CoV-2 [21]. HSP90 was reported to be significantly downregulated in the hesperidin-treated cells [22]. These findings raise the possibility that HT and HD functions as HSP70/90 inhibitors to cause the ACE2 and TMPRSS2 protein downregulation despite the induction of their mRNA expressions (Figure S3).

3.4. HT and HD Block the Cellular Entry of Vpp of SARS-CoV-2 Spike Protein (SARS-CoV-2-S)

Next, we verified the potential of HT and HD in suppressing the cellular entry of SARS-CoV-2. The VeroE6 monkey kidney cell line was infected with the Vpp of SARS-CoV-2-S followed by the pretreatments with HT and HD for 2 days due to their inhibitory effects on ACE2 and TMPRSS2 expressions. We found that treatments with HT and HD significantly impaired the infection of SARS-CoV-2-S with the Vpp (Figure 5B), but not VSVG pseudotyped lentivirus (Figure 5A). The globally uncontrolled transmission of SARS-CoV2 was due to the viral evolution. Starting from April 2020, the predominant strains of SARS-CoV2 were D614G (substitution of aspartate (D) to glycine (G) at site 614 in S protein) and 501Y.v2 (also called B.1.351; the simultaneous mutation of D614G and N501Y in the S protein) [23,24]. Compared to the wild-type S protein of SARS-CoV-2, both of these variants showed more robust binding activities to ACE2 to increase the efficacy of virus replication and transmission in host cells [25]. Treatments with HT and HD also dramatically diminished the cellular entry of the VPP of SARS-CoV-2-S, D614G and 501Y.v2 strains (Figure 5B,C), without affecting the viability of VeroE6 cells (Figure S2A,B).

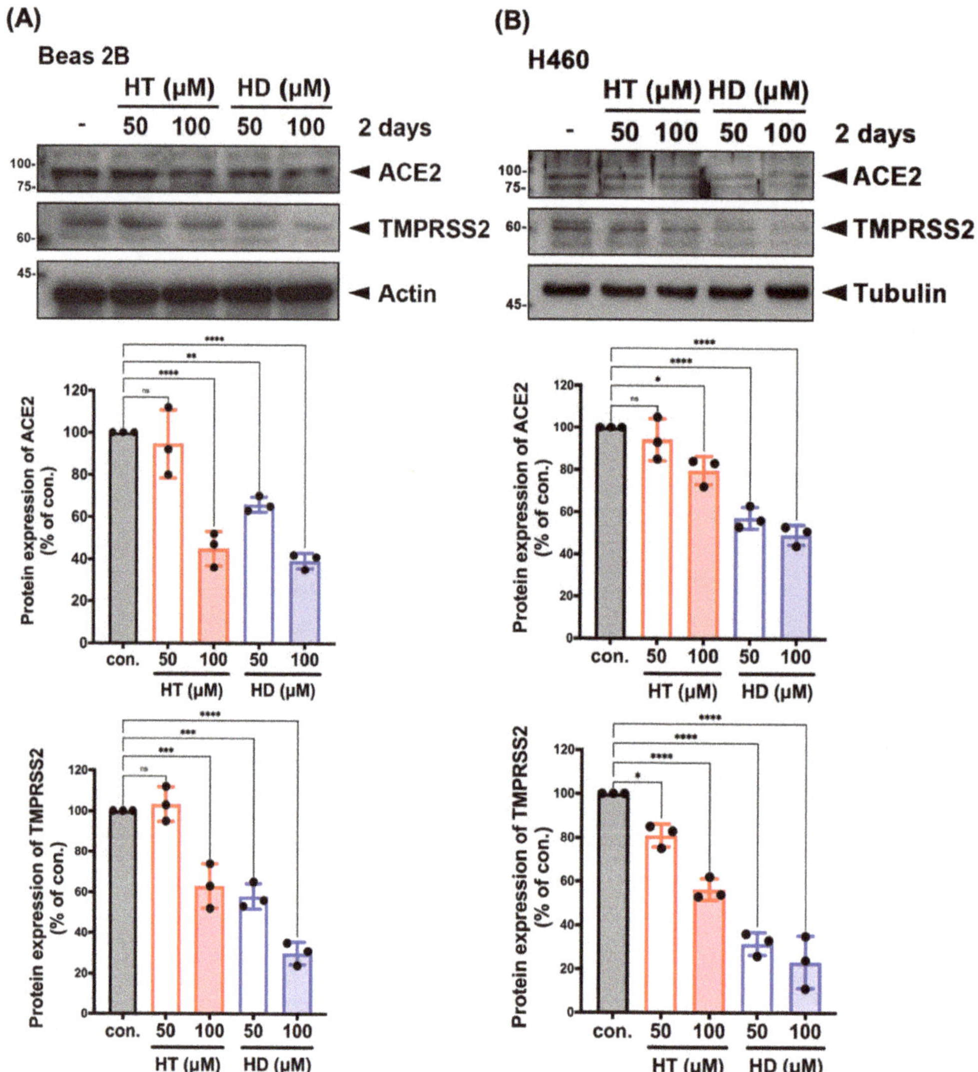

Figure 4. HT (Hesperetin) and HD (Hesperidin) suppressed the protein expressions of ACE2 and TMPRSS2 in normal and malignant lung cells. Beas 2B (**A**) and H460 (**B**) cell lines were treated with HT and HD in a dose-dependent manner for 2 days followed by the examination of protein expression in Western blotting with indicated antibodies. The quantitative results of ACE2 and TMPRSS2 expressions in Beas 2B and H460 were normalized with a level of actin and were shown below to Western blot images. • is shown as the mean of every independent experiment. Data are shown as mean ± SEM from 3 independent experiments with triplicates. * $p < 0.05$, ** $p < 0.01$, *** $p < 0.001$, and **** $p < 0.0001$. NS, no significance.

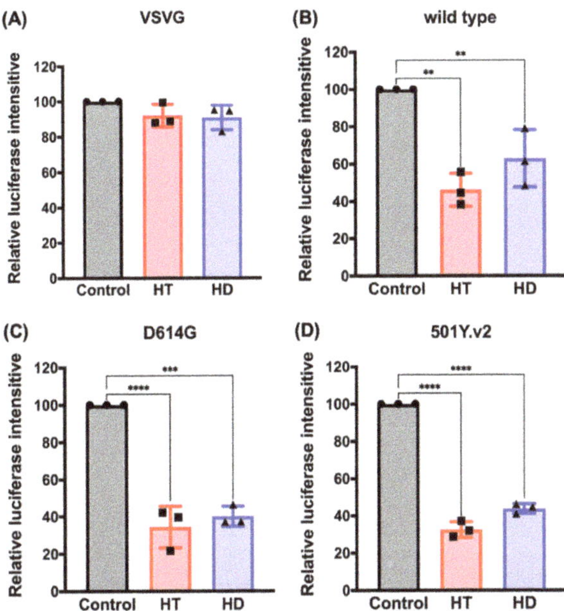

Figure 5. HT (Hesperetin) and HD (Hesperidin) impede SARS-CoV-2 pseudovirus into VeroE6 cell. The luciferase intensity of VSVG (**A**) or SARS-CoV-2 pseudovirus with S protein wild type (**B**), D614G strain (**C**), and 501Y.v2 stain (**D**) in VeroE6 cell line was measured after treatments with 100 μM of HT and HD for 2 days. •, ■, and ▲ are shown as the mean of every independent experiment. Data are shown as mean ± SEM from 3 independent experiments with triplicates. ** $p < 0.01$, *** $p < 0.001$, and **** $p < 0.0001$.

Furthermore, we treated VeroE6 cells with these compounds in the short term, before or after virus infection (as illustrated in Figure S4A), to demonstrate that the blockage of ACE/S protein interaction mediated the anti-SARS-CoV-2 infection activity of HT. As shown in Supplemental Figure S4C–E, the cell entry of the SARS-CoV-2 pseudovirus with wild types and variants of the S protein was reduced by adding HT (red) at 2 h pre-infection or during infection, but not post-infection, of pseudovirus. Interestingly, the treatments with HD (blue) in the short term did not suppress the infection of SARS-CoV-2 pseudovirus, which was consistent with its inability to suppress the interaction between ACE2 and S protein (Figure 3A) These results suggest HT and HD as potential agents against infection with SARS-CoV2 and its mutant strains.

4. Discussion

To date, the COVID-19 pandemic still cannot be controlled, even though many therapeutic strategies and vaccines have been developed. Similar to other RNA viruses, the genetic diversity of SARS-CoV-2 driven by high-random mutation and recombination enables this virus to increase the recognition of human cellular receptors, virus replication, and the higher rate of widespread infection [24,26,27]. However, these variants of SARS-CoV-2, such as the predominant strains of D614G and 501Y.v2, impair the neutralization by the vaccine-induced immunity [23,28]. In this study, we found that HT and HD are able to dramatically inhibit the cellular entry of Vpp of SARS-CoV-2 variants (Figure 5). HD has been considered as a favorable adjuvant for vaccines to prevent lung injury by promoting pro-inflammatory cytokines secretion and cell-autonomous immunity against influenza A (H1N1) infection [16,29,30]. Taken together, HT and HD may show benefits in fighting the threat of the COVID-19 pandemic.

There has been a wide prevalence of SARS-CoV-2 infection, and specific medicines for COVID-19 remain unavailable. Therefore, the components of SARS-CoV-2 and infection procedure are potential targets to screen for the pre-existing or marketed drugs which may possess preventive or therapeutic activity against SARS-CoV2 infection. Some potential medicines have been found to suppress SARS-CoV-2 infection in in vitro and animal studies [31]. However, few medicines have been proven to eradicate SARS-CoV-2 infection effectively in clinical trials. The use of chloroquine and hydroxychloquine, which inhibit the cellular entry, was revoked due to the high risk of mortality [32]. Camostat and Nafamostat, the synthetic inhibitors for TMPRSS2, also suppressed the cellular entry of SARS-CoV-2 but caused severe bleeding [33]. Our findings identified that HT and HD hindered the interaction between the S protein of SARS-CoV-2, hosted the cellular receptor ACE2 and downregulated the protein expression of ACE2 and TMPRSS2, thereby suppressing the infection with Vpp of SARS-CoV-2-S. In addition, the administration of HD at 500 mg/kg did not cause any abnormalities in the animal model, indicating a good safety profile. The median lethal dose of HD is 4837.5 mg/kg [34]. Therefore, HD could be considered as an effective and natural compound to fight SARS-CoV-2 infection.

Poncirus trifoliata (L.) *Raf.* (also called bitter orange fruit) belongs to the member of Rutaceae family and is closely related to *Citrus trifoliata*. This fruit contains various phytotherapeutic activities, which depend on its maturity, to alleviate symptoms in disorders. Poncirus fructus, the dry form of immature fruit of *Poncirus trifoliata* (L.) *Raf.*, is commonly known as a herbal medicine in East Asia for the dysfunction of the digestive system. The mature fruits demonstrated anti-cancer and anti-inflammatory activities [35]. Moreover, the seed extract from *Poncirus trifoliata* (L.) *Raf.* possessed the antiviral activity via the suppression of the cellular endocytosis of the oseltamivir-resistant influenza virus [36]. HD is one of the predominant phytochemicals found in *Poncirus trifoliata* (L.) *Raf* [37,38] and possesses the antioxidant, anti-inflammation, and anti-tumor properties [15]. These findings and our study support that HD isolated from *Poncirus trifoliata* (L.) *Raf.* could be considered as a potential agent to prevent SARS-CoV-2 infection.

Several review articles have predicted the anti-SARS-CoV-2 infection activity of the flavonoid family, including hesperidin [18,19]. Kandeil et al. cleanly demonstrated the inhibitory effect of hesperidin on the viral replication of SARS-CoV-2 at the early stage of virus infection [17]. Unfortunately, there is no direct evidence showing that the molecular mechanism of anti-SARS-CoV-2 activity is derived from hesperidin. It is also unknown whether hesperidin possesses anti-SARS-CoV-2 activity through the influence of the cellular components involved in SARS-CoV-2 infection. In this study, we demonstrated that hesperidin and its aglycone metabolite hesperetin repressed the protein expression of ACE2 and TMPRSS2 in lung cells, impeding the cell entry of the SARS-CoV-2 pseudo-virus. But hesperidin cannot directly decrease the activities of viral proteases, including PLpro and Mpro, in the enzyme activity assays (Figure 3C,D) even though the binding activity of hesperidin to these enzymes was predicted in the molecular docking analysis in the previous studies of [18,19], and this study (Figure 2G–L; Table 1). The reliability of protein structure and the environment of the binding site used for the ligand–protein complex docking assays would determine the prediction accuracy [39]. The molecular dynamic simulation would also be required to validate the predictions from molecular docking [39].

5. Conclusions

Currently, no specific therapy can significantly inhibit SARS-CoV-2 infection and help to prevent the COVID-19 pandemic in many parts of the world. Several vaccines were developed and approved by FDA to prevent SARS-CoV-2 infection and effectively suppressed the incidence of COVID-19 [40,41]. However, SARS-CoV-2 variants escaped from the inhibition by neutralizing antibodies [28]. Exploring the promising antiviral agents remains essential for the termination of SARS-CoV-2 spreading. Hesperidin is enriched in citrus fruits, which are common traditional medicines in Asia. In this study, we demonstrated that hesperidin and its aglycone, hesperetin, might provide benefits in fighting COVID-19 via blocking the binding of the S protein of SARS-CoV-2 to the human cellular receptor ACE2 and reducing the protein expression of ACE2 and TMPRSS2. These effects significantly suppress the cellular entry of the SARS-CoV-2 variant regardless of the mutation of the S protein. Therefore, hesperidin could be used as a potential prophylactic treatment against COVID-19.

Supplementary Materials: The following are available online at https://www.mdpi.com/article/10.3390/nu13082800/s1, Figure S1: the docking simulation of hesperetin in the pocket of human ACE2 bound to viral S protein. Figure S2: the effect of HT/HD on the viability of VeroE6 cells. Figure S3: the effect of HT and HD on the mRNA expressions of ACE2 and TMPRSS2. Figure S4: short term treatment with HT impedes the infection of SARS-CoV-2 pseudovirus into VeroE6 cells.

Author Contributions: F.-J.C. conceived the study, designed and performed the experiments, analyzed data, interpreted results, and wrote the manuscript; T.-K.H., C.-S.Y., D.-W.H. and Y.-C.S. performed experiments; C.-Y.T., Y.-C.W., C.-H.T. and Y.C. revised the manuscript; W.-C.H. and C.-Y.H. conceived and supervised the entire project, experimental designs, interpreted results, and wrote the manuscript. All authors have read and agreed to the published version of the manuscript.

Funding: This study was supported in part by grants from the Ministry of Science Technology, Taiwan (MOST 109-2320-B-039-014), China Medical University (CMU106-ASIA-18 and CMU107-TU-06), and China Medical University Hospital (DMR-109-212, DMR 109-213, CMUHCH DMR-110-003, and CMUHCH DMR-110-006). This work was also financially supported by the "Drug Development Center, China Medical University" from The Featured Areas Research Center Program within the framework of the Higher Education Sprout Project by the Ministry of Education (MOE) in Taiwan.

Institutional Review Board Statement: This study did not involve any humans or animals.

Acknowledgments: Experiments and data analysis were performed in part with the use of the Medical Research Core Facilities, Office of Research & Development at China Medical University, Taichung, Taiwan, R.O.C.

Conflicts of Interest: The authors declare no conflict of interest.

References

1. Del Rio, C.; Malani, P.N. COVID-19-New Insights on a Rapidly Changing Epidemic. *JAMA* **2020**, *323*, 1339–1340. [CrossRef]
2. Gao, Y.; Yan, L.; Huang, Y.; Liu, F.; Zhao, Y.; Cao, L.; Wang, T.; Sun, Q.; Ming, Z.; Zhang, L.; et al. Structure of the RNA-dependent RNA polymerase from COVID-19 virus. *Science* **2020**, *368*, 779–782. [CrossRef]
3. Chan, J.F.; Yuan, S.; Kok, K.H.; To, K.K.; Chu, H.; Yang, J.; Xing, F.; Liu, J.; Yip, C.C.; Poon, R.W.; et al. A familial cluster of pneumonia associated with the 2019 novel coronavirus indicating person-to-person transmission: A study of a family cluster. *Lancet* **2020**, *395*, 514–523. [CrossRef]
4. Walls, A.C.; Park, Y.J.; Tortorici, M.A.; Wall, A.; McGuire, A.T.; Veesler, D. Structure, Function, and Antigenicity of the SARS-CoV-2 Spike Glycoprotein. *Cell* **2020**, *181*, 281–292.e286. [CrossRef]
5. Hoffmann, M.; Kleine-Weber, H.; Schroeder, S.; Krüger, N.; Herrler, T.; Erichsen, S.; Schiergens, T.S.; Herrler, G.; Wu, N.H.; Nitsche, A.; et al. SARS-CoV-2 Cell Entry Depends on ACE2 and TMPRSS2 and Is Blocked by a Clinically Proven Protease Inhibitor. *Cell* **2020**, *181*, 271–280.e278. [CrossRef]
6. Dömling, A.; Gao, L. Chemistry and Biology of SARS-CoV-2. *Chem* **2020**, *6*, 1283–1295. [CrossRef]
7. Báez-Santos, Y.M.; St John, S.E.; Mesecar, A.D. The SARS-coronavirus papain-like protease: Structure, function and inhibition by designed antiviral compounds. *Antivir. Res.* **2015**, *115*, 21–38. [CrossRef]
8. Bouhaddou, M.; Memon, D.; Meyer, B.; White, K.M.; Rezelj, V.V.; Correa Marrero, M.; Polacco, B.J.; Melnyk, J.E.; Ulferts, S.; Kaake, R.M.; et al. The Global Phosphorylation Landscape of SARS-CoV-2 Infection. *Cell* **2020**, *182*, 685–712. [CrossRef]
9. Jan, J.T.; Cheng, T.R.; Juang, Y.P.; Ma, H.H.; Wu, Y.T.; Yang, W.B.; Cheng, C.W.; Chen, X.; Chou, T.H.; Shie, J.J.; et al. Identification of existing pharmaceuticals and herbal medicines as inhibitors of SARS-CoV-2 infection. *Proc. Natl. Acad. Sci. USA* **2021**, *118*, e2021579118. [CrossRef]
10. Jackson, L.A.; Anderson, E.J.; Rouphael, N.G.; Roberts, P.C.; Makhene, M.; Coler, R.N.; McCullough, M.P.; Chappell, J.D.; Denison, M.R.; Stevens, L.J.; et al. An mRNA Vaccine against SARS-CoV-2-Preliminary Report. *N. Engl. J. Med.* **2020**, *383*, 1920–1931. [CrossRef]
11. Iddir, M.; Brito, A.; Dingeo, G.; Fernandez Del Campo, S.S.; Samouda, H.; La Frano, M.R.; Bohn, T. Strengthening the Immune System and Reducing Inflammation and Oxidative Stress through Diet and Nutrition: Considerations during the COVID-19 Crisis. *Nutrients* **2020**, *12*, 1562. [CrossRef]
12. Manthey, J.A.; Grohmann, K. Flavonoids of the orange subfamily Aurantioideae. *Adv. Exp. Med. Biol.* **1998**, *439*, 85–101. [CrossRef]

13. González-Barrio, R.; Trindade, L.M.; Manzanares, P.; de Graaff, L.H.; Tomás-Barberán, F.A.; Espín, J.C. Production of bioavailable flavonoid glucosides in fruit juices and green tea by use of fungal alpha-L-rhamnosidases. *J. Agric. Food Chem.* **2004**, *52*, 6136–6142. [CrossRef]
14. Jin, M.J.; Kim, U.; Kim, I.S.; Kim, Y.; Kim, D.H.; Han, S.B.; Kim, D.H.; Kwon, O.S.; Yoo, H.H. Effects of gut microflora on pharmacokinetics of hesperidin: A study on non-antibiotic and pseudo-germ-free rats. *J. Toxicol. Environ. Health A* **2010**, *73*, 1441–1450. [CrossRef]
15. Li, C.; Schluesener, H. Health-promoting effects of the citrus flavanone hesperidin. *Crit. Rev. Food Sci. Nutr.* **2017**, *57*, 613–631. [CrossRef]
16. Ding, Z.; Sun, G.; Zhu, Z. Hesperidin attenuates influenza A virus (H1N1) induced lung injury in rats through its anti-inflammatory effect. *Antivir. Ther.* **2018**, *23*, 611–615. [CrossRef]
17. Kandeil, A.; Mostafa, A.; Kutkat, O.; Moatasim, Y.; Al-Karmalawy, A.A.; Rashad, A.A.; Kayed, A.E.; Kayed, A.E.; El-Shesheny, R.; Kayali, G.; et al. Bioactive Polyphenolic Compounds Showing Strong Antiviral Activities against Severe Acute Respiratory Syndrome Coronavirus 2. *Pathogens* **2021**, *10*, 758. [CrossRef]
18. Bellavite, P.; Donzelli, A. Hesperidin and SARS-CoV-2: New Light on the Healthy Function of Citrus Fruits. *Antioxidants* **2020**, *9*, 742. [CrossRef]
19. Haggag, Y.A.; El-Ashmawy, N.E.; Okasha, K.M. Is hesperidin essential for prophylaxis and treatment of COVID-19 Infection? *Med. Hypotheses* **2020**, *144*, 109957. [CrossRef]
20. Wang, Y.C.; Yang, W.H.; Yang, C.S.; Hou, M.H.; Tsai, C.L.; Chou, Y.Z.; Hung, M.C.; Chen, Y. Structural basis of SARS-CoV-2 main protease inhibition by a broad-spectrum anti-coronaviral drug. *Am. J. Cancer Res.* **2020**, *10*, 2535–2545.
21. Alreshidi, F.S.; Ginawi, I.A.; Hussain, M.A.; Arif, J.M. Piperaquine- and Aspirin- Mediated Protective Role of HSP70 and HSP90 as Modes to Strengthen the Natural Immunity against Potent SARS-CoV-2. *Biointerface Res. Appl. Chem.* **2021**, *11*, 12364–12379. [CrossRef]
22. Yumnam, S.; Venkatarame Gowda Saralamma, V.; Raha, S.; Lee, H.J.; Lee, W.S.; Kim, E.K.; Lee, S.J.; Heo, J.D.; Kim, G.S. Proteomic profiling of human HepG2 cells treated with hesperidin using antibody array. *Mol. Med. Rep.* **2017**, *16*, 5386–5392. [CrossRef]
23. Zhang, L.; Jackson, C.B.; Mou, H.; Ojha, A.; Peng, H.; Quinlan, B.D.; Rangarajan, E.S.; Pan, A.; Vanderheiden, A.; Suthar, M.S.; et al. SARS-CoV-2 spike-protein D614G mutation increases virion spike density and infectivity. *Nat. Commun.* **2020**, *11*, 6013. [CrossRef]
24. Tegally, H.; Wilkinson, E.; Giovanetti, M.; Iranzadeh, A.; Fonseca, V.; Giandhari, J.; Doolabh, D.; Pillay, S.; San, E.J.; Msomi, N.; et al. Emergence of a SARS-CoV-2 variant of concern with mutations in spike glycoprotein. *Nature* **2021**, *592*, 438–443. [CrossRef]
25. Mascola, J.R.; Graham, B.S.; Fauci, A.S. SARS-CoV-2 Viral Variants-Tackling a Moving Target. *JAMA* **2021**, *325*, 1261–1262. [CrossRef] [PubMed]
26. Baric, R.S. Emergence of a Highly Fit SARS-CoV-2 Variant. *N. Engl. J. Med.* **2020**, *383*, 2684–2686. [CrossRef]
27. Korber, B.; Fischer, W.M.; Gnanakaran, S.; Yoon, H.; Theiler, J.; Abfalterer, W.; Hengartner, N.; Giorgi, E.E.; Bhattacharya, T.; Foley, B.; et al. Tracking Changes in SARS-CoV-2 Spike: Evidence that D614G Increases Infectivity of the COVID-19 Virus. *Cell* **2020**, *182*, 812–827.e819. [CrossRef]
28. Abdool Karim, S.S.; de Oliveira, T. New SARS-CoV-2 Variants-Clinical, Public Health, and Vaccine Implications. *N. Engl. J. Med.* **2021**, *384*, 1866–1868. [CrossRef]
29. Dong, W.; Wei, X.; Zhang, F.; Hao, J.; Huang, F.; Zhang, C.; Liang, W. A dual character of flavonoids in influenza A virus replication and spread through modulating cell-autonomous immunity by MAPK signaling pathways. *Sci. Rep.* **2014**, *4*, 7237. [CrossRef]
30. Pappalardo, F.; Fichera, E.; Paparone, N.; Lombardo, A.; Pennisi, M.; Russo, G.; Leotta, M.; Pappalardo, F.; Pedretti, A.; De Fiore, F.; et al. A computational model to predict the immune system activation by citrus-derived vaccine adjuvants. *Bioinformatics* **2016**, *32*, 2672–2680. [CrossRef]
31. Hu, B.; Guo, H.; Zhou, P.; Shi, Z.L. Characteristics of SARS-CoV-2 and COVID-19. *Nat. Rev. Microbiol.* **2021**, *19*, 141–154. [CrossRef] [PubMed]
32. Geleris, J.; Sun, Y.; Platt, J.; Zucker, J.; Baldwin, M.; Hripcsak, G.; Labella, A.; Manson, D.K.; Kubin, C.; Barr, R.G.; et al. Observational Study of Hydroxychloroquine in Hospitalized Patients with Covid-19. *N. Engl. J. Med.* **2020**, *382*, 2411–2418. [CrossRef] [PubMed]
33. Hifumi, T.; Isokawa, S.; Otani, N.; Ishimatsu, S. Adverse events associated with nafamostat mesylate and favipiravir treatment in COVID-19 patients. *Crit. Care* **2020**, *24*, 497. [CrossRef] [PubMed]
34. Li, Y.; Kandhare, A.D.; Mukherjee, A.A.; Bodhankar, S.L. Acute and sub-chronic oral toxicity studies of hesperidin isolated from orange peel extract in Sprague Dawley rats. *Regul. Toxicol. Pharmacol.* **2019**, *105*, 77–85. [CrossRef]
35. Jang, Y.; Kim, E.K.; Shim, W.S. Phytotherapeutic effects of the fruits of *Poncirus trifoliata* (L.) Raf. on cancer, inflammation, and digestive dysfunction. *Phytother. Res.* **2018**, *32*, 616–624. [CrossRef]
36. Heo, Y.; Cho, Y.; Ju, K.S.; Cho, H.; Park, K.H.; Choi, H.; Yoon, J.K.; Moon, C.; Kim, Y.B. Antiviral activity of Poncirus trifoliata seed extract against oseltamivir-resistant influenza virus. *J. Microbiol.* **2018**, *56*, 586–592. [CrossRef] [PubMed]
37. Hussain, S.; Curk, F.; Dhuique-Mayer, C.; Urban, L.; Ollitrault, P.; Luro, F.; Morillon, R. Autotetraploid trifoliate orange (*Poncirus trifoliata*) rootstocks do not impact clementine quality but reduce fruit yields and highly modify rootstock/scion physiology. *Sci. Hortic.* **2012**, *134*, 100–107. [CrossRef]

38. Tundis, R.; Bonesi, M.; Sicari, V.; Pellicanò, T.M.; Tenuta, M.C.; Leporini, M.; Menichini, F.; Loizzo, M.R. *Poncirus trifoliata* (L.) Raf.: Chemical composition, antioxidant properties and hypoglycaemic activity via the inhibition of α-amylase and α-glucosidase enzymes. *J. Funct. Foods* **2016**, *25*, 477–485. [CrossRef]
39. Chen, Y.C. Beware of docking! *Trends Pharmacol. Sci.* **2015**, *36*, 78–95. [CrossRef]
40. Walsh, E.E.; Frenck, R.W., Jr.; Falsey, A.R.; Kitchin, N.; Absalon, J.; Gurtman, A.; Lockhart, S.; Neuzil, K.; Mulligan, M.J.; Bailey, R.; et al. Safety and Immunogenicity of Two RNA-Based Covid-19 Vaccine Candidates. *N. Engl. J. Med.* **2020**, *383*, 2439–2450. [CrossRef] [PubMed]
41. Baden, L.R.; El Sahly, H.M.; Essink, B.; Kotloff, K.; Frey, S.; Novak, R.; Diemert, D.; Spector, S.A.; Rouphael, N.; Creech, C.B.; et al. Efficacy and Safety of the mRNA-1273 SARS-CoV-2 Vaccine. *N. Engl. J. Med.* **2021**, *384*, 403–416. [CrossRef] [PubMed]

Article

Connecting Families to Food Resources amid the COVID-19 Pandemic: A Cross-Sectional Survey of Early Care and Education Providers in Two U.S. States

Lacy Stephens [1,*], Caroline Rains [2] and Sara E. Benjamin-Neelon [3]

[1] National Farm to School Network, P.M.B. #104, 8770 West Bryn Mawr Ave, Suite 1300, Chicago, IL 60631, USA
[2] Research Triangle Institute International, 3040 East Cornwallis Road, Research Triangle Park, NC 27709, USA; crains@rti.org
[3] Department of Health, Behavior and Society, Johns Hopkins University, 615 North Wolfe Street, Baltimore, MD 21205, USA; sara.neelon@jhu.edu
* Correspondence: lacy@farmtoschool.org; Tel.: +1-816-914-0597

Abstract: Early care and education (ECE) settings are important avenues for reaching young children and their families with food and nutrition resources, including through the U.S. federally funded Child and Adult Care Food Program (CACFP). Researchers conducted a cross-sectional survey of ECE providers in two U.S. states in November 2020 to identify approaches used to connect families with food and nutrition resources amid the COVID-19 pandemic. Logistic regression models were used to estimate odds of sites reporting no approaches and adjusted Poisson models were used to estimate the incidence rate ratio of the mean number of approaches, comparing sites that participate in CACFP to those that did not. A total of 589 ECE sites provided responses. Of those, 43% (n = 255) participated in CACFP. CACFP participating sites were more likely to report using any approaches to connecting families to food resources and significantly more likely to report offering "grab and go" meals, providing meal delivery, distributing food boxes to families, and recommending community food resources than non-CACFP sites. This study suggests that CACFP sites may have greater capacity to connect families to food resources amid emergencies than non-CACFP participating sites.

Keywords: CACFP; COVID-19; early care and education; food program

Citation: Stephens, L.; Rains, C.; Benjamin-Neelon, S.E. Connecting Families to Food Resources amid the COVID-19 Pandemic: A Cross-Sectional Survey of Early Care and Education Providers in Two U.S. States. *Nutrients* **2021**, *13*, 3137. https://doi.org/10.3390/nu13093137

Academic Editor: William B. Grant

Received: 30 July 2021
Accepted: 7 September 2021
Published: 9 September 2021

1. Introduction

Early care and education (ECE) sites, including child care centers, family child care homes, Head Starts, state or private preschools, nurseries, and childminders, are important settings for impacting nutrition and health behaviors in young children and reaching families with nutrition information [1–4]. Across "economically advanced" countries, 25% of children under the age of 3 years and 80% of children ages 3–6 are in some form of ECE [5]. In the United States (U.S.), 60% of children under the age of 5 years regularly spend time in non-parental care, and those children spend an average of approximately 25 h per week in care [6]. The Academy of Nutrition and Dietetics recommends that children in full-time care consume half to one-third of their daily calories in care, highlighting the importance of ECE sites as access point for healthy food for young children [7].

The Child and Adult Care Food Program (CACFP) is a federally funded, state-administered child nutrition program of the U.S. Department of Agriculture (USDA) that provides reimbursement to child care providers for eligible meals and snacks served to children in care [8]. Reimbursable meals and snacks must meet specific nutrition standards and meal patterns, which were most recently updated in 2010 U.S. legislation the Healthy Hunger Free Kids Act and implemented in 2017 [9]. CACFP participation is associated with more nutritious meals and better adherence to nutrition recommendations [10–14], greater capacity to connect with families [15], and reduced risk for household food insecurity for participating families [16]. However, while food offered at CACFP participating sites may be more nutritious, CACFP participation may not be associated with increased intake of nutritious foods by children [13]. Barriers to serving healthy foods and meeting healthy foods standard in ECE settings include cost of foods, child acceptance of foods, and provider time constraints for food purchasing and preparation [17–19]. One study suggests that CACFP participation may be associated with fewer ECE provider reported barriers to serving healthy foods [20].

Starting in the spring of 2020, the COVID-19 pandemic wreaked havoc on health and economic systems across the world, with reverberating impacts on families and children. Food insecurity in the U.S., particularly amongst families with children, increased dramatically. In November 2020, 28% of families with children reported being food insecure, nearly a threefold increase from pre-pandemic rates [21,22]. In the decade prior to the COVID-19 pandemic, household food insecurity rates in the U.S. had been steadily declining to a rate of approximately 10.5% in 2019 [23]. Globally, food insecurity has risen in recent years and the COVID-19 pandemic caused dramatic increases in acute food insecurity worldwide [24]. Families with children, particularly low-income families, were more likely to have lost income during the COVID-19 pandemic [21]. The Urban Institute reported 4 out of 10 parents with children under 6 lost employment and/or income in 2020 [25].

ECE programs and providers were severely impacted by the COVID-19 pandemic. Over half of ECE programs in the U.S. shut down at some point during the first 4 months of the pandemic and 18% remained closed as of November 2020 [26,27]. ECE site closures compounded family financial challenges and food insecurity by reducing already limited access to affordable child care, limiting parents' ability to work outside the home, and reducing access to meals and snacks served in ECE [21,28]. Even ECE sites that stayed open had limited capacity and resources to provide meals and nutrition resources [29].

As the impacts of the COVID-19 pandemic continue, the potential long-term impacts on child health are unclear. Early studies in school age children showed that the COVID-19 pandemic may have contributed to unhealthier eating habits, reduced physical activity, and increased screen time, all factors that may contribute to unhealthy weight gain [30,31]. ECE-based health and nutrition interventions may be increasingly important to mitigate potential long-term impacts of the COVID-19 pandemic on children's and families' access to sufficient and health-supporting food. Though initial data have been collected on CACFP participants' immediate responses and operational challenges due to the COVID-19 pandemic [32], the authors are not aware of any study yet to look at differences in food and nutrition response based on CACFP participation.

The purpose of this research was to assess ECE site participation in initiatives that connect children and families to food and nutrition resources amid the COVID-19 pandemic, comparing sites that did and did not participate in CACFP. Based on existing literature, the authors hypothesize that CACFP sites were more likely to participate in food- and nutrition-related initiatives than non-CACFP participating sites.

2. Materials and Methods

2.1. Study Design and Sample

Researchers conducted a cross-sectional survey of ECE providers from two U.S. states, Arizona and Pennsylvania, in the fall (November) of 2020. The target states were selected based on researcher relationships to these locales and access to complete lists of licensed ECE provider emails in the states. These lists were obtained from the respective states' ECE licensing agency websites. Researchers emailed a recruitment letter with survey link and consent information via Qualtrics to a total of 8171 recipients (2190 from Arizona and 5981 from Pennsylvania). Removing failed, bounced, and SPAM emails, a total of 7507 emails were delivered (1952 from Arizona and 5555 from Pennsylvania). The survey was open for responses for four weeks and two reminder emails were sent to non-respondents during the survey period. Participants who completed the survey were entered to win a random drawing for a $100 Amazon gift card. The survey tools and procedures were approved by Johns Hopkins Bloomberg School of Public Health Institutional Review Board.

2.2. Survey

The survey was designed to explore perceived impacts of the COVID-19 pandemic on ECE sites, including food- and nutrition-related activities and initiatives. The survey took approximately 10 min to complete and consisted of a total of 21 questions. The survey included 7 questions related to general program information, 5 questions on COVID-19 pandemic-related program model changes and perceived impact, 3 questions specific to CACFP and meal service, and 7 questions on activities, interests, and barriers related to serving and teaching about local food. Survey questions were primarily multiple choice with one Likert scale response and one open ended question. Questions were drawn from previous surveys related to food initiatives in ECE sites and the food-related COVID-19 pandemic response of ECE sites [32,33].

2.3. Analysis

To summarize descriptive characteristics of ECE sites, researchers calculated frequencies and percentages for categorical variables and means and standard deviations for continuous variables. The primary study outcome was reporting no approaches to connecting families to food and nutrition resources. Secondary study outcomes included mean number of approaches reported among sites reporting at least one approach and specific approaches used (offered "grab and go" meals, offered meal delivery, distributed food boxes, recommended community food resources, provided food from an on-site garden). Descriptive statistics were used to assess ECE site approaches to connecting families to food and nutrition resources amid the COVID-19 pandemic (March–November 2020). Unadjusted and adjusted logistic regression models were used to estimate odds of sites reporting no approaches to connecting children and families to food and nutrition resources since the onset of the COVID-19 pandemic and odds of sites reporting using specific approaches to connecting children and families to food and nutrition resources, comparing sites that participated in CACFP to those that did not. Unadjusted and adjusted Poisson models were also used to estimate the incidence rate ratio of the mean number of approaches used to connect families to food and nutrition resources among sites reporting at least one approach, comparing sites that participate in CACFP to those that do not. Sites reporting no approaches were removed from analysis comparing specific approaches and total number of approaches by CACFP status in order to assess differences only across those who reported approaches. This approach allows for more straightforward interpretation by key audiences (early childhood stakeholders and policy makers) and direct comparison to existing literature that uses the same analytical approach [20]. Models were adjusted for child enrollment (number of children enrolled in November 2020) and program type (family child care home, child care center/Head Start, or preschool); these potential confounders were selected *a priori* based on evidence that these factors differ between CACFP and non-CACFP participating sites [34]. State was included as an indicator in both adjusted and unadjusted models. Exploratory analysis was conducted to assess differences in approaches to connecting families to food and nutrition resources by site closure duration.

3. Results

3.1. Respondent Site Characteristics

A total of 589 ECE sites provided usable responses to the survey. This is a response rate of 8%. Of those respondents, 43% (n = 255) participated in CACFP and 57% (n = 334) did not participate in CACFP (Table 1). The distribution of program type across all respondents was 22% (n = 127) family or group child care home, 61% (n = 359) child care centers, 1% (n = 8) Head Start or Early Head Start, and 16% (n = 91) preschool programs (including state funded, private, and programs through K-12 schools). There were some differences in program type distribution between CACFP participating and non-participating sites, with more CACFP sites being family or group child care homes (33%, n = 83) than non-CACFP participating sites (13%, n = 44) and fewer CACFP sites being preschool programs (5%, n = 13) than non-CACFP participating sites (23%, n = 78). Approximately 21% of responding sites were from Arizona (n = 124) and 79% were from Pennsylvania (n = 465).

Table 1. Respondent site characteristics reported by CACFP participation status.

	Participating in CACFP	Not Participating in CACFP	Total
Program Type [a]		N (%)	
Family or Group Child Care Home	83 (33)	44 (13)	127 (22)
Child Care Center	148 (59)	211 (63)	359 (61)
Head Start and/or Early Head Start	8 (3)	0 (0)	8 (1)
Preschool (state, private, or through a K-12 district)	13 (5)	78 (23)	91 (16)
State		N (%)	
Arizona	43 (17)	81 (24)	124 (21)
Pennsylvania	212 (83)	253 (76)	465 (79)
Child Enrollment		**Mean (SD)**	
February 2020	65 (86)	69 (92)	67 (90)
November 2020	43 (64)	45 (47)	44 (55)

[a] Three centers participating in CACFP and one center not participating in CACFP did not indicate program type. CACFP = Child and Adult Care Food Program; N = number; SD = standard deviation.

Sites reported a mean number of 67 (SD = 90) children enrolled in February 2020 (pre-COVID-19 enrollment) and 44 (SD = 55) children enrolled in November 2020. Mean enrollment in both February 2020 and November 2020 was similar across CACFP participating and non-participating sites. Mean site enrollment decreased 34% between February and November across all sites.

3.2. Impacts of the COVID-19 Pandemic on ECE Sites Operations

Respondents were asked to report how severely, overall, the COVID-19 crisis impacted their organizations. More than half of sites (56%, n = 331) reported significant impacts from the COVID-19 pandemic. Most of the remaining sites reported moderate impacts (42%, n = 246), with just 2% (n = 9) reporting no impacts (Table 2). The reported impact of the COVID-19 pandemic on sites did not vary significantly across CACFP and non-CACFP participating sites. In regard to COVID-19 pandemic-related site closures, most sites reported closure between one month and less than three months (39%, n = 226) or between three months and less than five months (23%, n = 136). Fewer sites reported never closing (18%, n = 105), closing between one day and less than one month (10%, n = 60), and closing for 5 months or longer (10%, n = 59). The reported site closure duration was largely the same across CACFP and non-CACFP participating sites with the exception of sites having never closed and sites closed for five months or longer. Approximately one quarter (24%, n = 62) of CACFP participating sites never closed, while 13% (n = 43) of non CACFP participating sites never closed. For sites closed for five months or longer, approximately 7% (n = 19) participated in CACFP and 12% (n = 40) did not participate in CACFP. The most frequently reported changes to service models due to the COVID-19 pandemic were reducing hours or days open (47%, n = 279), limiting the number of children served (46%, n = 274), and offering virtual education for children and families (43%, n = 257). CACFP and non-CACFP participating sites were similar in most reported changes in service models, but more non-CACFP participating sites limited the number of children served (50%, n = 168) compared to CACFP participating sites (41%, n = 104).

Table 2. Reported impacts of COVID-19 on site function and service models by CACFP participation status.

	Participating in CACFP	Not Participating in CACFP	Total
Impact of COVID-19 on Site [a]		N (%)	
None	3 (1)	6 (2)	9 (2)
Moderate	108 (43)	138 (41)	246 (42)
Significant	142 (56)	189 (57)	331 (56)
Site Closure Duration [b]		N (%)	
Never closed	62 (24)	43 (13)	105 (18)
1 day to <1 month	30 (12)	30 (9)	60 (10)
Between 1 month and <3 months	91 (36)	135 (41)	226 (39)
Between 3 months and <5 months	52 (20)	84 (25)	136 (23)
5 months or longer	19 (7)	40 (12)	59 (10)
Changes in service models due to COVID-19		N (%)	
Offered virtual education for children and families	109 (43)	143 (43)	252 (43)
Reduced hours or days open	114 (45)	16 (49)	130 (22)
Limited services to children of essential workers only	45 (18)	53 (16)	98 (17)
Limited the number of children served	104 (41)	168 (50)	272 (46)

[a] Two centers participating in CACFP and one center not participating in CACFP did not indicate impact of COVID-19 on site. [b] One center participating in CACFP and two centers not participating in CACFP did not indicate site closure duration. CACFP = Child and Adult Care Food Program; N = number.

3.3. Approaches to Connecting Families to Food and Nutrition Resources

Overall, just under half of all respondents (48%, n = 281) reported no approaches to connecting families and children to food and nutrition resources amid the COVID-19 pandemic (Table 3). Of CACFP participating sites, 42% (n = 106) reported no approaches and 52% (n = 175) of non-CACFP participating sites reported no approaches. For sites reporting such approaches, the most frequently reported initiative was providing recommendations for community food resources (including food pantries and Supplemental Nutrition Assistance Program (SNAP)) (38%, n = 222). Less frequently reported approaches include offering "grab and go" (non-congregate meals) (12%, n = 68), distributing food boxes to families (11%, n = 67), offering meal delivery (6%, n = 35), and providing food from on-site gardens for families (4%, n = 23). These results were similar across sites participating in CACFP and sites not participating in CACFP, although a slightly larger proportion of sites participating in CACFP reported conducting each approach compared to sites not participating in CACFP. The mean total number of approaches used by sites that reported at least one approach was 1.5 (SD = 0.8) overall, 1.6 (SD = 0.9) for CACFP participating sites, and 1.4 (SD = 0.7) for non-CACFP participating sites.

Table 3. Reported approaches to connecting families to food and nutrition resources by CACFP participation status.

	Participating in CACFP	Not Participating in CACFP	Total
No approaches to connecting families to food and nutrition resources	106 (42)	175 (52)	281 (48)
Specific approaches to connecting families to food and nutrition resources		N (%)	
Offered "grab and go" meals	42 (16)	26 (8)	68 (12)
Offered meal delivery	23 (9)	12 (4)	35 (6)
Distributed food boxes	39 (15)	28 (8)	67 (11)
Provided food from on-site garden	14 (5)	9 (3)	23 (4)
Recommended community food resources	107 (42)	115 (34)	222 (38)
Total number of approaches to connecting families to food and nutrition resources reported [a]—Mean (SD)	1.6 (0.9)	1.4 (0.7)	1.5 (0.8)

[a] Among sites reporting at least one approach. CACFP = Child and Adult Care Food Program; N = number; SD = standard deviation.

3.4. Approaches to Connecting Families to Food Resources by CACFP Status

After adjusting for site enrollment and program type, CACFP participating sites were significantly more likely to offer "grab and go" (non-congregate) meals (OR = 3.9, 95% CI [2.1, 7.1]; $p < 0.001$), offer meal delivery (OR = 4.4, 95% CI [1.9, 10.0]; $p < 0.001$), distribute food boxes to families (OR = 2.6, 95% CI [1.5, 4.5]; $p = 0.001$), and to recommend community food resources (OR = 1.5, 95% CI [1.0, 2.1]; $p = 0.04$) than non-CACFP participating respondents (Table 4). CACFP participating sites were less likely to report using no approaches to connecting children and families to food during the COVID-19 pandemic (OR = 0.6, 95% CI [0.4, 0.8]; $p = 0.003$) (Table 5). Disaggregated by site closure, the difference was only significant for sites closed three months or longer (OR = 0.4, 95% CI [0.2, 0.7]; $p = 0.002$). The mean number of farm to ECE activities among respondents reporting at least one activity was not significantly different between CACFP and non-CACFP sites except for sites that closed 3 months or longer (IRR = 1.5, 95% CI [1.1, 2.0]; $p = 0.02$).

Table 4. Unadjusted and adjusted [a] odds ratios for specific approaches reported by CACFP compared to non-CACFP sites [b].

Specific Approaches to Connecting Families to Food and Nutrition Resources	Unadjusted Odds Ratios (95% CI)	*p*-Value	Adjusted Odds Ratios (95% CI)	*p*-Value
Offered "grab and go" meals	2.6 (1.5, 4.4)	<0.001	3.9 (2.1, 7.1)	<0.001
Offered meal delivery	2.7 (1.3, 5.6)	0.007	4.4 (1.9, 10.0)	<0.001
Distributed food boxes	2.0 (1.2, 3.3)	0.01	2.6 (1.5, 4.5)	0.001
Recommended community food resources	1.4 (1.0, 1.9)	0.08	1.5 (1.0, 2.1)	0.04
Provided food from on-site garden	2.1 (0.9, 4.9)	0.10	1.6 (0.6, 3.9)	0.35

[a] Adjusted for number of children enrolled and site type. [b] Among sites reporting at least one approach. CACFP = Child and Adult Care Food Program.

Table 5. Adjusted [a] odds ratios and incidence rate ratios for approaches connecting families to food and nutrition resources [b] reported by CACFP compared to non-CACFP sites, overall and by length of closure during the COVID-19 pandemic.

	Odds Ratios (95% CI)	***p*-Value**
No approaches to connecting families to food and nutrition resources [b]	0.6 (0.4, 0.8)	0.003
Never closed (n = 105)	0.9 (0.4, 2.3)	0.87
Closed 1 day to <3 months (n = 286)	0.7 (0.4, 1.1)	0.13
Closed ≥3 months (n = 195)	0.4 (0.2, 0.7)	0.002
	Incidence Rate Ratios (95% CI)	***p*-Value**
Mean number of approaches connecting families to food and nutrition resources [b]	1.2 (1.0, 1.5)	0.06
Never closed (n = 105)	0.7 (0.4, 1.3)	0.26
Closed 1 day to <3 months (n = 286)	1.2 (0.9, 1.5)	0.32
Closed ≥3 months (n = 195)	1.5 (1.1, 2.0)	0.02

[a] Adjusted for number of children enrolled and site type. [b] Among sites reporting at least one approach. CACFP = Child and Adult Care Food Program; CI = confidence interval.

4. Discussion

This study presents results from a cross-sectional survey of 589 responding ECE sites from two U.S. states aimed at exploring approaches to providing food and nutrition resources for children and families during the COVID-19 pandemic. Of responding sites, 43% participated in CACFP and 57% did not participate in CACFP. Most sites, regardless of CACFP status, reported that they were "significantly" impacted by the COVID-19 pandemic and were most likely to have closed for between one month and less than three month due to the COVID-19 pandemic. When it comes to connecting families to food and nutrition resources, CACFP participating sites were more likely to report using any approaches and significantly more likely to report offering "grab and go" meals, providing meal delivery, distributing food boxes to families, and providing recommendations for community food resources. Among sites closed for greater than three months, the mean number of approaches to connecting families to food and nutrition resources was significantly higher in CACFP participating sites than non-CACFP participating sites. Non-CACFP sites were significantly more likely to report using no approaches to connect families to food and nutrition resources.

The primary findings confirm the researchers' initial hypothesis that CACFP sites would be more likely to provide food and nutrition resources to families during the COVID-19 pandemic. Previous literature indicates that CACFP participating sites were more likely to provide family nutrition education and engagement than non-CACFP participating sites [15,35]. Non-CACFP sites may experience more barriers to communicating with parents about nutrition [15]. An existing precedent of communication with parents about food and nutrition may have better set the stage for CACFP participating sites to provide resources during the COVID-19 pandemic. Existing barriers to reaching parents reported in non-CACFP participating sites were likely exacerbated in the midst of the COVID-19 pandemic. In previous literature, Head Start sites, which are required by U.S. federal law to participate in CACFP, were even more likely to provide parents with nutrition education than non-Head Start CACFP sites and non-CACFP participating sites, potentially due to the Head Start policies that require family engagement and supplemental training provided to Head Start staff [35]. Further research is required to better understand how Head Start response may have differed from other program types and how existing nutrition and family engagement policy may have influenced food and nutrition practices amid the COVID-19 pandemic.

CACFP participating sites were also more likely to serve children and families with food insecurity than non-CACFP participating sites [16,36]. CACFP participating sites may have been responding to a greater need for continued food access experienced by the families they serve. CACFP has been shown to be an important potential pathway to supporting family food security by increasing the quality and quantity of food offered in the ECE setting and thus reducing food expenses for families [16,34]. As food insecurity rates increased suddenly and dramatically during the COVID-19 pandemic, especially for families with young children, this additional support likely became even more important for families [25]. Simply by continuing CACFP meal operations, sites that remained fully open to provide in-person care for children were providing a form of food resource to families. However, this study aimed to assess activities that went beyond providing meals as per usual. Initiatives that CACFP sites were significantly more likely to offer, including "grab and go" meals, family food boxes, and recommendations for community food resources, were likely aimed at supporting families who may have been experiencing new, ongoing, or deepened food insecurity. This may also account for the even greater significance of CACFP sites closed for three months or longer to provide food and nutrition supportive approaches, as the sites aimed to meet ongoing food security challenges. CACFP participating sites' capacity to provide resources may be due to increased funds available through the CACFP reimbursement. Previous literature points to CACFP participating sites reporting fewer barriers to serving healthier foods and being less likely to report cost and lack of staff knowledge as barriers [20]. Without knowledge and cost as a barrier, CACFP participating sites may have had greater capacity to offer food and nutrition services and resources to families.

There were differences in site closure status between CACFP and non-CACFP participating sites, particularly across sites that never closed. This may suggest CACFP participating sites may have had more capacity to remain open to serve children and families. The approaches to connecting families to food and nutrition resources explored in this survey are applicable to sites that were open for in-person care, fully or partially, or just operational, that is they may have had staff working, but were not providing in-person care or learning. There is significant nuance and fluidity in this distinction because sites were frequently shifting closure status, changing capacity guidelines, and adapting to staffing issues and limitations. The study findings suggest that CACFP participating sites may have had more capacity to remain operational and offer services like "grab and go" meals and meal delivery that would be even more relevant during site closure. Because the survey did not ask specifically about periods of operation, additional research would be necessary to further explore this nuance.

Though CACFP sites were more likely to offer approaches to connecting families with food and nutrition resources, the relatively low number of CACFP sites that continued to offer meals through "grab and go" (16%, n = 42) or delivery (9%, n = 23) reflects the limited capacity for most ECE sites to continue to provide food and nutrition resources to families. While the USDA was granted ability to issue waivers for meal time flexibility and non-congregate feeding (allowing for "grab and go" and meal delivery) in CACFP programs through the Families First Coronavirus Response Act on March 18, 2020, delays in issuing the waivers and providing guidance for CACFP programs left many providers unsure if they would be reimbursed for meals and snacks served outside of the traditional CACFP model [29,37]. Overall CACFP program participation decreased dramatically throughout the COVID-19 pandemic [38]. From March through September of 2020, 480 million fewer CACFP meals were served than in the same time period in 2019, a 41% decrease [38]. According to Nutrition and Obesity Policy Research and Evaluation Network (NOPREN) COVID-19 Early Childcare and Education Working Group, the root of this breakdown in capacity for child care sites to continue to meet the nutritional needs of children and families is threefold: loss of income and reimbursement left child care sites unable to pay staff; child care providers, disproportionately within the "high-risk" category, closed sites or limited initiatives to protect their own health; and CACFP reimbursement is insufficient to cover food costs and administrative costs, let alone additional costs of transitioning service models [29].

Despite closures and limited capacity, ECE sites remained an important pathway to connect children and families to food resources. During the COVID-19 pandemic, food banks and emergency food assistance systems were severely strained. In many communities, food pantries could not meet the demand of the dramatically increased numbers of families experiencing food insecurity [39]. Additionally, families with young children were not included in some of the federal nutrition programs aimed at addressing childhood hunger during the COVID-19 pandemic. Specifically, children under six years of age were initially not eligible for Pandemic Electronic Benefits Transfer (P-EBT), a program that provides funding to families to purchase foods when children miss school meals due to school closure [40]. ECE sites must be one of a spectrum of resources to support families experiencing food insecurity. Based on existing relationships with caregivers and parents, ECE sites are especially well positioned to help families access and navigate other food assistance programs.

However, even outside of the COVID-19 pandemic, CACFP programming itself may not be reaching children and families that could be significantly benefitting from the program. Due to the eligibility limitations, lower-income children living in high-income areas may not have access to CACFP [36]. Lack of awareness, administrative burden and insufficient reimbursement may inhibit eligible ECE sites from choosing to participate in the program [11,34]. Additional research and qualitative exploration would provide deeper insight into the barriers that ECE providers faced in continuing to offer meals and food resources to children during the COVID-19 pandemic and how those barriers may or may not be alleviated with CACFP participation. For sites that are not eligible for CACFP, voluntary initiatives that prioritize nutrition and build relationships across families, community partners, and local food systems stakeholders can support continued access to food resources for families. This may include farm to child care initiatives that engage children and families or regularly assessing family food security and referring families to community food resources and programs [7]. While additional exploration is needed, state child care policies that include health and nutrition guidance may be one pathway to supporting healthier nutrition environments in ECE settings [7,41].

This study provides novel insight into the food and nutrition responses of ECE sites during the COVID-19 pandemic. However, it does have several limitations. First, the geographic scope of this study was limited to two U.S. states, Arizona and Pennsylvania. These states provide geographic and policy diversity, but it is not possible to generalize the results due to this limited geographic reach. Second, all site information is self-reported and not confirmed through any other sources. The individual completing the survey may not have been fully informed of all activities at a site and thus activities may have been underreported. Alternately, approaches to connecting families to food and nutrition resources may have been overreported due to perceived social desirability of those activities. Third, the low overall response rate (7.8%) may limit generalizability of survey results. Prior studies similar in nature have garnered response rates around 50% [11,42,43]. The low response rate in this survey is likely due to the impacts of the COVID-19 pandemic itself. When the survey was distributed (November 2020) there were still significant fluctuations in closure status for ECE sites and many ECE sites had already permanently closed but may not have yet been removed from official state lists. Sites that were opened experienced extremely limited staff capacity while also having to pivot program models and meet new health and safety protocols, which may have limited their capacity to respond to the survey. Future studies may consider telephone outreach and options to complete surveys verbally (via phone) or in written mail format to increase response rate. A final limitation is the lack of economic status information collected from survey respondents. The populations served by CACFP and non-CACFP participating sites could differ significantly in economic status and future studies should control for that difference.

5. Conclusions

No studies that the authors are aware of have examined ECEE sites approaches to providing families with food and nutrition resources during the COVID-19 pandemic by CACFP status. This study suggests that CACFP sites may have greater capacity to continue to connect families to food and nutrition resources in the midst of emergencies than non-CACFP participating sites. However, the small number of ECE providers, CACFP participating and non-participating, that continued to provide resources to families suggests that the CACFP program in its current form and ECE systems in general may be insufficient to fully support ECE sites in meeting the needs of families, particularly in a crisis. Further research is needed to better understand the barriers and facilitators to accessing and implementing CACFP and thus identify opportunities to increase reach of and participation in the program. Exploration of policy changes that would strengthen CACFP and its function as a pathway to increase food security for families is also needed. We are only now starting to see the long-term impacts of the COVID-19 pandemic, the ensuing economic crisis, and the federal response on the ECE system. Amid the COVID-19 pandemic, ECE sites emerged as an important connector to community food and nutrition resources for families. Proportionally fewer sites reported providing direct food resources, but for families able to access it, this was likely an especially vital source of food as the emergency food system was overwhelmed by demand. Though emergency funding and policy flexibilities sought to bolster ECE sites' capacity to continue to serve children and families, longer-term investments and policy change may be required to rebuild ECE sites' long-term resilience and capacities. With sufficient funding and capacity development, ECE sites can be one important entity in the network of resources ensuring comprehensive food security for children. The impacts of the COVID-19 pandemic on the ECE industry and the way that those impacts reverberate for the families of young children offer an opportunity to understand and elevate the current and potential role that ECE sites can play in supporting families and children in accessing food and nutrition resources.

Author Contributions: Conceptualization, L.S., C.R. and S.E.B.-N.; methodology, L.S., C.R. and S.E.B.-N.; formal analysis, C.R.; investigation, L.S.; data curation, C.R. writing—original draft preparation, L.S. writing—review and editing, C.R. and S.E.B.-N.; visualization, C.R. and S.E.B.-N.; supervision, S.E.B.-N.; project administration, L.S.; funding acquisition, L.S., C.R. and S.E.B.-N. All authors have read and agreed to the published version of the manuscript.

Funding: This research was funded by the Bloomberg American Health Initiative Collaboration Award Funding, Johns Hopkins University and the Johns Hopkins Bloomberg School of Public Health Lerner Center for Public Health Promotion.

Institutional Review Board Statement: This study was deemed exempt from institutional review by the Johns Hopkins Bloomberg School of Public Health Institutional Review Board.

Informed Consent Statement: Informed consent was obtained from all subjects involved in this study.

Conflicts of Interest: The authors declare no conflict of interest.

References

1. Gubbels, J.S.; Kremers, S.P.J.; Stafleu, A.; Dagnelie, P.C.; Vries, N.K.D.; Thijs, C. Child-care environment and dietary intake of 2- and 3-year-old children. *J. Hum. Nutr. Diet.* **2010**, *23*, 97–101. [CrossRef]
2. *Institute of Medicine Early Childhood Obesity Prevention Policies*; The National Academies Press: Washington, DC, USA, 2011; ISBN 978-0-309-21024-9.
3. Larson, N.; Ward, D.S.; Neelon, S.B.; Story, M. What role can child-care settings play in obesity prevention? A review of the evidence and call for research efforts. *J. Am. Diet. Assoc.* **2011**, *111*, 1343–1362. [CrossRef]
4. Story, M.; Kaphingst, K.M.; French, S. The role of child care settings in obesity prevention. *Future Child.* **2006**, *16*, 143–168. [CrossRef]
5. UNICEF. *The Child Care Transition, Innocenti Report Card 8*; UNICEF Innocenti Research Centre: Florence, Italy, 2008; Available online: https://www.unicef-irc.org/publications/pdf/rc8_eng.pdf (accessed on 12 June 2021).
6. National Center on Education Statistics National Household Education Surveys Program (NHES). Available online: https://nces.ed.gov/nhes/tables/ECPP_HoursPerWeek_Care.asp (accessed on 7 February 2021).

7. Benjamin-Neelon, S.E. Position of the Academy of Nutrition and Dietetics: Benchmarks for nutrition in child care. *J. Acad. Nutr. Diet.* **2018**, *118*, 1291–1300. [CrossRef] [PubMed]
8. United States Department of Agriculture Child and Adult Care Food Program (CACFP), Food and Nutrition Service. Available online: https://www.fns.usda.gov/cacfp/child-and-adult-care-food-program (accessed on 27 February 2021).
9. United States Department of Agriculture, Food and Nutrition Service. *Child and Adult Care Food Program: Meal Pattern Revisions Related to the Healthy, Hunger-Free Kids Act of 2010*; Final Rule; Federal Register: Washington, DC, USA, 2016; Volume 81, p. 79. Available online: https://www.federalregister.gov/documents/2016/04/25/2016-09412/child-and-adult-care-food-program-meal-pattern-revisions-related-to-the-healthy-hunger-free-kids-act (accessed on 27 February 2021).
10. Loth, K.A.; Shanafelt, A.; Davey, C.S.; O'Meara, J.; Johnson-Reed, J.; Larson, N.; Nanney, S. Does adherence to child care nutrition and physical activity best practices differ by child care provider's participation in support programs and training? *Child. Youth Serv. Rev.* **2019**, *105*, 104417. [CrossRef]
11. Andreyeva, T.; Henderson, K.E. Center-reported adherence to nutrition standards of the Child and Adult Care Food Program. *Child. Obes.* **2018**, *14*, 421–428. [CrossRef] [PubMed]
12. Erinosho, T.; Vaughn, A.; Hales, D.; Mazzucca, S.; Gizlice, Z.; Ward, D. Participation in the Child and Adult Care Food Program is associated with healthier nutrition environments at family child care homes in Mississippi. *J. Nutr. Educ. Behav.* **2018**, *50*, 441–450. [CrossRef] [PubMed]
13. Hasnin, S.; Dev, D.A.; Tovar, A. Participation in the CACFP ensures availability but not intake of nutritious foods at lunch in preschool children in child-care centers. *J. Acad. Nutr. Diet.* **2020**, *120*, 1722–1729.e1. [CrossRef] [PubMed]
14. Gurzo, K.; Lee, D.L.; Ritchie, K.; Yoshida, S.; Vitale, E.H.; Hecht, K.; Ritchie, L.D. Child care sites participating in the federal Child and Adult Care Food Program provide more nutritious foods and beverages. *J. Nutr. Educ. Behav.* **2020**, *52*, 697–704. [CrossRef]
15. Dev, D.A.; Byrd-Williams, C.; Ramsay, S.; McBride, B.; Srivastava, D.; Murriel, A.; Arcan, C.; Adachi-Mejia, A.M. Engaging parents to promote children's nutrition and health. *Am. J. Health Promot.* **2017**, *31*, 153–162. [CrossRef]
16. Heflin, C.; Arteaga, I.; Gable, S. The Child and Adult Care Food Program and food insecurity. *Soc. Serv. Rev.* **2015**, *89*, 77–98. [CrossRef]
17. Carroll, J.D.; Demment, M.M.; Stiles, S.B.; Devine, C.M.; Dollahite, J.S.; Sobal, J.; Olson, C.M. Overcoming barriers to vegetable consumption by preschool children: A child care center buying club. *J. Hunger Environ. Nutr.* **2011**, *6*, 153–165. [CrossRef]
18. Nanney, M.S.; LaRowe, T.L.; Davey, C.; Frost, N.; Arcan, C.; O'Meara, J. Obesity prevention in early child care settings: A bistate (Minnesota and Wisconsin) assessment of best practices, implementation difficulty, and barriers. *Health Educ. Behav.* **2017**, *44*, 23–31. [CrossRef]
19. Zaltz, D.A.; Pate, R.R.; O'Neill, J.R.; Neelon, B.; Benjamin-Neelon, S.E. Barriers and facilitators to compliance with a state healthy eating policy in early care and education centers. *Child. Obes.* **2018**, *14*, 349–357. [CrossRef]
20. Zaltz, D.A.; Hecht, A.A.; Pate, R.R.; Neelon, B.; O'Neill, J.R.; Benjamin-Neelon, S.E. Participation in the Child and Adult Care Food Program is associated with fewer barriers to serving healthier foods in early care and education. *BMC Public Health* **2020**, *20*, 856. [CrossRef] [PubMed]
21. Bauer, L. Hungry at Thanksgiving: A Fall 2020 Update on Food Insecurity in the U.S. *Brookings*. 2020. Available online: https://www.brookings.edu/blog/up-front/2020/11/23/hungry-at-thanksgiving-a-fall-2020-update-on-food-insecurity-in-the-u-s/ (accessed on 14 July 2021).
22. Schanzenbach, D.W.; Pitts, A. *How Much Has Food Insecurity Risen? Evidence from the Census Household Pulse Survey*; Institute for Policy Rapid Research Report: Evanston, IL, USA, 2020; Available online: https://www.ipr.northwestern.edu/documents/reports/ipr-rapid-research-reports-pulse-hh-data-10-june-2020.pdf (accessed on 14 July 2021).
23. Coleman-Jensen, A.; Rabbitt, M.P.; Gregory, C.A.; Singh, A. Household Food Security in the United States in 2019. In *U.S. Department of Agriculture Economic Research Report*; 2019; 47p. Available online: https://www.ers.usda.gov/webdocs/publications/99282/err-275.pdf?v=1127.7 (accessed on 12 June 2021).
24. Food Security and COVID-19. Available online: https://www.worldbank.org/en/topic/agriculture/brief/food-security-and-covid-19 (accessed on 20 June 2021).
25. Waxman, E.; Gupta, P.; Gonzalez, D. Six Months into the Pandemic, 40 Percent of Parents with Young Children Have Experienced Economic Fallout. Urban Institute 2020. Available online: https://www.urban.org/research/publication/six-months-pandemic-40-percent-parents-young-children-have-experienced-economic-fallout (accessed on 12 June 2021).
26. Gilliam, W.S.; Malik, A.A.; Shafiq, M.; Klotz, M.; Reyes, C.; Humphries, J.E.; Murray, T.; Elharake, J.A.; Wilkinson, D.; Omer, S.B. COVID-19 transmission in US child care programs. *Pediatrics* **2021**, *147*, e202003197. [CrossRef] [PubMed]
27. *Piecing Together Solutions: The Importance of Childcare to U.S. Families and Businesses*; U.S. Chamber of Commerce Foundation: Washington, DC, USA, 2020; Available online: https://www.uschamberfoundation.org/sites/default/files/EarlyEd_Minis_Report6_121420_Final.pdf (accessed on 12 June 2021).
28. Esposito, S.; Principi, N. School closure during the Coronavirus Disease 2019 (COVID-19) Pandemic: An effective intervention at the global level? *JAMA Pediatrics* **2020**, *174*, 921–922. [CrossRef] [PubMed]
29. Bauer, K.W.; Chriqui, J.F.; Andreyeva, T.; Kenney, E.L.; Stage, V.C.; Dev, D.; Lessard, L.; Cotwright, C.J.; Tovar, A. A safety net unraveling: Feeding young children during COVID-19. *Am. J. Public Health* **2021**, *111*, 116–120. [CrossRef]
30. Cuschieri, S.; Grech, S. COVID-19: A one-way ticket to a global childhood obesity crisis? *J. Diabetes Metab. Disord.* **2020**, *19*, 1–4. [CrossRef]

31. Workman, J. How much may COVID-19 school closures increase childhood obesity? *Obesity* **2020**, *28*, 1787. [CrossRef]
32. National CACFP Sponsors Association. *COVID Impact on the CACFP*; National CACFP Sponsors Association: Round Rock, TX, USA, 2020; Available online: https://growthzonesitesprod.azureedge.net/wp-content/uploads/sites/2039/2021/01/Impact_of_COVID-19_on_the_CACFP_June_2020_cacfp.org_.pdf (accessed on 12 January 2021).
33. Shedd, M.K.; Stephens, L.; Matts, C.; Laney, J. *Results From the 2018 National Farm to Early Care and Education Survey*; National Farm to School Network: Chicago, IL, USA, 2018; Available online: www.farmtoschool.org/Resources/results-of-the-2018-national-farm-toearly-care-and-education-survey.pdf (accessed on 12 January 2021).
34. Korenman, S.; Abner, K.S.; Kaestner, R.; Gordon, R.A. The Child and Adult Care Food Program and the nutrition of preschoolers. *Early Child. Res. Q.* **2013**, *28*, 325–336. [CrossRef] [PubMed]
35. Dev, D.A.; McBride, B.A. STRONG Kids Research Team Academy of Nutrition and Dietetics Benchmarks for Nutrition in Child Care 2011: Are child-care providers across contexts meeting recommendations? *J. Acad. Nutr. Diet.* **2013**, *113*, 1346–1353. [CrossRef]
36. Gordon, R.; Kaestner, R.; Korenman, S.; Abner, K. *The Child and Adult Care Food Program: Who Is Served and What Are Their Nutritional Outcomes?* National Bureau of Economic Research: Cambridge, MA, USA, 2010. [CrossRef]
37. *Families First Coronavirus Response Act*; Public Law No 116–127; U.S. Government Publishing Office: Washington, DC, USA, 2020.
38. Food Research and Action Center. *CACFP during COVID-19*; Food Research and Action Center: Wasshington, DC, USA, 2021; Available online: https://frac.org/wp-content/uploads/CACFP-Program-Brief-March-2021.pdf (accessed on 23 May 2021).
39. Kulish, N. Never seen anything like it': Cars line up for miles at food banks. *The New York Times*, 8 April 2020. Available online: https://www.nytimes.com/2020/04/08/business/economy/coronavirus-food-banks.html(accessed on 28 August 2021).
40. United States Department of Agriculture, Food and Nutrition ServiceState Guidance on Coronavirus P-EBT. 26 August 2021. Available online: https://www.fns.usda.gov/snap/state-guidance-coronavirus-pandemic-ebt-pebt (accessed on 28 August 2021).
41. Kenney, E.L.; Mozaffarian, R.S.; Frost, N.; Looby, A.A.; Cradock, A.L. Opportunities to promote healthy weight through child care llicensing regulations: Trends in the United States, 2016–2020. *J. Acad. Nutr. Diet.* **2021**, *121*, 1763–1774.e2. [CrossRef] [PubMed]
42. Benjamin-Neelon, S.E.; Hecht, A.A.; Burgoine, T.; Adams, J. Perceived barriers to fruit and vegetable gardens in early years settings in england: Results from a cross-sectional survey of nurseries. *Nutrients* **2019**, *11*, 2925. [CrossRef] [PubMed]
43. Williams, B.D.; Sisson, S.B.; Padasas, I.O.; Dev, D.A. Food program participation influences nutrition practices in early care and education settings. *J. Nutr. Educ. Behav.* **2021**, *53*, 299–308. [CrossRef]

Review

Changes in Dietary Patterns and Clinical Health Outcomes in Different Countries during the SARS-CoV-2 Pandemic

Roxana Filip [1,2], Liliana Anchidin-Norocel [1,*], Roxana Gheorghita [1,3], Wesley K. Savage [1,3] and Mihai Dimian [3,4]

1 Faculty of Medicine and Biological Sciences, Stefan cel Mare University of Suceava, 720229 Suceava, Romania; roxana_filip@yahoo.com (R.F.); roxana.puscaselu@usm.ro (R.G.); wesley.savage@usm.ro (W.K.S.)
2 Regional County Emergency Hospital, 720224 Suceava, Romania
3 Integrated Center for Research, Development and Innovation in Advanced Materials, Nanotechnologies, and Distributed Systems for Fabrication and Control, Stefan cel Mare University of Suceava, 720229 Suceava, Romania; dimian@usm.ro
4 Department of Computers, Electronics and Automation, Stefan cel Mare University of Suceava, 720229 Suceava, Romania
* Correspondence: liliana.norocel@usm.ro

Abstract: Coronavirus disease 2019 (COVID-19), caused by severe acute respiratory syndrome coronavirus 2 (SARS-CoV-2), has led to an excess in community mortality across the globe. We review recent evidence on the clinical pathology of COVID-19, comorbidity factors, immune response to SARS-CoV-2 infection, and factors influencing infection outcomes. The latter specifically includes diet and lifestyle factors during pandemic restrictions. We also cover the possibility of SARS-CoV-2 transmission through food products and the food chain, as well as virus persistence on different surfaces and in different environmental conditions, which were major public concerns during the initial days of the pandemic, but have since waned in public attention. We discuss useful measures to avoid the risk of SARS-CoV-2 spread through food, and approaches that may reduce the risk of contamination with the highly contagious virus. While hygienic protocols are required in food supply sectors, cleaning, disinfection, avoidance of cross-contamination across food categories, and foodstuffs at different stages of the manufacturing process are still particularly relevant because the virus persists at length on inert materials such as food packaging. Moreover, personal hygiene (frequent washing and disinfection), wearing gloves, and proper use of masks, clothes, and footwear dedicated to maintaining hygiene, provide on-site protections for food sector employees as well as supply chain intermediates and consumers. Finally, we emphasize the importance of following a healthy diet and maintaining a lifestyle that promotes physical well-being and supports healthy immune system function, especially when government movement restrictions ("lockdowns") are implemented.

Keywords: pandemic; prevention; food hygiene; bioactive compounds; diet; lifestyle

Citation: Filip, R.; Anchidin-Norocel, L.; Gheorghita, R.; Savage, W.K.; Dimian, M. Changes in Dietary Patterns and Clinical Health Outcomes in Different Countries during the SARS-CoV-2 Pandemic. *Nutrients* **2021**, *13*, 3612. https://doi.org/10.3390/nu13103612

Academic Editors: William B. Grant and Ronan Lordan

Received: 31 August 2021
Accepted: 12 October 2021
Published: 15 October 2021

1. Introduction

Since early 2020, the novel severe acute respiratory syndrome coronavirus 2 (SARS-CoV-2) causing the COVID-19 pandemic has led to health care systems around the world facing insufficient resources [1]. As new SARS-CoV-2 variants arise, individual immunodeficiencies and co-morbidity factors, diet and lifestyle factors, vaccine effectiveness, and vaccination rates will determine infection rates and persistence of the pandemic, and as we have seen with the delta variant, health care systems everywhere struggle to treat the number of severe COVID-19 cases, in many situations because of compromised immune systems due to poor diets and lifestyle habits. The appearance of COVID-19 revealed shortcomings with the capacity for public health systems to respond early to a novel pathogen and prevent a global pandemic, and moreover, it highlighted an emerging public health crisis of comorbidities due to diet and lifestyles that exacerbate COVID-19 progression and outcome. While scientific research and public health systems need to adapt quickly to combat emerging pathogens, as we see in the current pandemic where medical science produced the fastest development of a vaccine that has and continues to prevent countless COVID-19 cases requiring hospitalization and reduced deaths due to co-morbidities, it also became abundantly clear that the general state of public health can determine the outcome of infection and the magnitude of excess mortality [2,3].

Despite efforts to contain the spread of SARS-CoV-2, this new, contagious virus is challenging social welfare, public health, and the medical and scientific research community through its unpredictable spread via new variants, long-term effects of infection (i.e., "long covid" and organ injury), and differing outcomes across demographic groups. Initial measures were focused on containing the spread of SARS-CoV-2, through promoting mask-wearing, hand-sanitizing, and stay-at-home measures, all of which are similar to measures implemented to slow the Spanish flu pandemic in 1917–1918 [4]. However, these are only measures of containment and do little to address the problem of infection outcomes, which are exacerbated by the modern diet, nutrition, and lifestyle behaviors, all of which have been linked to prolonged infection, admission to intensive care units, and excess death rates usually linked to co-morbidity factors.

We know from previous outbreaks of viruses similar to SARS-CoV-2 that higher pathogenicity is generally correlated with lower transmissibility, and vice versa. As far as we know now, SARS-CoV-2 has lower pathogenicity and moderate transmissibility compared to MERS-CoV, avian SARS-CoV-1, Ebola, and H7N9 [2]. However, in the many areas of the world that lack sufficient medical resources to treat severe covid cases, infections are resulting in greater mortality due to COVID-19 than in areas with higher vaccination rates and greater availability of health care. These outcomes may be mitigated by dietary strategies aimed at maximizing dietary approaches that benefit infection outcomes and minimizing lifestyle behaviors that contribute to excess mortality. Further, as in the early stages of the pandemic, the World Health Organization (WHO) produced guidelines on water, sanitation, hygiene (WASH), and waste management approaches relevant to viruses, including coronaviruses [5]. While global vaccination and boosters are the main axes of containing the virus, and thereby mitigating transmission rates, delays in access to and deployment of the vaccine in parts of the world will ensure that SARS-CoV-2 variants arise via mutations, and continue to spread across populations. This means other measures remain crucial to minimizing viral spread; as we saw earlier in the pandemic before vaccination rollouts, public health measures, such as focus on using proper hygiene, social distancing, mask use, and population lockdowns, had the effect of slowing down the pandemic, but not without physical, social, and mental health consequences. This had unforeseen consequences for population health, namely that people may not have had access to healthy foods and instead were faced with limited choices that affected what they consumed, how often, and their ability to exercise or adopt positive lifestyle behaviors.

How SARS-CoV-2 presented itself to the human populations is still not known, and will likely be difficult to track because of rapidly evolving genomes (i.e., mutation), genetic recombination (as yet unknown), and switching from reservoir to novel host species. It took more than a decade after the SARS-CoV-1 outbreak in China in 2002 before it was reported that this coronavirus originated in bats and spread to palm civets, then infecting humans [6,7]. In the case of the MERS coronavirus outbreak, a clear transmission route is not exactly known, but it is theorized to also have originated in bats, with camels perhaps the intermediate host leading to transmission to humans [8]. To date, no wild animal tested in the region of the outbreak (i.e., "ground zero") in Wuhan, China has been identified as transmitting the SARS-CoV-2 virus to humans, but given that bats are known reservoirs for SARS-like coronaviruses, a natural origin is plausible, and likely involves contact with an intermediate host as in the case for MERS and SARS-CoV-1 [8,9]. The salient point of this is that we will likely continue to host SARS-CoV-2 in the human population, and aside from the impossible task of chasing spike protein mutations with designer vaccines, our best chance at reducing the impact of it is through population-level hygiene, food chain hygiene, and diet and healthy lifestyle awareness.

The SARS-CoV-2 virus reached pandemic proportions in short time due to modifications in the receptor binding domain (RBD) that enhances viral binding to the angiotensin converting enzyme 2 (ACE2) receptors, with particular affinity to pulmonary tissues (pneumocytes) where the virus can have more severe disease outcomes [10]. More concerning is that ACE2 is also highly expressed on adipose cells, which suggests that weight gain via fat deposition (through unhealthy dietary habits) can pose a serious problem with COVID-19 recovery by being a viral reservoir in overweight people. This may explain some of the evidence that obesity is a significant co-morbidity factor with COVID-19 hospitalization and mortality, and further evidence that dietary habits during the pandemic may continue to exacerbate severe disease outcomes. Transmission can also happen directly from the reservoir host to humans without receptor-binding domain adaptations [11]. The bat coronavirus that is currently in circulation possesses spike proteins that facilitate human infection. Interspecies transmission from animals to humans is possible by the high plasticity in receptor binding and the possibility of viral antigenic "make-up" by mutation and recombination. The in vitro and in vivo studies of isolated virus demonstrate that there is a potential risk for the re-emergence of coronavirus infection from viruses that are currently circulating in bat populations in nature [12]. Given these realities, and the fact that vaccines are chasing a moving target, the host immune response to combat infection has re-emerged as an important focus on preventative health.

Recent published articles address changes in food hygiene, lifestyle, and diet from different regions of the world during the COVID-19 pandemic, but given the recent emergence of SARS-CoV-2, changes are ongoing and much more data are being gathered. Ultimately, understanding how the pandemic is affecting mental, social, and physical health in the global human population relative to pre-pandemic conditions is important for defining the broader health impacts of COVID-19 beyond direct clinical disease pathology [13–16].

A healthy diet, based on plant-healthy fats and proteins, together with regular exercise and sunlight exposure, is of paramount importance to help prevent viral infection by strengthening the immune system. However, sedentarism, unease, and tediousness caused by social isolation could lead to changes and worsening of lifestyle patterns while also promoting binge-eating, which is worsened by limited access to healthy food rich in natural vitamins, minerals, and antioxidants [17].

The aim of this review is to highlight the latest evidence on how the clinical pathology of COVID-19, including comorbidities and immune response due to lifestyle behaviors and diet, may exacerbate outcomes and prolong the severity of the novel SARS-CoV-2 pandemic.

2. Clinical Pathology of SARS-CoV-2 Infection and COVID-19 Syndrome

Severe acute respiratory syndrome (SARS-CoV-2) causes COVID-19, a disease that presents a complex of syndromes, including severe specific contagious pneumonia (SSCP) and Wuhan pneumonia [18]. While this coronavirus has less severe pathogenesis compared to SARS-CoV-1, it is highly transmissible, demonstrated repeatedly by the rapidly and continuously increasing number of COVID-19 cases since it emerged in December 2019. The mean incubation time of SARS-CoV-2 in familial clusters is reportedly 3 to 6 days [19,20]. Similar to MERS and SARS-CoV-1, the severity of COVID-19 is higher in age groups above 50 years [21,22]. Since the onset, while the per-capita mortality rate is lower than recorded in outbreaks of SARS-CoV-1 and MERS, the high rate of transmission means that specific demographics are particularly impacted by COVID-19 [23]. Information obtained from outbreaks in Thailand, South Korea, China, and Japan confirm that COVID-19 patients usually had mild manifestations compared to those with SARS-CoV-1 and MERS, and with a much larger sample size owing to the high rate of transmission. Regardless of the SARS-CoV type, the primary barrier against SARS-CoV-2 is the immune system, and the first line of defense is mast cells in the submucosa of the respiratory tract and nasal cavity [24].

Severe interstitial inflammation of the lungs is caused by invasion of pulmonary parenchyma by SARS-CoV-2 [25]. In radiology, the characteristic image is "ground glass "opacity of the lungs. The lesion initially involves a single lobe but later expands to other lobes [26]. Lung tissue biopsies of COVID-19 patients reveal diffuse alveolar damage, desquamation of pneumocytes, hyaline membrane formation, and cellular fibromyxoid exudates indicative of acute respiratory distress syndrome [27]. Hematological findings show lymphocytopenia, both with and without leukocyte abnormalities, and the degree of lymphocytopenia is positively associated with disease severity [26].

3. COVID-19 and Comorbidities

Cardiovascular disease, obesity, hypertension, diabetes, and other pre-existing conditions are highly correlated with the severity of COVID-19 infection and cause excess deaths via co-morbidities (Figure 1) [28]. The clinical manifestation and relevance of specific co-morbidities due to COVID-19 infection is heterogenous [29]. In a large study of 460 general practices in England, of the 4300 COVID-19 patients with hypertension, about 20% died within 1 month of infection. Of note, the authors did not find any correlation between COVID-19 diagnosis or hospitalization and blood pressure control [30].

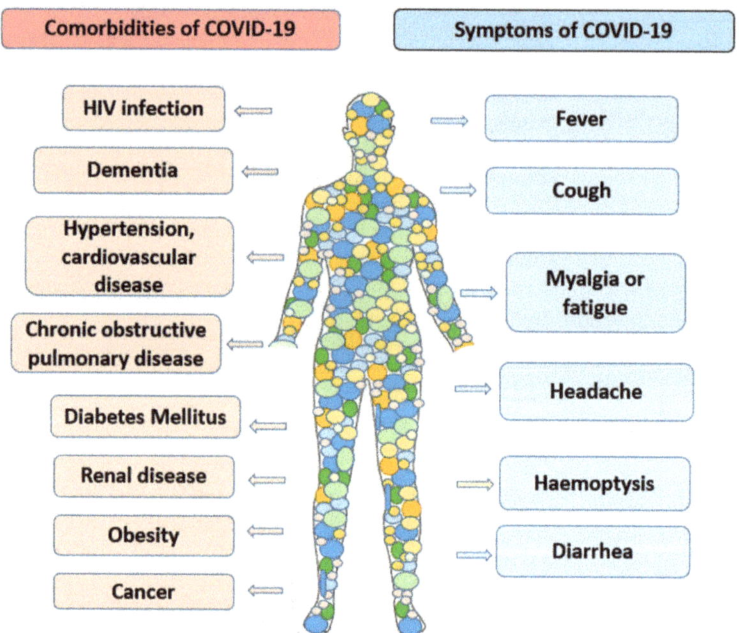

Figure 1. The main comorbidities and symptoms of COVID-19.

Patients with underlying cardiovascular disease and pre-existing blood vessel damage, such as atherosclerosis, may be at higher risk for severe disease. In addition to respiratory infection and inflammation, the SARS-CoV-2 coronavirus can directly and indirectly affect the renal system and cardiovascular tissue, which cause organ and tissue damage to the kidneys, heart, and blood vessels, and exacerbate inflammation that induces cytokine storms.

Similar to increased COVID-19 severity in patients with cardiovascular disease, especially hypertension, many studies show greater severity of infection in diabetics [31–34]. Current data indicate that diabetes in COVID-19 patients is correlated with a two-fold increase in mortality as well as severity of COVID-19. A meta-analysis of 30 studies and 16,003 patients conducted by Kumar et al. [35] suggests that diabetes and COVID-19 infection are significantly correlated with mortality [36]. The method for influencing the relationship is rigorous glucose monitoring and consideration of drug interactions.

As treatment for COVID-19, there are several pharmaceutical drugs options, such as lopinavir and steroids, which have a risk of hyperglycemia. In contrast, hydroxychloroquine may improve glycemic control in diabetic patients with decompensated refractory treatment [37,38]. It remains uncertain which treatment is suitable and works best for COVID-19 disease, and if treatment of diabetic patients should be different from those without diabetes. It is also uncertain whether specific diabetes drugs, such as DPP4 inhibitors, increase or decrease the susceptibility or severity of coronavirus infection. Isolated reports of new-onset diabetes in COVID-19 cases may suggest that coronavirus infection is directly cytotoxic to pancreatic islet β cells. Careful investigation [39] indicated that interaction of the coronavirus and diabetes is mediated by systemic inflammation and/or metabolic changes in adipose tissue, muscle, or liver, and not by direct infection of pancreatic cells.

Chronic obstructive pulmonary disease (COPD), a complex disease related to airway and/or alveolar abnormalities, is a lung dysfunction that is manifested by limited airflow mainly caused by exposure to harmful gases and particles over a long period of time (e.g., tobacco smoke). A meta-analysis of 15 studies that examined 2400 confirmed COVID-19 cases suggested that patients with COPD were at higher risk of more severe disease outcomes, with 60% higher mortality [40]. Multivariate logistic regression models identifying risk factors for positive SARS-CoV-2 tests in smokers were inconclusive [41]. While it cannot be concluded that smoking enhances SARS-CoV-2 infection, smokers are more likely to present a cough that can signal pulmonary distress and possible advance of COVID-19 infection; however, more testing is required to determine whether this population is directly susceptible to the virus by pulmonary infection, or by suppressed immune system function [42–44].

HIV infection serves as a model of cellular immune deficiency and it seems that antiretroviral therapy is thought to have various effects against the new coronavirus [45]. Given the fact that HIV positive patients may be at higher risk from other infectious diseases such as sexually transmitted diseases, these percentages were so low that some experts have already speculated on potential protective factors [46].

There are still debates about the effects of antiretroviral therapy against SARS-CoV-2. Regarding lopinavir, there is currently concrete evidence that it does not work. Regarding tenofovir alafenamide, there are some chemical similarities with remdesivir and it has been shown to bind to SARS-CoV-2 RNA polymerase with high binding affinity, being suggested as a potential treatment for COVID-19 [47]. However, the most serious concern about HIV is the collateral damage induced by COVID-19 [48]. The story of immunosuppression is uncertain, and the available data are insufficient to draw any conclusion. Despite the large lack of data, numerous views and guidelines have been published on how to manage immunosuppressed patients (who may be more susceptible to COVID-19 infection) and the development of severe cases [49–53].

A big challenge of the pandemic is to offer continuous care for cancer patients. Cancer patients are more vulnerable due to their underlying disease and immunosuppressed condition, and may therefore be at increased risk of developing severe complications due to the coronavirus. In fact, COVID-19 triage and management may leave some vital activities uncovered, such as treatment administration or surgery, and also a fragile immune system. It is well known that suboptimal synchronization and delayed oncological treatment can lead to disease progression, leading to reduced survival outcomes. There are various recommendations to minimize the exposure of cancer patients to COVID-19 without compromising the oncological outcome of radiation for breast cancer [54], hematopoietic cell transplant [55], and leukemia treatment [56].4. Immune Response to SARS-CoV-2.

In the ongoing SARS-CoV-2 pandemic, the consequences of infection range from asymptomatic to mild to moderate symptoms in most affected COVID-19 cases, but also can have a rapid and progressive disease that damages organs and leads to early deaths, in some cases as soon as 14–21 days from onset of infection. Since the start of the pandemic, facts have been complicated surrounding whether the virus can continue to be transmitted by asymptomatic individuals [57,58], and certainly by those with upper respiratory tract symptoms, or interstitial pneumonia that can progress rapidly to respiratory failure and acute respiratory distress syndrome, in which mechanical ventilation and admission to an intensive care unit and culminating in multiorgan failure [59–61]. Disease spread is correlated with longer viral shedding periods, encountered especially in asymptomatic patients [62]. After viral contamination, an effective adaptive immune response able to neutralize new antigens can be expected to develop in 14–21 days [63].

Antiviral innate immunity consists of coagulant factor, and components of the complement and fibrinolytic systems, soluble proteins that recognize glycans on cell surfaces, interferons, chemokines, and naturally occurring antibodies (mainly IgM but also IgA and IgG). The cellular components are natural killer cells and other innate lymphoid cells but also gamma delta T cells, which generally limit the spread of viral infection [64]. The viral spike protein preferentially binds to the angiotensin converting enzyme 2 receptor (ACE-2), prevalent in cells in the mammalian respiratory tract. Glycosylation of the viral surface can affect some aspects of virus biology, such as cell tropism, stability of protein components, camouflage of recognized antigens by neutralizing antibodies, and recognition by immune mechanisms.

Antiglycan antibodies are naturally identified in serum, i.e., they are identified in the absence of previous immunization, similar to natural ABO antibodies. Like ABO antibodies, they belong to the IgM class. Natural IgM concentrations appear to reflect some of the clinical severity patterns in COVID-19 [65]; they decrease significantly with age (>40 years) and are found in lower concentrations in people with blood type A. A protective role of high anti-A antibody titers described for SARS-CoV-1 [66] has been suggested for the SARS-CoV-2 [67].

Mannose binding lectin (MBL) is one of the components of the complement system in innate immunity, which recognizes mannose residues in the membrane of a variety of microorganisms, and acts as a soluble pattern recognition receptor (PRR). This recognition component activates the complement system, induces inflammation, and improves phagocytosis. MBL can bind to coronavirus, conducting to C4 deposition in the virus and in experimental models, decreases the capacity for infection [68]. The existence of mannose-rich glycans in the S1 region of SARS-CoV-2 has led to the hypothesis that glycan recognition and binding to MBL may inhibit the S1–ACE2 interaction [69]. However, with age, serum MBL levels decrease [70].

The first line of the innate immune response against viral infections is represented by type I interferons. They induce viral resistance in both infected cells and neighboring cells by interfering with cellular and viral replication. In MERS and SARS-CoV-1 infection, delayed production of interferon I favors the accumulation of inflammatory monocyte–macrophages [62,71].

The key diet-related changes in the developmental process of disease progression in humans include increased production of reactive oxygen species, oxidative stress, development of hyperinsulinemia, insulin resistance, low-grade inflammation, and an abnormal activation of the sympathetic nervous system and the renin–angiotensin system. Further, diet plays an important role in epigenetic changes and fetal programming that may have large effects on immune system efficiency. This suggested pathomechanism also explains the close relationship between obesity and the wide range of comorbidities, such as type 2 diabetes mellitus, cardiovascular disease, etc., and diseases of similar etiopathology. Changing lifestyle behaviors in accordance with human genetic makeup, including diet and physical activity, may help prevent or limit the development of these diseases [72]. COVID-19 poses a serious challenge to health-care systems worldwide, with an enormous impact on health conditions and loss of life at a remarkable scale. Notably, obesity and related comorbidities are strictly related with worse clinical outcomes of COVID-19 disease.

Recently, there is a growing interest in the clinical use of ketogenic diets, particularly in the context of severe obesity with related metabolic complications that are ameliorated through ketogenesis. Ketogenic diets have proven effective for a rapid reduction in fat mass, preserving lean mass and providing an adequate nutritional status. In particular, the physiological increase in plasma levels of ketone bodies exerts important anti-inflammatory and immunomodulating effects, which may prevent infection and potential adverse outcomes of COVID-19 disease [73].

3.1. Nutrients and Food Bioactive Components Involved in Immune System Stimulation

The immune system is one of the most important defense mechanism of body against disease, and the survival of humans is dependent on this system of fighting against viruses or pathogenic microorganisms [74]. There are studies that have indicated that some nutrients can have effects on immune functions through cell activation and modification of both production of signaling molecules and gene expression. Several micronutrients, such as vitamins and minerals, have essential roles in both adaptive and innate immune responses, and micronutrient homeostasis is central to the maintenance of a healthy immune system (Table 1). The efficacy of micronutrients in infections can be influenced by different factors, such as the dose, duration of administration, type of pathogen, genetics, age, lifestyle, and nutritional and immunological status [75].

Table 1. Micronutrients and their role in COVID-19.

Vitamins and Minerals	Food Sources	Actions	Role	Reference
Vitamin A	Meat, poultry, fish, dairy, eggs	Shows efficiency for immune function and resistance to infection	Immunomodulatory, Anti-inflammatory	[76]
Vitamin B1 (thiamine)	Meat, poultry, fish, whole-grains, brown rice dried beans, soybeans, nuts,	Eliminates the SARS-CoV-2 virus by triggering cell-mediated and antibody-mediated immunity	Supports immune response	[77]
Vitamin B2 (riboflavin)	Calf liver, fish, nuts, wild rice, dark green leafy vegetables, mushrooms, certain fruits and legumes, beer, yeast, milk, cheese, egg,	Reduces number of pathogens in the blood plasma of COVID-19 patients, and reduce the risk of transfusion–transmission of COVID-19.	Supports immune response	[77,78]
Vitamin B3 (niacin)	Meat, liver, beans	Reduces viral infection & stimulates defense mechanisms Reduces neutrophil infiltration in patients with ventilator-induced lung injury	Supports immune response	[77]
Vitamin B6 (pyridoxine)	Cereal grains, vegetables (carrots, spinach, peas), milk, potatoes, eggs, cheese, fish, liver, meat,	Relieves COVID-19 symptoms by improving immune response, supporting endothelial integrity, preventing hyper-coagulability & reducing pro-inflammatory cytokines	Supports immune response	[77]
Vitamin B12 (cyanocobalamin/ cobalamin)	Meat, milk, egg, fish, and shellfish	Essential role in improved immune system function	Supports immune response	[77]
Vitamin C	Citrus fruits kiwi, tomato, pineapple, kale, spinach, beef liver, milk, cabbage, broccoli, chicken, oysters, strawberries	Reduces symptom duration Reduces mortality Prevents COVID-19 progression Decreases risk of respiratory failure requiring a ventilator Reduces death rate & dependency on ventilator	Antioxidant, immunomodulatory	[79]
Vitamin D	Wild mushroom, fungi, fortified, bread fortified orange juice, milk, eggs, cheese, yogurt, fortified margarine	Improves prognosis in older patients Prevents respiratory infections	Antioxidant, immunomodulatory	[80]

Table 1. *Cont.*

Vitamins and Minerals	Food Sources	Actions	Role	Reference
Vitamin E	Vegetable oils, Nuts, seeds, avocado, green leafy vegetables, mango, salmon fortified cereals	Antioxidant defense Role in immune response & to reduce viral pathogenicity	Antioxidant, immunomodulatory	[81]
Selenium	Whole grains, nuts, mushrooms, dairy products, poultry, cereals, red meat, seafood	Essential for protection against viral infection	Antioxidant, ROS balance in inflammation, immune-cell function	[79]
Zinc	Fortified breakfast cereal, nuts, beans, poultry, red meat, whole grains, crustaceans, mollusks	Reduces inflammatory reaction Increases ventilator-free days Organ failure-free days Acute inflammation-free days	Antioxidant, anti-inflammatory, reduces ROS in viral infection	[80]

There are studies that emphasize a significant relationship between immunity, diet, and disease susceptibility. Nutritional deficits in macro- and micro-nutrients can affect the immune system and resistance to infection. Various functional food plants, such as pepper, garlic, turmeric, and onion, may have immunomodulatory and antiviral properties (Table 2) [75].

Table 2. Biological activities of foods bioactive compounds.

Food Source	Compounds	Biological Activities	Reference
Onion	Quercetin, thiosulfinates, & anthocyanins	Antioxidant	[82]
Citrus fruits	Hesperidin	Antioxidant, anti-inflammatory, antiviral	[83]
Garlic	Diallyl disulphide, alliin, polyphenols, proteins	Antioxidant, antiviral	[84]
Honey	p-coumaric acid, ellagic acid	Antimicrobial, antiviral	[85]
Tea Plant	Gallic acid, theaflavin-3,3′-digallate, quercetin, catechins	Antioxidant, antiviral, immunomodulatory	[86]
Cranberry	Myricetin	Antiviral	[87]
Barberry	Berbamine, berberine	Anticancer	[82]
Turmeric	Curcumin	Anti-inflammatory	[83]
Soybean	Flavonoids, Isoflavones, phytosterols, saponins & organic acid	Antioxidant	[82]
Banana	Bananin	Antiviral	[88]
Long pepper, black pepper	Piperine	Anticancer	[82]
Plum	Anthocyanins, protocatechuic acid	Antioxidant	[82]
Grapes, berries	Quercetin	Antioxidant, anti-inflammatory	[89]
Kale	Kaempferol	Anti-inflammatory	[90]
Avocado, pistachio, almond	B-sitosterol	Anti-inflammatory	[89]
Mango	Flavonoids, xanthones, phenolic acids, triterpenes	Antioxidant, antiviral	[82]
Nuts, seeds	Stigmasterol	Antiviral	[89]
Red grape	Resveratrol	Anti-inflammatory	[89]

3.2. The Effects of Isolation, Quarantine, and Lockdowns on Dietary Health

Measures taken by governments around the world to contain the spread of COVID-19 have had measurable impacts on health and habits of people everywhere. A general consequence of quarantine is a change in lifestyle: reduced physical activity and unhealthy dietary choices (Table 3) [91]. Access to healthy foods, such as fresh produce, has been limited, and people have been in lockdowns which prevent outdoor movement and access to sunlight and clean air. Those who have acquired the disease or have been quarantined have often been deprived of sources of higher-nutrition foods [92]. In addition, containment measures restricting free movement and creating physical lockdowns have been detrimental to healthy lifestyles. Regulations aimed at containing viral spread had differing outcomes on different demographics groups, communities, and socioeconomic groups. In particular, failures in infrastructures and supply chain resilience during the early and mid-stages of the pandemic disrupted food availability. For less economically fortunate populations, the breakdown in supply chains resulted in limited access to fresh produce, and instead forced a switch to processed foods that have a longer shelf life. Notably, the connections for supply chain deliveries to communities that are further away from distribution centers/ports/hubs/farms, or that would otherwise depend on imports for healthier foods, had reduced availability of healthy food options. This exacerbated public health issues that have been being largely overlooked by media and governments throughout the pandemic. In contrast, in more economically well-to-do communities, fresh imports and local produce options were available due, in part, to demand, economic liquidity of the communities, and proximity to distribution hubs in those areas. It is well known that a diet rich in fresh produce and whole foods is necessary for healthy immune function, and is thus preferable to a high-calorie processed food diet that increases the risk of developing and even aggravating autoimmune problems and chronic diseases [93]. Recent work reported by Fernández-Quintela et al. in 2020 [94] found that two particular omega-3 fatty acids, eicosapentaenoic acid (EPA) and docosahexaenoic acid (DHA), are both effective at inactivating entrapped SARS-CoV-2 viruses by modulating optimal lipid conditions to reduce viral replication. Both EPA and DHA can also inhibit cyclooxygenase enzymes (COX), which inhibits prostaglandin production, reducing tissue inflammation. Quarantine isolation and lockdown interventions for COVID-19 created, especially for older adults, a severe sense of social isolation and loneliness with potentially serious mental and physical health consequences. The impact was disproportionately amplified in those with pre-existing mental illness, who often suffered from loneliness and social isolation prior to the enhanced distancing from others imposed by the COVID-19 pandemic public health measures. Older adults are also more vulnerable to social isolation and loneliness as they are functionally very dependent on family members or support by community services [95].

3.2.1. Spain

A recent survey of 1036 individuals in Spain reported that, during the pandemic, people consumed more fresh produce as well as fish than before [96]. Another study of 1073 persons reported decreased consumption of poultry and mammal, as well as rice and pasta [97]. Further, a larger study of 7514 individuals reported that people generally consumed a Mediterranean diet more than they did before the onset of the pandemic [98]. These results suggest that people in Spain sought foods that are part of a healthy diet and lifestyle, which may reduce the incidence of severe COVID-19 outcomes.

3.2.2. Italy

Italy was one of the first countries most affected by COVID-19, when hospitalization cases rapidly overwhelmed public healthcare capacity. Early in the pandemic the public faced lockdowns, and that coupled with panic, led to fresh food shortages. In contrast to findings from populations in Spain, a study of 1519 people in Italy reported that the diet of the average person increased in the consumption of frozen foods and foods made with refined sugars [99]. Grant et al., 2021, collected data from 2678 people and observed that many improved the quality of their diet by increasing the consumption of fruits, vegetables, legumes, nuts, fish, or shellfish. However, unfavorable dietary changes were also reported; people consumed an excess of sweets, pastries, and comfort foods than they reportedly did before the pandemic [100]. In another study, 3533 individuals reported a decreased intake of fresh fish, processed sugar foods and baked goods, food delivery, alcohol intake, and an increase in homemade recipes (pizza, sweets, and bread), vegetables, cereals, white meat, and hot beverages consumption [101]. Increased alcohol consumption was observed in a study of 1383 participants, who, according to their responses, also chose foods high in carbohydrates, such as potatoes, cereals, fruits, leading to weight gain amongst the respondents; conversely, this same group reduced their consumption of dairy products, vegetables, and red meat. Anxiety, fear, stress, or moments of boredom have encouraged over 40% of people in Italy to eat foods high in refined processed sugars and oils, leading to weight gain and possible side effects related to COVID-19 infection and disease severity [102]. However, these variations in the weekly frequency of food consumption did not alter the adherence score to the Mediterranean diet, which remained at medium-high values [103]. In this sense, Italians have been trained to transform their green spaces into food gardens, especially taking into account the benefits of eating fresh fruits and vegetables. The television programs followed the training of small farmers, the purpose being not only to obtain their own crops, but also to focus their attention on recreational activities to maintain mental health [104].

3.2.3. France

A study of 938 individuals showed increased intake of fresh produce, legumes, and seafood. Consumption of refined sugars, processed meats, sweet drinks and alcoholic beverages also increased, leading to a decrease in nutritional quality of the average diet [105]. A total of 11,391 participants surveyed in the first 8–13 days after home confinement measures were implemented revealed increased consumption of caloric and salty foods (28.4%), alcohol (24.8%), tobacco (35.6%) and even cannabis (31.2%) [106]. Almost a quarter of French people engaged in behaviors that contributed to poor health outcomes, in many cases by stress management through eating more and unhealthy foods, and lack of physical activity due to confinement indoors [107].

A study of 498 parents of children aged 3.0–12.3 years presented no changes in eating behavior for other reasons than a change in eating habits. A significant decrease was observed for rules and limits around unhealthy foods, setting, and on scheduled meals. Children increased their consumption of high carbohydrate sources and processed foods, including sweets/chocolates, fruit juices, soft drinks, chips/crackers, ice cream, pastries/sweet cakes, dessert cream, milk, yogurt/cheese/quark, fresh and dried fruits, and a significant decrease in the consumption of compote/fruit in syrup [108].

3.2.4. Greece

Unlike other countries heavily affected by COVID-19, such as Spain or Italy, the Greek population was not so emotionally involved, and the signs of anxiety and depression were less obvious among adults. Thus, the emotional eating of unhealthy foods was not so high [109].

Nearly one in three of 2258 participants reported that they changed their dietary habits during the pandemic towards a healthier diet rich in fruits, vegetables, salads, green vegetables, cereals, legumes, and olive oil, and consumed less meat, especially processed meat [110]. Similarly, another study of 741 individuals report a prudent dietary pattern containing of fruits, vegetables, fish, and rice [111].

3.2.5. Denmark

The data presented by Giacalone, et al., 2020, based on a questionnaire with a number of 2462 subjects, suggest that the pandemic affected the lifestyle and eating habits of some adults living in Denmark. The main findings include the fact that they ate more frequently. During this period, survey respondents reported they consumed more processed, canned, frozen, or ready-to-eat food, and reduced consumption of bread, alcohol, and dairy [112]. Moreover, unhealthy eating habits were observed, such as the increased intake of pastries and carbonated beverages [113].

3.2.6. Poland

People in Poland reported positive dietary changes, consuming less red meat, processed flours and baked goods, prepared foods, fast-food, canned meat products, as well as energy drinks and refined sugars. However, increased consumption of alcoholic beverages and sweets, both of which are poor dietary choices, contributed to unhealthy weight gain [114]. Increased BMI was associated with less frequent consumption of vegetables and fruits during quarantine, and higher adherence to meat, dairy, and fast-foods. Increased alcohol consumption was reported in 14.6% of study participants, with a tendency to drink more among regular alcohol consumers [115].

3.2.7. China

In China, the general frequency of intake of fresh vegetables, fruits, soybean products, and dairy decreased during the lockdown. Average weekly consumption of rice decreased, but there were increases by younger age classes in wheat products, other staple foods, fresh vegetables, fresh fruit, preserved vegetables eggs, fish, and dairy products. Furthermore, the frequency of sugar-sweetened beverage consumption had decreased, while the frequency of other beverages had increased [116].

3.2.8. United States

In general, people shifted their diet away from healthy animal proteins, fruits, and vegetables, reportedly due to increased cost because of supply chain issues related to the pandemic. Local supply of food was disrupted, which in turn affected local economies that led to social, mental, and physical health changes. The municipal authorities developed programs to support and protect food security during the pandemic, but especially in the post-pandemic period. In the U.S., people reported experiencing greater stress, anxiety, and boredom, which led to overeating and weight gain. People ate more, and in particular they ate more processed snacks and comfort foods, which are rich in additives and fats, processed trans-fats, high salt, and sugar [117]. According to statistical analysis, the population consumed mainly red and processed meats, fast food, sweets, and refined cereals during the pandemic, and with the return of the US economy, prices fell and facilitated access to vegetables, oils, nuts, and lean proteins [118].

Table 3. Dietary behavior in different countries on COVID-19.

	Colombia [119]	Poland [120]	Italy [100]	France [121]	Saudi Arabia [122]	Germany [123]	China [124]	Spain [125]
Ate more								
Increase	45%	-	-	-	-	40%	31%	64%
Decrease	20%	-	-	-	-	21%	17%	36%
As before	35%	-	-	-	-	39%%	52%	-
Weight gain								
Yes	22%	40%	37%	38%	38%	31%	-	13%
No	38%	10%	52%	18%	26%	16%	-	47%
Unknown	40%	50%	11%	44%	36%	53%	-	40%
Fast food								
Increase	21%	8%	-	-	-	-	-	5%
Decrease	34%	37%	-	-	-	-	-	35%
As before	45%	55%	-	-	-	-	-	60%
Snacking								
Increase	48%	30%	33%	24%	45%	-	25%	37%
Decrease	22%	10%	11%	18%	19%	-	37%	16%
As before	30%	60%	56%	58%	36%	-	38%	47%
Meals out of home								
Increase	-	-	-	-	-	7%	-	21%
Decrease	-	-	-	-	-	80%	-	50%
As before	-	-	-	-	-	13%	-	29%
Alcohol intake								
Increase	7%	18%	16%	12%	-	-	26%	11%
Decrease	18%	11%	13%	12%	-	-	39%	57%
As before	22%	71%	71%	39%	-	-	35%	32%
Never	53%	-	-	23%	-	-	-	-
Water intake								
Increase	36%	-	20%	-	57%	-	-	-
Decrease	26%	-	8%	-	7%	-	-	-
As before	38%	-	72%	-	36%	-	-	-
Physical activity								
Increase	23%	-	36%	26	27%	-	-	16%
Decrease	48%	-	11%	50%	52%	-	-	59%
As before	18%	-	16%	24%	21%	-	-	19%
Never	11%	-	37%	-	-	-	-	6%
Home cooking								
Increase	66%	-	-	-	73%	96%	65%	45%
Decrease	5%	-	-	-	4%	4%	9%	4%
As before	23%	-	-	-	23%	-	26%	51%
Never	6%	-	-	-	-	-	-	-

- Data not available. *N*, number of respondents: Colombia = 2745, Poland = 407, Italy = 2678, Saudi Arabia = 2255, Germany = 1964, China = 1994, Spain = 7514, France = 4005.

Regarding food hygiene, it is suggested that consumers living in communities with COVID-19 cases have higher food safety knowledge scores, disinfect cooking surfaces more, pay more attention to food safety information, and have more timely access to food safety news. So, people with COVID-19 pandemic-related information tend to have higher food safety knowledge and practice food safety behavior [126,127].

3.3. Importance of Lifestyle in Prevention of COVID-19

Since early in the pandemic, the most effective measures that reduce transmission of SARS-CoV-2 and prevent COVID-19 spread have been physical distancing and proper use of face masks that have multiple layers of tightly woven, breathable fabric, a nose wire, and a thickness that block lights when held up to bright light source [128] (Figure 2).

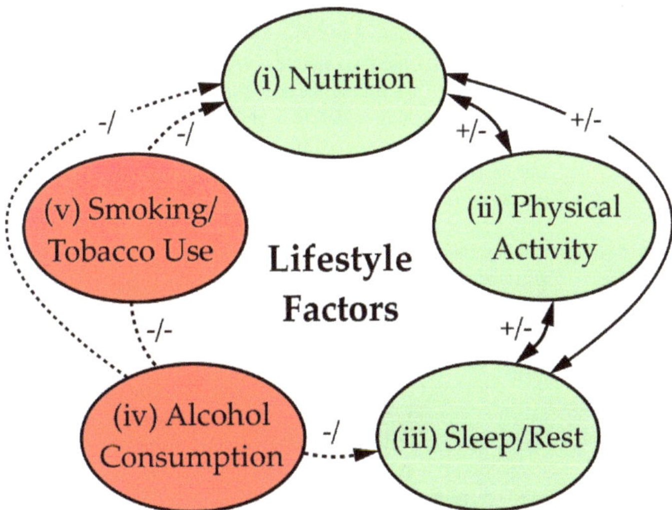

Figure 2. Lifestyle factors that affect clinical health outcomes in COVD-19 infections. Dashed lines indicate negative effects, whereas solid lines connect factors known to improve health outcomes; arrows denote interaction direction(s). Red indicates negative lifestyle factors; green indicates positive factors. (i) Nutrition is essential in supporting immune system function and is affected by other factors positively/negatively (+/−) [129]. Poor quality rest or lack of physical activity can limit the benefit of nutrition as a health factor (−), and the inverse is true (+). (ii) Physical activity and regular exercise help mitigate disease effects and are related to nutrition and rest similarly [130]. (iii) Quality sleep and rest contribute to healthy outcomes, and affect and are affected by nutrition and activity. Note that inadequate sleep can induce stress/anxiety that can exacerbate health outcomes (−). (iv) Alcohol consumption leads to organ stress and dysfunction, depresses the immune system response to viral and bacterial infections, and negatively impacts sleep quality (−/) [131]. (v) Tobacco use has well-known detrimental effects on the immune system and leads to many clinical health problems that exacerbate infectious disease outcomes (−/) [132]. Smoking and alcohol consumption are often correlated and thus have negative interactions (−/−).

Lifestyles have changed substantially due to isolation and distancing measures (Figure 2), as people are more sedentary, and the lack of physical activity is correlated with poor dietary changes and unhealthy weight gain, both of which contribute to severity of COVID-19 outcomes. Indeed, obesity and being in poorer health with a less nutritious diet is associated with greater severity of COVID-19 cases requiring hospitalization. Other problematic outcomes of social isolation measures implemented during the pandemic are changes in smoking and sleep habits. Several studies reported associations between sleep disorders and obesity due to increased secretion of pro-inflammatory cytokines by increasing visceral adipose that can contribute to altered sleep-wake rhythms [88,133–135].

3.4. Food Hygiene in COVID-19 Pandemic

Public health and food safety authorities have found no evidence that SARS-CoV-2 spreads via food [136]. The only transmission path involving food is the packaging, which could be contaminated with coronavirus [137–139]. Thus, handling or consumption of contaminated food packaging carries similar risks as other surfaces known to transfer coronavirus [140].

The COVID-19 pandemic has caused temporal food shortages due to supply chains changes, labor shortages [141–144], training personnel in hygiene, food safety, incident management, recreating business models regarding packaging, and other unanticipated impacts [145]. Furthermore, lockdown measures enacted at regional and national levels, such as the closure of universities, schools, workplaces, restaurants, public events, so-called non-essential businesses, and travel restrictions [146], changed the way people purchased food, where they ate, what they ate, and how their food was prepared [112,147]. Some of these changes may be latent symptoms of post-COVID lifestyles.

The plan for preventing the transmission of the coronavirus includes control requirements for food facility disinfection, sanitation, cleaning, monitoring and screening of workers for COVID-19, education programs to prevent the spread of SARS-CoV-2, and management of sick employees [148].

Inactivation methods (thermal and non-thermal) are effective at minimizing pathogens and viruses in the food sector [149]. For the inactivation of foodborne viruses (e.g., hepatitis A and norovirus) on food matrices or liquids, different thermal treatments have been used [150], such as dry (hot air oven) and humid heat (autoclave) which are very effective methods for inactivating both viruses and bacteria [151,152].

As is specified in research studies, cold-chain food contributes to contamination because coronavirus is stable at 4 °C on poultry, meat, fish, and swine skin, for 3 weeks [137]. Thus, the possibility of transmission through food chain is very high in the frozen food. Therefore, risk management approaches should be adopted to inspect potentially infected foods, especially cold-chain foods (Figure 3) [89].

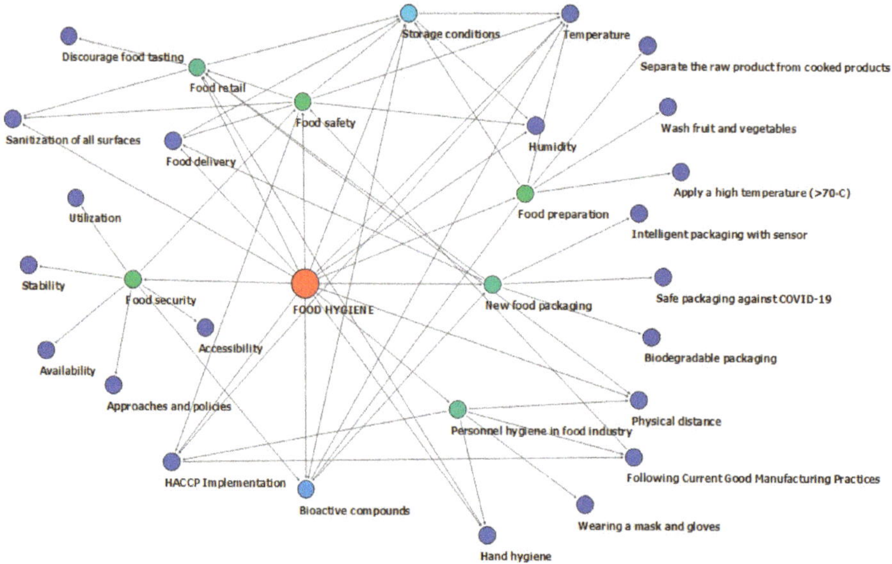

Figure 3. Network connections to food hygiene in food supply systems.

3.4.1. Food Security

Food security includes availability, accessibility, stability, and utilization of food, at all times for all people. Food must be in sufficient quantities, be safe and nutritious to afford people a healthy life [153].

Availability. What foods are available has demonstrably affected the nutritional habits of consumers during the pandemic. The available foods need to be good sources of necessary macro- and micro-nutrients to ensure public health by minimizing severity of COVID-19 cases, and ensuring that basic nutritional requirements are being met is crucial [154].

Accessibility. The pandemic has highlighted how much food accessibility can be affected by interruptions in food supply distribution and logistics, leading to rising prices and lack of food options. Restrictions on food logistics are likely to increase transaction costs, and therefore food prices could adversely affect access to healthy food and contribute to food insecurity, obesity, and malnutrition [155]. Reduced access to healthy foods leads to higher consumption of preserved and ultra-processed, which, combined with reduced physical activity, leads to obesity and other diet-related diseases. The ability to deliver whole foods faster and at reasonable costs is a difficult task during the pandemic [156].

Utilization. Eating food through a proper diet, drinking water, sanitation, and health care to achieve a state of nutritional well-being in which all physiological needs are met [157].

Stability. Food stability emphasizes that all humans should have access to enough food all the time, regardless of any unforeseen risk (such as a pandemic), which could prevent people from accessing food [158,159].

3.4.2. Food Safety in COVID-19 Pandemic

The food industry has Food Safety Management Systems (FSMS) based on the Hazard Analysis and Critical Control Point (HACCP) principles for preventing food contamination and manage food safety risks. FSMSs contain good hygiene practices, zoning of processing areas, storage, supplier control, personnel hygiene, cleaning and sanitation, and fitness to work distribution and transport—all the basic conditions and activities necessary to maintain a hygienic food processing environment [160].

3.4.3. Personnel Hygiene in Food Industry in COVID-19 Pandemic

Cold air conditions in food factories make it particularly difficult to prevent transmission of COVID-19 because the virus is stable over longer periods, and can be moved on aerial particulates by recirculated air systems [161].

Food industry workers are in some cases tested for SARS-CoV-2 to eliminate the potential risk of food contamination [139]. Regular hand washing is crucial in the food sector as well as in all the industries. Similar viruses are spread by droplets when an infected person coughs or sneezes. The WHO recommends measures applicable to the food industry, such as frequent washing of hands with soap and water or alcohol-based disinfectants; maintaining physical distance; and avoiding contact of the hands with the eyes, nose, and mouth. In addition to these practices, the mobility of food industry staff, such as air transport, should be monitored [143,162].

3.4.4. Food Retail in COVID-19 Pandemic

COVID-19 has significantly changed the retail customer experience through changes in availability and increases in pricing. People have to visit more retailers to find specific items, and are often unable to find the types of food and beverages they need to maintain a healthy diet, which can have negative impacts on emotional and mental well-being [163].

Retail workers are often at risk of exposure due to the nature of the workplace that involves interactions with unknown individuals, and consequently we have seen SARS-CoV-2 cluster infections in retail environments and other settings where there is higher traffic of retail individuals that means less control over containment [164].

Measures taken during the pandemic include physical distancing, providing sanitation stations to clean shared shopping equipment, and by regulating the number of customers inside the premises. Physical barriers, such as plexiglass, were deployed to separate food from any risk of danger and to protect staff at cashier point [140].

Food preparation includes measures such as separating the raw product from the cooked product to prevent cross-contamination, as well as washing, rinsing, and sanitizing surfaces or utensils in contact with food and beverage equipment after use. At the same time, practicing good hand hygiene before eating and washing fruits and vegetables with drinking water before consumption are crucial. For cooking food, it is recommended to apply a high temperature (>70 °C) [165].

3.4.5. Food Delivery Include

Customers are more interested in delivery hygiene and food safety with the COVID-19 pandemic because they could be infected with COVID-19 if they contact contaminated food and infected delivery personnel [166]. Contactless delivery in many countries is practiced through the "leave delivery at the door" option or workers leave materials two meters away for customers.

It is also recommended to use face masks and gloves as well as keep physical distance. Employees receiving and delivering must wash or disinfect their hands and implement appropriate hygiene and hygiene protocols. Another measure to avoid contamination is to use an electronic wallet or credit card payment method. It is also recommended that consumers throw away the packaging as soon as possible and wash their hands immediately afterwards [167].

3.4.6. Importance of Smart Packaging with Antiviral Properties

Due to the sensitivity of food security during the pandemic, and faced with the state of global panic on the means of transmission of the virus, the food industry, throughout the world, has been obliged, to also have emergency plans in place. The results of the field survey made it clear that consumers are fearful of the health impacts of COVID-19 on their lives, and may want packaging to be safe, sustainable, and meet their expectations [168].

COVID-19 has had a major impact on consumer choice and eating habits. A review article suggests that during the COVID-19 pandemic, consumers and policy makers responded to an increased perception of the risk to food safety by increasing their dependence on disposable plastic packaging. Attitudes towards food packaging have major implications for both food and environmental policy, thus feeding the need for smart, biodegradable, and safe packaging [169].

The stability of coronavirus on food packaging has led to the development of materials based on biopolymers with antiviral properties (Figure 4). The use of these materials has shown high efficacy against human norovirus and hepatitis A virus. Some research studies show that the release of ions from the surfaces of copper or copper alloy can help inactivate HuCoV-229E.

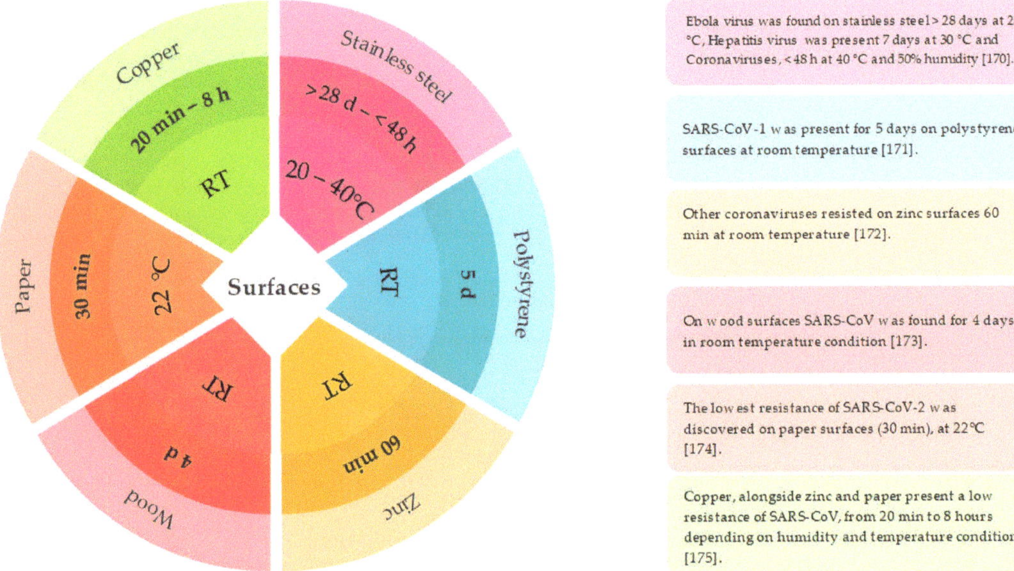

Figure 4. Capacity of SARS-CoV-2 and other viruses to persist on the package surfaces. RT, room temperature: 20–26 °C [170–175].

The development of biopolymers with antiviral properties and their applications in the food packaging industry remains an open field of research. Recently, it has been reported that the use of nanomaterial-based packaging or films containing zinc, copper, and silver nanoparticles can inhibit SARS-CoV-2, prevent contamination of food packaging surfaces, and thus diminish its transmission [138,176].

4. Conclusions

Social limits and movement restrictions introduced on the public during the pandemic have had unforeseen health outcomes that reinforce the importance of maintaining a healthy lifestyle through diet, exercise, and stress management. Because the COVID-19 pandemic changed how and when people could access provisions, what kinds of provisions, and how much were available at any one time, diets in many populations and demographic groups became less healthy compared to pre-pandemic life, notably that people consumed more calories of lower nutritional quality, which can exacerbate COVID-19 outcomes. While not all populations experienced an unhealthy diet shift, it is concerning because, since the start of the pandemic, there have been telling signs that diet patterns led to higher BMI, which, along with obesity, are known to worsen outcomes from COVID-19, with more severe infection requiring intervention via hospitalization, intensive care, and possibly a ventilator, or lead to a fatal outcome. In some populations, the severity of COVID-19 infection seems to be moderated by the consumption of specific micro- and macronutrients, such as those found in the Mediterranean diet. However, an unfortunate outcome of the pandemic has been that, during lockdowns, the slide in adherence to healthy behaviors (i.e., healthy eating, restful sleep, stress management, physical activity, avoidance of risky substances such as smoking and alcohol, and healthy relationships) which can prevent, treat, and even reverse disease, may have been overlooked.

Practicing food hygiene and a healthy diet containing fresh vegetables and fruits are key components of a healthy lifestyle essential for maintaining a properly functioning and efficient immune system to defend against infection and disease. While there is no evidence that the coronavirus can spread directly via foods, packages can be contaminated with

SARS-CoV-2, and could transmit the virus; therefore, there may be a need for packaging with bioactive compounds that neutralize infectious contaminants to reduce transmission risks. Essentially, with the likelihood that SARS-CoV-19 continues to move about the population, we want to avoid future outbreaks that consume health care infrastructures.

Author Contributions: Conceptualization, R.F., L.A.-N., R.G., W.K.S. and M.D.; resources, R.F., L.A.-N., R.G., W.K.S. and M.D.; data curation, L.A.-N., R.G., W.K.S.; Writing—Original Draft preparation, R.F., L.A.-N., R.G., W.K.S., M.D.; Writing—Review and Editing, R.F., L.A.-N., R.G., W.K.S., M.D.; visualization, L.A.-N., R.G.; supervision, M.D., W.K.S.; funding acquisition, R.F., W.K.S. All authors have contributed equally to this work. All authors have read and agreed to the published version of the manuscript.

Funding: This research received no external funding.

Institutional Review Board Statement: Not applicable.

Informed Consent Statement: Not applicable.

Data Availability Statement: The data presented in this study are available upon request from the corresponding author.

Acknowledgments: This work was supported by Romania National Council for Higher Education Funding, CNFIS, project number CNFIS-FDI-2021-0357.

Conflicts of Interest: The authors declare no conflict of interest.

References

1. Wu, Y.-C.; Chen, C.-S.; Chan, Y.-J. The outbreak of COVID-19: An overview. *J. Chin. Med. Assoc.* **2020**, *83*, 217–220. [CrossRef]
2. Ng, W.H.; Tipih, T.; Makoah, N.A.; Vermeulen, J.-G.; Goedhals, D.; Sempa, J.B.; Burt, F.J.; Taylor, A.; Mahalingam, S. Comorbidities in SARS-CoV-2 Patients: A Systematic Review and Meta-Analysis. *mBio* **2021**, *12*, 1–12. [CrossRef] [PubMed]
3. Fathi, M.; Vakili, K.; Sayehmiri, F.; Mohamadkhani, A.; Hajiesmaeili, M.; Rezaei-Tavirani, M.; Eilami, O. The prognostic value of comorbidity for the severity of COVID-19: A systematic review and meta-analysis study. *PLoS ONE* **2021**, *16*, e0246190. [CrossRef]
4. Bertuzzi, S.; DiRita, V.J. After COVID-19 We Will Need a New Research System. We Need to Start Planning Now. *mBio* **2020**, *11*, 1–4. [CrossRef] [PubMed]
5. Water, Sanitation, Hygiene, and Waste Management for SARS-CoV-2, the Virus that Causes COVID-19. Available online: https://www.who.int/publications/i/item/WHO-2019-nCoV-IPC-WASH-2020.4 (accessed on 31 August 2021).
6. Cyranoski, D. Bat cave solves mystery of deadly SARS virus—And suggests new outbreak could occur. *Nature* **2017**, *552*, 15–16. [CrossRef]
7. Xiao-Shuang, Z.; Zeng, L.; Yang, X.-L.; Ge, X.-Y.; Zhang, W.; Lin-Fa, W.; Xie, J.-Z.; Dong-Sheng, L.; Zhang, Y.-Z.; Wang, N.; et al. Discovery of a rich gene pool of bat SARS-related coronaviruses provides new insights into the origin of SARS coronavirus. *PLOS Pathog.* **2017**, *13*, e1006698. [CrossRef]
8. Lau, S.K.P.; Lee, P.; Tsang, A.K.L.; Yip, C.C.Y.; Tse, H.; Lee, R.A.; So, L.-Y.; Lau, Y.L.; Chan, K.-H.; Woo, P.C.Y.; et al. Molecular Epidemiology of Human Coronavirus OC43 Reveals Evolution of Different Genotypes over Time and Recent Emergence of a Novel Genotype due to Natural Recombination. *J. Virol.* **2011**, *85*, 11325–11337. [CrossRef]
9. Li, W.; Shi, Z.; Yu, M.; Ren, W.; Smith, C.; Epstein, J.H.; Wang, H.; Crameri, G.; Hu, Z.; Zhang, H.; et al. Bats Are Natural Reservoirs of SARS-Like Coronaviruses. *Science* **2005**, *310*, 676–679. [CrossRef]
10. Mulay, A.; Konda, B.; Garcia, G.; Yao, C.; Beil, S.; Villalba, J.M.; Koziol, C.; Sen, C.; Purkayastha, A.; Kolls, J.K.; et al. SARS-CoV-2 infection of primary human lung epithelium for COVID-19 modeling and drug discovery. *Cell Rep.* **2021**, *35*, 109055. [CrossRef] [PubMed]
11. Gryseels, S.; De Bruyn, L.; Gyselings, R.; Calvignac-Spencer, S.; Leendertz, F.H.; Leirs, H. Risk of human-to-wildlife transmission of SARS-CoV-2. *Mammal Rev.* **2020**, *51*, 272–292. [CrossRef]
12. Menachery, V.D.; Yount, B.L.; Debbink, K.; Agnihothram, S.; Gralinski, L.E.; Plante, J.A.; Graham, R.L.; Scobey, T.; Ge, X.-Y.; Donaldson, E.F.; et al. A SARS-like cluster of circulating bat coronaviruses shows potential for human emergence. *Nat. Med.* **2015**, *21*, 1508–1513. [CrossRef] [PubMed]
13. De Backer, C.; Teunissen, L.; Cuykx, I.; Decorte, P.; Pabian, S.; Gerritsen, S.; Matthys, C.; Al Sabbah, H.; Van Royen, K.; The Corona Cooking Survey Study Group. An Evaluation of the COVID-19 Pandemic and Perceived Social Distancing Policies in Relation to Planning, Selecting, and Preparing Healthy Meals: An Observational Study in 38 Countries Worldwide. *Front. Nutr.* **2021**, *7*, 621726. [CrossRef]
14. Czarniecka-Skubina, E.; Pielak, M.; Sałek, P.; Głuchowski, A.; Kobus-Cisowska, J.; Owczarek, T. Use of Food Services by Consumers in the SARS-CoV-2 Pandemic. How the Eating Habits of Consumers Changed in View of the New Disease Risk Factors? *Nutrients* **2021**, *13*, 2760. [CrossRef]

15. Memon, S.; Pawase, V.; Pavase, T.; Soomro, M. Investigation of COVID-19 Impact on the Food and Beverages Industry: China and India Perspective. *Foods* **2021**, *10*, 1069. [CrossRef]
16. Kassir, R. Risk of COVID-19 for patients with obesity. *Obes. Rev.* **2020**, *21*, e13034. [CrossRef]
17. Celorio-Sardà, R.; Comas-Basté, O.; Latorre-Moratalla, M.; Zerón-Rugerio, M.; Urpi-Sarda, M.; Illán-Villanueva, M.; Farran-Codina, A.; Izquierdo-Pulido, M.; Vidal-Carou, M. Effect of COVID-19 Lockdown on Dietary Habits and Lifestyle of Food Science Students and Professionals from Spain. *Nutrients* **2021**, *13*, 1494. [CrossRef]
18. Overview of The 2019 Novel #Coronavirus (2019-nCoV): The #Pathogen of Severe Specific Contagious #Pneumonia (SSCP) (J Chin Med Assoc., Abstract)—ETIDIoH. Available online: https://etidioh.wordpress.com/2020/02/13/overview-of-the-2019-novel-coronavirus-2019-ncov-the-pathogen-of-severe-specific-contagious-pneumonia-sscp-j-chin-med-assoc-abstract/ (accessed on 31 July 2021).
19. Chan, J.F.-W.; Yuan, S.; Kok, K.-H.; To, K.K.-W.; Chu, H.; Yang, J.; Xing, F.; Liu, J.; Yip, C.C.-Y.; Poon, R.W.-S.; et al. A familial cluster of pneumonia associated with the 2019 novel coronavirus indicating person-to-person transmission: A study of a family cluster. *Lancet* **2020**, *395*, 514–523. [CrossRef]
20. Backer, J.A.; Klinkenberg, D.; Wallinga, J. Incubation period of 2019 novel coronavirus (2019-nCoV) infections among travellers from Wuhan, China, 20–28 January 2020. *Eurosurveillance* **2020**, *25*, 2000062. [CrossRef] [PubMed]
21. Assiri, A.; Al-Tawfiq, J.A.; Al-Rabeeah, A.A.; Al-Rabiah, F.A.; Al-Hajjar, S.; Al-Barrak, A.; Flemban, H.; Al-Nassir, W.N.; Balkhy, H.H.; Al-Hakeem, R.F.; et al. Epidemiological, demographic, and clinical characteristics of 47 cases of Middle East respiratory syndrome coronavirus disease from Saudi Arabia: A descriptive study. *Lancet Infect. Dis.* **2013**, *13*, 752–761. [CrossRef]
22. Cockrell, A.S.; Yount, B.L.; Scobey, T.; Jensen, K.; Douglas, M.; Beall, A.; Tang, X.-C.; Marasco, W.A.; Heise, M.T.; Baric, R.S. A mouse model for MERS coronavirus-induced acute respiratory distress syndrome. *Nat. Microbiol.* **2016**, *2*, 16226. [CrossRef] [PubMed]
23. Mahase, E. Coronavirus: Covid-19 has killed more people than SARS and MERS combined, despite lower case fatality rate. *BMJ* **2020**, *368*, m641. [CrossRef]
24. Sas, B. Mast cells contribute to coronavirus-induced inflammation: New anti-inflammatory strategy. *J. Biol. Regul. Homeost. Agents* **2020**, *33*, 9–14. [CrossRef]
25. Duzgun, S.A.; Durhan, G.; Demirkazik, F.B.; Akpinar, M.G.; Ariyurek, O.M. COVID-19 pneumonia: The great radiological mimicker. *Insights Imaging* **2020**, *11*, 118. [CrossRef]
26. Xie, C.; Ng, M.-Y.; Ding, J.; Leung, S.T.; Lo, C.S.Y.; Wong, H.Y.F.; Vardhanabhuti, V. Discrimination of pulmonary ground-glass opacity changes in COVID-19 and non-COVID-19 patients using CT radiomics analysis. *Eur. J. Radiol. Open* **2020**, *7*, 100271. [CrossRef]
27. Xu, Z.; Shi, L.; Wang, Y.; Zhang, J.; Huang, L.; Zhang, C.; Liu, S.; Zhao, P.; Liu, H.; Zhu, L.; et al. Pathological findings of COVID-19 associated with acute respiratory distress syndrome. *Lancet Respir. Med.* **2020**, *8*, 420–422. [CrossRef]
28. Guan, W.-J.; Liang, W.-H.; Zhao, Y.; Liang, H.-R.; Chen, Z.-S.; Li, Y.-M.; Liu, X.-Q.; Chen, R.-C.; Tang, C.-L.; Wang, T.; et al. Comorbidity and its impact on 1590 patients with COVID-19 in China: A nationwide analysis. *Eur. Respir. J.* **2020**, *55*, 2000547. [CrossRef]
29. Sheppard, J.P.; Nicholson, B.D.; Lee, J.; McGagh, D.; Sherlock, J.; Koshiaris, C.; Oke, J.; Jones, N.R.; Hinton, W.; Armitage, L.; et al. Association Between Blood Pressure Control and Coronavirus Disease 2019 Outcomes in 45 418 Symptomatic Patients with Hypertension. *Hypertension* **2021**, *77*, 846–855. [CrossRef]
30. Schiffrin, E.L.; Flack, J.M.; Ito, S.; Muntner, P.; Webb, R.C. Hypertension and COVID-19. *Am. J. Hypertens.* **2020**, *33*, 373–374. [CrossRef]
31. Mazucanti, C.H.; Egan, J.M. SARS-CoV-2 disease severity and diabetes: Why the connection and what is to be done? *Immun. Ageing* **2020**, *17*, 21. [CrossRef] [PubMed]
32. Tabrizi, F.M.; Rasmi, Y.; Hosseinzadeh, E.; Rezaei, S.; Balvardi, M.; Kouchari, M.R.; Ebrahimi, G. Diabetes is associated with increased mortality and disease severity in hospitalized patients with covid-19. *EXCLI J.* **2021**, *20*, 444–453. [CrossRef]
33. Chávez-Reyes, J.; Escárcega-González, C.E.; Chavira-Suárez, E.; León-Buitimea, A.; Vázquez-León, P.; Morones-Ramírez, J.R.; Villalón, C.M.; Quintanar-Stephano, A.; Marichal-Cancino, B.A. Susceptibility for Some Infectious Diseases in Patients with Diabetes: The Key Role of Glycemia. *Front. Public Health* **2021**, *9*, 559595. [CrossRef]
34. Erener, S. Diabetes, infection risk and COVID-19. *Mol. Metab.* **2020**, *39*, 101044. [CrossRef]
35. Kumar, A.; Arora, A.; Sharma, P.; Anikhindi, S.A.; Bansal, N.; Singla, V.; Khare, S.; Srivastava, A. Is diabetes mellitus associated with mortality and severity of COVID-19? A meta-analysis. *Diabetes Metab. Syndr. Clin. Res. Rev.* **2020**, *14*, 535–545. [CrossRef]
36. Hussain, A.; Bhowmik, B.; Moreira, N.C.D.V. COVID-19 and diabetes: Knowledge in progress. *Diabetes Res. Clin. Pract.* **2020**, *162*, 108142. [CrossRef]
37. Gerstein, H.C.; Thorpe, K.E.; Taylor, D.W.; Haynes, R.B. The effectiveness of hydroxychloroquine in patients with type 2 diabetes mellitus who are refractory to sulfonylureas—A randomized trial. *Diabetes Res. Clin. Pract.* **2002**, *55*, 209–219. [CrossRef]
38. Rekedal, L.R.; Massarotti, E.; Garg, R.; Bhatia, R.; Gleeson, T.; Lu, B.; Solomon, D.H. Changes in glycosylated hemoglobin after initiation of hydroxychloroquine or methotrexate treatment in diabetes patients with rheumatic diseases. *Arthritis Rheum.* **2010**, *62*, 3569–3573. [CrossRef]

39. Coate, K.C.; Cha, J.; Shrestha, S.; Wang, W.; Gonçalves, L.M.; Almaça, J.; Kapp, M.E.; Fasolino, M.; Morgan, A.; Dai, C.; et al. SARS-CoV-2 Cell Entry Factors ACE2 and TMPRSS2 Are Expressed in the Microvasculature and Ducts of Human Pancreas but Are Not Enriched in β Cells. *Cell Metab.* **2020**, *32*, 1028–1040.e4. [CrossRef]
40. Alqahtani, J.S.; Oyelade, T.; Aldhahir, A.M.; Alghamdi, S.M.; Almehmadi, M.; Alqahtani, A.S.; Quaderi, S.; Mandal, S.; Hurst, J.R. Prevalence, Severity and Mortality associated with COPD and Smoking in patients with COVID-19: A Rapid Systematic Review and Meta-Analysis. *PLoS ONE* **2020**, *15*, e0233147. [CrossRef]
41. de Lusignan, S.; Dorward, J.; Correa, A.; Jones, N.; Akinyemi, O.; Amirthalingam, G.; Andrews, N.; Byford, R.; Dabrera, G.; Elliot, A.; et al. Risk factors for SARS-CoV-2 among patients in the Oxford Royal College of General Practitioners Research and Surveillance Centre primary care network: A cross-sectional study. *Lancet Infect. Dis.* **2020**, *20*, 1034–1042. [CrossRef]
42. Increased Susceptibility to Infection in Smokers—Tobacco in Australia. Available online: https://www.tobaccoinaustralia.org.au/chapter-3-health-effects/3-9-increased-susceptibility-to-infection-in-smoke (accessed on 8 October 2021).
43. Grundy, E.J.; Suddek, T.; Filippidis, F.T.; Majeed, A.; Coronini-Cronberg, S. Smoking, SARS-CoV-2 and COVID-19: A review of reviews considering implications for public health policy and practice. *Tobacco Induced Dis.* **2020**, *18*. [CrossRef] [PubMed]
44. Paleiron, N.; Mayet, A.; Marbac, V.; Perisse, A.; Barazzutti, H.; Brocq, F.-X.; Janvier, F.; Dautzenberg, B.; Bylicki, O. Impact of Tobacco Smoking on the Risk of COVID-19: A Large Scale Retrospective Cohort Study. *Nicotine Tob. Res.* **2021**, *23*, 1398–1404. [CrossRef]
45. Jiang, H.; Zhou, Y.; Tang, W. Maintaining HIV care during the COVID-19 pandemic. *Lancet HIV* **2020**, *7*, e308–e309. [CrossRef]
46. Inciarte, A.; Gonzalez-Cordon, A.; Rojas, J.; Torres, B.; De Lazzari, E.; De La Mora, L.; Martinez-Rebollar, M.; Laguno, M.; Callau, P.; Gonzalez-Navarro, A.; et al. Clinical characteristics, risk factors, and incidence of symptomatic coronavirus disease 2019 in a large cohort of adults living with HIV: A single-center, prospective observational study. *AIDS* **2020**, *34*, 1775–1780. [CrossRef] [PubMed]
47. Elfiky, A.A. Ribavirin, Remdesivir, Sofosbuvir, Galidesivir, and Tenofovir against SARS-CoV-2 RNA dependent RNA polymerase (RdRp): A molecular docking study. *Life Sci.* **2020**, *253*, 117592. [CrossRef]
48. Etienne, N.; Karmochkine, M.; Slama, L.; Pavie, J.; Batisse, D.; Usubillaga, R.; Letembet, V.-A.; Brazille, P.; Canouï, E.; Slama, D.; et al. HIV infection and COVID-19. *AIDS* **2020**, *34*, 1771–1774. [CrossRef] [PubMed]
49. Bousquet, J.; Akdis, C.; Jutel, M.; Bachert, C.; Klimek, L.; Agache, I.; Ansotegui, I.J.; Bedbrook, A.; Bosnic-Anticevich, S.; Canonica, G.W.; et al. Intranasal corticosteroids in allergic rhinitis in COVID-19 infected patients: An ARIA-EAACI statement. *Allergy* **2020**, *75*. [CrossRef] [PubMed]
50. Conforti, C.; Giuffrida, R.; Dianzani, C.; Di Meo, N.; Zalaudek, I. COVID-19 and psoriasis: Is it time to limit treatment with immunosuppressants? A call for action. *Dermatol. Ther.* **2020**, *33*, e13298. [CrossRef]
51. Torres, T.; Puig, L. Managing Cutaneous Immune-Mediated Diseases During the COVID-19 Pandemic. *Am. J. Clin. Dermatol.* **2020**, *21*, 307–311. [CrossRef]
52. Favalli, E.G.; Ingegnoli, F.; De Lucia, O.; Cincinelli, G.; Cimaz, R.; Caporali, R. COVID-19 infection and rheumatoid arthritis: Faraway, so close! *Autoimmun. Rev.* **2020**, *19*, 102523. [CrossRef]
53. Kennedy, N.A.; Jones, G.-R.; Lamb, C.A.; Appleby, R.; Arnott, I.; Beattie, R.M.; Bloom, S.; Brooks, A.J.; Cooney, R.; Dart, R.J.; et al. British Society of Gastroenterology guidance for management of inflammatory bowel disease during the COVID-19 pandemic. *Gut* **2020**, *69*, 984–990. [CrossRef]
54. Coles, C.E.; Aristei, C.; Bliss, J.; Boersma, L.; Brunt, A.M.; Chatterjee, S.; Hanna, G.; Jagsi, R.; Person, O.K.; Kirby, A.; et al. International guidelines on radiation therapy for breast cancer during the COVID-19 pandemic. *Clin. Oncol.* **2020**, *32*, 279. [CrossRef] [PubMed]
55. Dholaria, B.; Savani, B.N. How do we plan hematopoietic cell transplant and cellular therapy with the looming COVID-19 threat? *Br. J. Haematol.* **2020**, *189*, 239–240. [CrossRef] [PubMed]
56. Zeidan, A.M.; Boddu, P.C.; Patnaik, M.M.; Bewersdorf, J.P.; Stahl, M.; Rampal, R.K.; Shallis, R.; Steensma, D.P.; Savona, M.R.; Sekeres, M.A.; et al. Special considerations in the management of adult patients with acute leukaemias and myeloid neoplasms in the COVID-19 era: Recommendations from a panel of international experts. *Lancet Haematol.* **2020**, *7*, e601–e612. [CrossRef]
57. Lu, X.; Zhang, L.; Du, H.; Zhang, J.; Li, Y.Y.; Qu, J.; Zhang, W.; Wang, Y.; Bao, S.; Li, Y.; et al. SARS-CoV-2 Infection in Children. *N. Engl. J. Med.* **2020**, *382*, 1663–1665. [CrossRef]
58. Hoehl, S.; Rabenau, H.; Berger, A.; Kortenbusch, M.; Cinatl, J.; Bojkova, D.; Behrens, P.; Böddinghaus, B.; Götsch, U.; Naujoks, F.; et al. Evidence of SARS-CoV-2 Infection in Returning Travelers from Wuhan, China. *N. Engl. J. Med.* **2020**, *382*, 1278–1280. [CrossRef]
59. Grasselli, G.; Zangrillo, A.; Zanella, A.; Antonelli, M.; Cabrini, L.; Castelli, A.; Cereda, D.; Coluccello, A.; Foti, G.; Fumagalli, R.; et al. Baseline Characteristics and Outcomes of 1591 Patients Infected with SARS-CoV-2 Admitted to ICUs of the Lombardy Region, Italy. *JAMA* **2020**, *323*, 1574. [CrossRef]
60. Lippi, G.; Mattiuzzi, C.; Sanchis-Gomar, F.; Henry, B.M. Clinical and demographic characteristics of patients dying from COVID-19 in Italy vs China. *J. Med Virol.* **2020**, *92*, 1759–1760. [CrossRef]
61. Zhu, N.; Zhang, D.; Wang, W.; Li, X.; Yang, B.; Song, J.; Zhao, X.; Huang, B.; Shi, W.; Lu, R.; et al. A Novel Coronavirus from Patients with Pneumonia in China, 2019. *N. Engl. J. Med.* **2020**, *382*, 727–733. [CrossRef]
62. Long, Q.-X.; Tang, X.-J.; Shi, Q.-L.; Li, Q.; Deng, H.-J.; Yuan, J.; Hu, J.-L.; Xu, W.; Zhang, Y.; Lv, F.-J.; et al. Clinical and immunological assessment of asymptomatic SARS-CoV-2 infections. *Nat. Med.* **2020**, *26*, 1200–1204. [CrossRef]

63. Okba, N.M.; Müller, M.A.; Li, W.; Wang, C.; GeurtsvanKessel, C.H.; Corman, V.M.; Lamers, M.M.; Sikkema, R.S.; de Bruin, E.; Chandler, F.D.; et al. Severe Acute Respiratory Syndrome Coronavirus 2−Specific Antibody Responses in Coronavirus Disease Patients. *Emerg. Infect. Dis.* **2020**, *26*, 1478–1488. [CrossRef]
64. Boechat, J.L.; Chora, I.; Morais, A.; Delgado, L. The immune response to SARS-CoV-2 and COVID-19 immunopathology—Current perspectives. *Pulmonology* **2021**, *27*, 423–437. [CrossRef] [PubMed]
65. Matricardi, P.M.; Negro, R.W.D.; Nisini, R. The first, holistic immunological model of COVID-19: Implications for prevention, diagnosis, and public health measures. *Pediatr. Allergy Immunol.* **2020**, *31*, 454–470. [CrossRef] [PubMed]
66. Cheng, Y.; Cheng, Y.; Cheng, G.; Chui, C.H.; Lau, F.Y.; Chan, P.K.S.; Ng, M.H.L.; Sung, J.J.Y.; Wong, R.S.M. ABO Blood Group and Susceptibility to Severe Acute Respiratory Syndrome. *JAMA* **2005**, *293*, 1447–1451. [CrossRef]
67. Zhao, J.; Yang, Y.; Huang, H.; Li, D.; Gu, D.; Lu, X.; Zhang, Z.; Liu, L.; Liu, T.; Liu, Y.; et al. Relationship Between the ABO Blood Group and the Coronavirus Disease 2019 (COVID-19) Susceptibility. *Clin. Infect. Dis.* **2020**, *73*, 328–331. [CrossRef] [PubMed]
68. Zhou, Y.; Lu, K.; Pfefferle, S.; Bertram, S.; Glowacka, I.; Drosten, C.; Pöhlmann, S.; Simmons, G. A Single Asparagine-Linked Glycosylation Site of the Severe Acute Respiratory Syndrome Coronavirus Spike Glycoprotein Facilitates Inhibition by Mannose-Binding Lectin through Multiple Mechanisms. *J. Virol.* **2010**, *84*, 8753–8764. [CrossRef] [PubMed]
69. Watanabe, Y.; Allen, J.D.; Wrapp, D.; McLellan, J.S.; Crispin, M. Site-specific glycan analysis of the SARS-CoV-2 spike. *Science* **2020**, *369*, 330–333. [CrossRef]
70. Tomaiuolo, R.; Ruocco, A.; Salapete, C.; Carru, C.; Baggio, G.; Franceschi, C.; Zinellu, A.; Vaupel, J.; Bellia, C.; Sasso, B.L.; et al. Activity of mannose-binding lectin in centenarians. *Aging Cell* **2012**, *11*, 394–400. [CrossRef]
71. Boudewijns, R.; Thibaut, H.J.; Kaptein, S.J.F.; Li, R.; Vergote, V.; Seldeslachts, L.; Van Weyenbergh, J.; De Keyzer, C.; Bervoets, L.; Sharma, S.; et al. STAT2 signaling restricts viral dissemination but drives severe pneumonia in SARS-CoV-2 infected hamsters. *Nat. Commun.* **2020**, *11*, 5838. [CrossRef]
72. Kopp, W. Diet-induced hyperinsulinemia as a key factor in the etiology of both benign prostatic hyperplasia and essential hypertension? *Nutr. Metab. Insights* **2018**, *11*, 1178638818773072. [CrossRef]
73. Kopp, W. How Western Diet And Lifestyle Drive The Pandemic Of Obesity And Civilization Diseases. *Diabetes Metab. Syndr. Obesity Targets Ther.* **2019**, *12*, 2221–2236. [CrossRef]
74. Paoli, A.; Gorini, S.; Caprio, M. The dark side of the spoon—Glucose, ketones and COVID-19: A possible role for ketogenic diet? *J. Transl. Med.* **2020**, *18*, 441. [CrossRef]
75. Clemente-Suárez, V.; Ramos-Campo, D.; Mielgo-Ayuso, J.; Dalamitros, A.; Nikolaidis, P.; Hormeño-Holgado, A.; Tornero-Aguilera, J. Nutrition in the Actual COVID-19 Pandemic. A Narrative Review. *Nutrients* **2021**, *13*, 1924. [CrossRef] [PubMed]
76. Lange, K.W. Food science and COVID-19. *Food Sci. Hum. Wellness* **2020**, *10*, 1–5. [CrossRef]
77. Dhok, A.; Butola, L.K.; Anjankar, A.; Shinde, A.D.R.; Kute, P.; Jha, R.K. Role of Vitamins and Minerals in Improving Immunity during Covid-19 Pandemic—A Review. *J. Evol. Med. Dent. Sci.* **2020**, *9*, 2296–2300. [CrossRef]
78. Kumar, P.; Kumar, M.; Bedi, O.; Gupta, M.; Kumar, S.; Jaiswal, G.; Rahi, V.; Yedke, N.G.; Bijalwan, A.; Sharma, S.; et al. Role of vitamins and minerals as immunity boosters in COVID-19. *Inflammopharmacology* **2021**, *29*, 1001–1016. [CrossRef] [PubMed]
79. Suwannasom, N.; Kao, I.; Pruß, A.; Georgieva, R.; Bäumler, H. Riboflavin: The Health Benefits of a Forgotten Natural Vitamin. *Int. J. Mol. Sci.* **2020**, *21*, 950. [CrossRef] [PubMed]
80. Cámara, M.; Sánchez-Mata, M.; Fernández-Ruiz, V.; Cámara, R.; Cebadera, E.; Domínguez, L. A Review of the Role of Micronutrients and Bioactive Compounds on Immune System Supporting to Fight against the COVID-19 Disease. *Foods* **2021**, *10*, 1088. [CrossRef]
81. Di Matteo, G.; Spano, M.; Grosso, M.; Salvo, A.; Ingallina, C.; Russo, M.; Ritieni, A.; Mannina, L. Food and COVID-19: Preventive/Co-therapeutic Strategies Explored by Current Clinical Trials and in Silico Studies. *Foods* **2020**, *9*, 1036. [CrossRef] [PubMed]
82. Shakoor, H.; Feehan, J.; Al Dhaheri, A.S.; Ali, H.I.; Platat, C.; Ismail, L.C.; Apostolopoulos, V.; Stojanovska, L. Immune-boosting role of vitamins D, C, E, zinc, selenium and omega-3 fatty acids: Could they help against COVID-19? *Maturitas* **2020**, *143*, 1–9. [CrossRef]
83. Yang, F.; Zhang, Y.; Tariq, A.; Jiang, X.; Ahmed, Z.; Zhihao, Z.; Idrees, M.; Azizullah, A.; Adnan, M.; Bussmann, R.W. Food as medicine: A possible preventive measure against coronavirus disease (COVID-19). *Phytotherapy Res.* **2020**, *34*, 3124–3136. [CrossRef]
84. Kandeil, A.; Mostafa, A.; Kutkat, O.; Moatasim, Y.; Al-Karmalawy, A.A.; Rashad, A.A.; Kayed, A.E.; Kayed, A.E.; El-Shesheny, R.; Kayali, G.; et al. Bioactive Polyphenolic Compounds Showing Strong Antiviral Activities against Severe Acute Respiratory Syndrome Coronavirus 2. *Pathogens* **2021**, *10*, 758. [CrossRef]
85. Khubber, S.; Hashemifesharaki, R.; Mohammadi, M.; Gharibzahedi, S.M.T. Garlic (*Allium sativum* L.): A potential unique therapeutic food rich in organosulfur and flavonoid compounds to fight with COVID-19. *Nutr. J.* **2020**, *19*, 124. [CrossRef]
86. Shaldam, M.A.; Yahya, G.; Mohamed, N.H.; Abdel-Daim, M.M.; Al Naggar, Y. In silico screening of potent bioactive compounds from honeybee products against COVID-19 target enzymes. *Environ. Sci. Pollut. Res.* **2021**, *28*, 40507–40514. [CrossRef]
87. Chojnacka, K.; Witek-Krowiak, A.; Skrzypczak, D.; Mikula, K.; Młynarz, P. Phytochemicals containing biologically active polyphenols as an effective agent against Covid-19-inducing coronavirus. *J. Funct. Foods* **2020**, *73*, 104146. [CrossRef] [PubMed]
88. Bhushan, I.; Sharma, M.; Mehta, M.; Badyal, S.; Sharma, V.; Sharma, I.; Singh, H.; Sistla, S. Bioactive compounds and probiotics—A ray of hope in COVID-19 management. *Food Sci. Hum. Wellness* **2021**, *10*, 131–140. [CrossRef]

89. Phytochemicals Candidates as Promising Preventives and/or Curatives for COVID-19 Infection: A Brief Review. Available online: https://www.heighpubs.org/hjbm/ibm-aid1019.php (accessed on 31 July 2021).
90. Li, S.; Cheng, C.-S.; Zhang, C.; Tang, G.-Y.; Tan, H.-Y.; Chen, H.-Y.; Wang, N.; Lai, A.Y.-K.; Feng, Y. Edible and Herbal Plants for the Prevention and Management of COVID-19. *Front. Pharmacol.* **2021**, *12*, 656103. [CrossRef] [PubMed]
91. Tallei, T.E.; Tumilaar, S.G.; Niode, N.J.; Fatimawali; Kepel, B.J.; Idroes, R.; Effendi, Y.; Alam Sakib, S.; Bin Emran, T. Potential of Plant Bioactive Compounds as SARS-CoV-2 Main Protease (Mpro) and Spike (S) Glycoprotein Inhibitors: A Molecular Docking Study. *Scientifica* **2020**, *2020*, 6307457. [CrossRef] [PubMed]
92. Mattioli, A.V.; Puviani, M.B.; Nasi, M.; Farinetti, A. COVID-19 pandemic: The effects of quarantine on cardiovascular risk. *Eur. J. Clin. Nutr.* **2020**, *74*, 852–855. [CrossRef] [PubMed]
93. Bennett, G.; Young, E.; Butler, I.; Coe, S. The Impact of Lockdown During the COVID-19 Outbreak on Dietary Habits in Various Population Groups: A Scoping Review. *Front. Nutr.* **2021**, *8*, 626432. [CrossRef] [PubMed]
94. Fernández-Quintela, A.; Milton-Laskibar, I.; Trepiana, J.; Gómez-Zorita, S.; Kajarabille, N.; Léniz, A.; González, M.; Portillo, M.P. Key Aspects in Nutritional Management of COVID-19 Patients. *J. Clin. Med.* **2020**, *9*, 2589. [CrossRef] [PubMed]
95. Hwang, T.-J.; Rabheru, K.; Peisah, C.; Reichman, W.; Ikeda, M. Loneliness and social isolation during the COVID-19 pandemic. *Int. Psychogeriatr.* **2020**, *32*, 1217–1220. [CrossRef]
96. Pérez-Rodrigo, C.; Citores, M.G.; Bárbara, G.H.; Ruiz-Litago, F.; Sáenz, L.C.; Arija, V.; López-Sobaler, A.; de Victoria, E.M.; Ortega, R.; Partearroyo, T.; et al. Patterns of Change in Dietary Habits and Physical Activity during Lockdown in Spain due to the COVID-19 Pandemic. *Nutrients* **2021**, *13*, 300. [CrossRef]
97. Sánchez-Sánchez, E.; Ramírez-Vargas, G.; Avellaneda-López, Y.; Orellana-Pecino, J.I.; García-Marín, E.; Díaz-Jimenez, J. Eating Habits and Physical Activity of the Spanish Population during the COVID-19 Pandemic Period. *Nutrients* **2020**, *12*, 2826. [CrossRef]
98. Molina-Montes, E.; Uzhova, I.; Verardo, V.; Artacho, R.; García-Villanova, B.; Guerra-Hernández, E.J.; Kapsokefalou, M.; Malisova, O.; Vlassopoulos, A.; Katidi, A.; et al. Impact of COVID-19 confinement on eating behaviours across 16 European countries: The COVIDiet cross-national study. *Food Qual. Prefer.* **2021**, *93*, 104231.
99. Izzo, L.; Santonastaso, A.; Cotticelli, G.; Federico, A.; Pacifico, S.; Castaldo, L.; Colao, A.; Ritieni, A. An Italian Survey on Dietary Habits and Changes during the COVID-19 Lockdown. *Nutrients* **2021**, *13*, 1197. [CrossRef]
100. Grant, F.; Scalvedi, M.; Scognamiglio, U.; Turrini, A.; Rossi, L. Eating Habits during the COVID-19 Lockdown in Italy: The Nutritional and Lifestyle Side Effects of the Pandemic. *Nutrients* **2021**, *13*, 2279. [CrossRef]
101. Di Renzo, L.; Gualtieri, P.; Pivari, F.; Soldati, L.; Attinà, A.; Cinelli, G.; Leggeri, C.; Caparello, G.; Barrea, L.; Scerbo, F.; et al. Eating habits and lifestyle changes during COVID-19 lockdown: An Italian survey. *J. Transl. Med.* **2020**, *18*, 1–15. [CrossRef] [PubMed]
102. Scarmozzino, F.; Visioli, F. Covid-19 and the Subsequent Lockdown Modified Dietary Habits of Almost Half the Population in an Italian Sample. *Foods* **2020**, *9*, 675. [CrossRef] [PubMed]
103. Mascherini, G.; Catelan, D.; Pellegrini-Giampietro, D.E.; Petri, C.; Scaletti, C.; Gulisano, M. Changes in physical activity levels, eating habits and psychological well-being during the Italian COVID-19 pandemic lockdown: Impact of socio-demographic factors on the Florentine academic population. *PLoS ONE* **2021**, *16*, e0252395. [CrossRef] [PubMed]
104. Sofo, A.; Sofo, A. Converting Home Spaces into Food Gardens at the Time of Covid-19 Quarantine: All the Benefits of Plants in this Difficult and Unprecedented Period. *Hum. Ecol.* **2020**, *48*, 131–139. [CrossRef] [PubMed]
105. Marty, L.; de Lauzon-Guillain, B.; Labesse, M.; Nicklaus, S. Food choice motives and the nutritional quality of diet during the COVID-19 lockdown in France. *Appetite* **2020**, *157*, 105005. [CrossRef]
106. Rolland, B.; Haesebaert, F.; Zante, E.; Benyamina, A.; Haesebaert, J.; Franck, N. Global Changes and Factors of Increase in Caloric/Salty Food Intake, Screen Use, and Substance Use during the Early COVID-19 Containment Phase in the General Population in France: Survey Study. *JMIR Public Health Surveill.* **2020**, *6*, e19630. [CrossRef]
107. Cherikh, F.; Frey, S.; Bel, C.; Attanasi, G.; Alifano, M.; Iannelli, A. Behavioral Food Addiction During Lockdown: Time for Awareness, Time to Prepare the Aftermath. *Obes. Surg.* **2020**, *30*, 3585–3587. [CrossRef] [PubMed]
108. Philippe, K.; Chabanet, C.; Issanchou, S.; Monnery-Patris, S. Child eating behaviors, parental feeding practices and food shopping motivations during the COVID-19 lockdown in France: (How) did they change? *Appetite* **2021**, *161*, 105132. [CrossRef] [PubMed]
109. Papandreou, C.; Arija, V.; Aretouli, E.; Tsilidis, K.K.; Bulló, M. Comparing eating behaviours, and symptoms of depression and anxiety between Spain and Greece during the COVID-19 outbreak: Cross-sectional analysis of two different confinement strategies. *Eur. Eat. Disord. Rev.* **2020**, *28*. [CrossRef]
110. Panagiotakos, D.; Kosti, R.I.; Pitsavos, C. How will the way we live look different in the wake of the COVID-19 pandemic? A nutrition survey in Greece. *Nutr. Health* **2021**. [CrossRef] [PubMed]
111. Grigoriou, E.; Trovas, G.; Papaioannou, N.; Dontas, I.; Makris, K.; Apostolou-Karampelis, K.; Dedoussis, G. Dietary Patterns of Greek Adults and Their Associations with Serum Vitamin D Levels and Heel Quantitative Ultrasound Parameters for Bone Health. *Nutrients* **2020**, *12*, 123. [CrossRef]
112. Giacalone, D.; Frøst, M.B.; Rodríguez-Pérez, C. Reported Changes in Dietary Habits During the COVID-19 Lockdown in the Danish Population: The Danish COVIDiet Study. *Front. Nutr.* **2020**, *7*, 592112. [CrossRef]
113. Janssen, M.; Chang, B.P.I.; Hristov, H.; Pravst, I.; Profeta, A.; Millard, J. Changes in Food Consumption during the COVID-19 Pandemic: Analysis of Consumer Survey Data From the First Lockdown Period in Denmark, Germany, and Slovenia. *Front. Nutr.* **2021**, *8*, 635859. [CrossRef]

114. Błaszczyk-Bębenek, E.; Jagielski, P.; Bolesławska, I.; Jagielska, A.; Nitsch-Osuch, A.; Kawalec, P. Nutrition Behaviors in Polish Adults before and during COVID-19 Lockdown. *Nutrients* **2020**, *12*, 3084. [CrossRef]
115. Sidor, A.; Rzymski, P. Dietary Choices and Habits during COVID-19 Lockdown: Experience from Poland. *Nutrients* **2020**, *12*, 1657. [CrossRef]
116. Yu, B.; Zhang, D.; Yu, W.; Luo, M.; Yang, S.; Jia, P. Impacts of lockdown on dietary patterns among youths in China: The COVID-19 Impact on Lifestyle Change Survey. *Public Health Nutr.* **2021**, *24*, 3221–3232. [CrossRef]
117. Naja, F.; Hamadeh, R. Nutrition amid the COVID-19 pandemic: A multi-level framework for action. *Eur. J. Clin. Nutr.* **2020**, *74*, 1117–1121. [CrossRef]
118. Bin Zarah, A.; Enriquez-Marulanda, J.; Andrade, J.M. Relationship between Dietary Habits, Food Attitudes and Food Security Status among Adults Living within the United States Three Months Post-Mandated Quarantine: A Cross-Sectional Study. *Nutrients* **2020**, *12*, 3468. [CrossRef] [PubMed]
119. Pertuz-Cruz, S.L.; Molina-Montes, E.; Rodríguez-Pérez, C.; Guerra-Hernández, E.J.; de Rangel, O.P.C.; Artacho, R.; Verardo, V.; Ruiz-Lopez, M.D.; García-Villanova, B. Exploring Dietary Behavior Changes Due to the COVID-19 Confinement in Colombia: A National and Regional Survey Study. *Front. Nutr.* **2021**, *8*, 644800. [CrossRef] [PubMed]
120. Skotnicka, M.; Karwowska, K.; Kłobukowski, F.; Wasilewska, E.; Małgorzewicz, S. Dietary Habits before and during the COVID-19 Epidemic in Selected European Countries. *Nutrients* **2021**, *13*, 1690. [CrossRef] [PubMed]
121. Bakaloudi, D.R.; Jeyakumar, D.T.; Jayawardena, R.; Chourdakis, M. The impact of COVID-19 lockdown on snacking habits, fast-food and alcohol consumption: A systematic review of the evidence. *Clin. Nutr.* **2021**. [CrossRef] [PubMed]
122. Bakhsh, M.A.; Khawandanah, J.; Naaman, R.K.; Alashmali, S. The impact of COVID-19 quarantine on dietary habits and physical activity in Saudi Arabia: A cross-sectional study. *BMC Public Health* **2021**, *21*, 1487. [CrossRef]
123. Huber, B.C.; Steffen, J.; Schlichtiger, J.; Brunner, S. Altered nutrition behavior during COVID-19 pandemic lockdown in young adults. *Eur. J. Nutr.* **2020**, *60*, 2593–2602. [CrossRef] [PubMed]
124. Li, S.; Jiang, W.; Zheng, C.; Shao, D.; Liu, Y.; Huang, S.; Han, J.; Ding, J.; Tao, Y.; Li, M. Oral delivery of bacteria: Basic principles and biomedical applications. *J. Control. Release* **2020**, *327*, 801–833. [CrossRef] [PubMed]
125. Serrem, K.; Illés, C.B.; Serrem, C.; Atubukha, B.; Dunay, A. Food safety and sanitation challenges of public university students in a developing country. *Food Sci Nutr.* **2021**, *9*, 4287–4297.
126. Rodríguez-Pérez, C.; Molina-Montes, E.; Verardo, V.; Artacho, R.; García-Villanova, B.; Guerra-Hernández, E.J.; Ruíz-López, M.D. Changes in Dietary Behaviours during the COVID-19 Outbreak Confinement in the Spanish COVIDiet Study. *Nutrients* **2020**, *12*, 1730. [CrossRef]
127. Min, S.; Xiang, C.; Zhang, X.-H. Impacts of the COVID-19 pandemic on consumers' food safety knowledge and behavior in China. *J. Integr. Agric.* **2020**, *19*, 2926–2936. [CrossRef]
128. Lange, K.W.; Nakamura, Y. Lifestyle factors in the prevention of COVID-19. *Glob. Health J.* **2020**, *4*, 146–152. [CrossRef]
129. Mentella, M.; Scaldaferri, F.; Gasbarrini, A.; Miggiano, G. The Role of Nutrition in the COVID-19 Pandemic. *Nutrients* **2021**, *13*, 1093. [CrossRef]
130. Woods, J.A.; Hutchinson, N.T.; Powers, S.K.; Roberts, W.O.; Gomez-Cabrera, M.C.; Radak, Z.; Berkes, I.; Boros, A.; Boldogh, I.; Leeuwenburgh, C.; et al. The COVID-19 pandemic and physical activity. *Sports Med. Health Sci.* **2020**, *2*, 55–64. [CrossRef] [PubMed]
131. Chick, J. Alcohol and COVID-19. *Alcohol Alcohol.* **2020**, *55*, 341–342. [CrossRef] [PubMed]
132. Vardavas, C.I.; Nikitara, K. COVID-19 and smoking: A systematic review of the evidence. *Tob. Induc. Dis.* **2020**, *18*, 20. [CrossRef] [PubMed]
133. Al-Rashed, F.; Sindhu, S.; Al Madhoun, A.; Alghaith, A.; Azim, R.; Al-Mulla, F.; Ahmad, R. Short Sleep Duration and Its Association with Obesity and Other Metabolic Risk Factors in Kuwaiti Urban Adults. *Nat. Sci. Sleep* **2021**, *13*, 1225. [CrossRef]
134. Silva, E.D.S.M.E.; Ono, B.H.V.S.; Souza, J.C. Sleep and immunity in times of COVID-19. *Revista da Associação Médica Brasileira* **2020**, *66*, 143–147. [CrossRef]
135. Fidancı, I.; Aksoy, H.; Taci, D.Y.; Fidancı, I.; Başer, D.A.; Cankurtaran, M. Evaluation of the effect of the COVID-19 pandemic on sleep disorders and nutrition in children. *Int. J. Clin. Pract.* **2021**, e14170. [CrossRef]
136. Kingsbury, J.; Lake, R.; Hewitt, J.; Smit, E.; King, N. Potential for foodborne transmission of Covid-19: Literature review update. *Update* **2020**, 1–78. Available online: https://mia.co.nz/assets/Covid-19/2021-01-14-NZFSSRC-COVID-19-foodborne-transmission-final-Kingsbury.pdf (accessed on 31 August 2021).
137. Han, J.; Zhang, X.; He, S.; Jia, P. Can the coronavirus disease be transmitted from food? A review of evidence, risks, policies and knowledge gaps. *Environ. Chem. Lett.* **2020**, *19*, 5–16. [CrossRef] [PubMed]
138. Olaimat, A.N.; Shahbaz, H.M.; Fatima, N.; Munir, S.; Holley, R.A. Food Safety During and After the Era of COVID-19 Pandemic. *Front. Microbiol.* **2020**, *11*, 1854. [CrossRef] [PubMed]
139. Ceylan, Z.; Meral, R.; Cetinkaya, T. Relevance of SARS-CoV-2 in food safety and food hygiene: Potential preventive measures, suggestions and nanotechnological approaches. *VirusDisease* **2020**, *31*, 154–160. [CrossRef]
140. Ceniti, C.; Tilocca, B.; Britti, D.; Santoro, A.; Costanzo, N. Food Safety Concerns in "COVID-19 Era". *Microbiol. Res.* **2021**, *12*, 6. [CrossRef]
141. Mardones, F.O.; Rich, K.M.; Boden, L.A.; Moreno-Switt, A.I.; Caipo, M.L.; Zimin-Veselkoff, N.; Alateeqi, A.M.; Baltenweck, I. The COVID-19 Pandemic and Global Food Security. *Front. Vet. Sci.* **2020**, *7*, 578508. [CrossRef]

142. Chowdhury, T.; Sarkar, A.; Paul, S.K.; Moktadir, A. A case study on strategies to deal with the impacts of COVID-19 pandemic in the food and beverage industry. *Oper. Manag. Res.* **2020**, 1–13. [CrossRef]
143. Nakat, Z.; Bou-Mitri, C. COVID-19 and the food industry: Readiness assessment. *Food Control* **2020**, *121*, 107661. [CrossRef]
144. Hossain, S.T. Impacts of COVID-19 on the Agri-food Sector: Food Security Policies of Asian Productivity Organization Members. *J. Agric. Sci. Sri Lanka* **2020**, *15*, 116. [CrossRef]
145. Telukdarie, A.; Munsamy, M.; Mohlala, P. Analysis of the Impact of COVID-19 on the Food and Beverages Manufacturing Sector. *Sustainability* **2020**, *12*, 9331. [CrossRef]
146. Haider, N.; Osman, A.Y.; Gadzekpo, A.; Akipede, G.O.; Asogun, D.; Ansumana, R.; Lessells, R.J.; Khan, P.; Hamid, M.M.A.; Yeboah-Manu, D.; et al. Lockdown measures in response to COVID-19 in nine sub-Saharan African countries. *BMJ Glob. Health* **2020**, *5*, e003319. [CrossRef]
147. Madeira, A.; Palrão, T.; Mendes, A. The Impact of Pandemic Crisis on the Restaurant Business. *Sustainability* **2020**, *13*, 40. [CrossRef]
148. Aday, S.; Aday, M.S. Impact of COVID-19 on the food supply chain. *Food Qual. Saf.* **2020**, *4*, 167–180. [CrossRef]
149. Pexara, A.; Govaris, A. Foodborne Viruses and Innovative Non-Thermal Food-Processing Technologies. *Foods* **2020**, *9*, 1520. [CrossRef] [PubMed]
150. Espinosa, M.F.; Sancho, A.N.; Mendoza, L.M.; Mota, C.R.; Verbyla, M.E. Systematic review and meta-analysis of time-temperature pathogen inactivation. *Int. J. Hyg. Environ. Health* **2020**, *230*, 113595. [CrossRef] [PubMed]
151. Anelich, L.E.C.M.; Lues, R.; Farber, J.M.; Parreira, V.R. SARS-CoV-2 and Risk to Food Safety. *Front. Nutr.* **2020**, *7*, 580551. [CrossRef] [PubMed]
152. Gamble, A.; Fischer, R.J.; Morris, D.H.; Yinda, C.K.; Munster, V.J.; Lloyd-Smith, J.O. Heat-Treated Virus Inactivation Rate Depends Strongly on Treatment Procedure: Illustration with SARS-CoV-2. *Appl. Environ. Microbiol.* **2021**, *87*, AEM0031421. [CrossRef]
153. Lugo-Morin, D.R. Global Food Security in a Pandemic: The Case of the New Coronavirus (COVID-19). *World* **2020**, *1*, 13. [CrossRef]
154. Bakalis, S.; Valdramidis, V.P.; Argyropoulos, D.; Ahrne, L.; Chen, J.; Cullen, P.; Cummins, E.; Datta, A.K.; Emmanouilidis, C.; Foster, T.; et al. Perspectives from CO+RE: How COVID-19 changed our food systems and food security paradigms. *Curr. Res. Food Sci.* **2020**, *3*, 166–172. [CrossRef]
155. Jafri, A.; Mathe, N.; Aglago, E.K.; Konyole, S.O.; Ouedraogo, M.; Audain, K.; Zongo, U.; Laar, A.K.; Johnson, J.; Sanou, D. Food availability, accessibility and dietary practices during the COVID-19 pandemic: A multi-country survey. *Public Health Nutr.* **2021**, *24*, 1798–1805. [CrossRef] [PubMed]
156. Rejeb, A.; Rejeb, K.; Keogh, J.G. Covid-19 and the food chain? Impacts and future research trends. *Logforum* **2020**, *16*, 475–485. [CrossRef]
157. Mouloudj, K.; Bouarar, A.C.; Fechit, H. The Impact of COVID-19 Pandemic on Food Security. *Les Cah. du CREAD* **2020**, *36*, 159–184. Available online: https://www.asjp.cerist.dz/en/article/120912 (accessed on 31 July 2021).
158. Nosratabadi, S.; Khazami, N.; Abdallah, M.; Lackner, Z.; Band, S.S.; Mosavi, A.; Mako, C. Social Capital Contributions to Food Security: A Comprehensive Literature Review. *Foods* **2020**, *9*, 1650. [CrossRef]
159. Chiwona-Karltun, L.; Amuakwa-Mensah, F.; Wamala-Larsson, C.; Amuakwa-Mensah, S.; Abu Hatab, A.; Made, N.; Taremwa, N.K.; Melyoki, L.; Rutashobya, L.K.; Madonsela, T.; et al. COVID-19: From health crises to food security anxiety and policy implications. *Ambio* **2021**, *50*, 794–811. [CrossRef]
160. COVID-19 and Food Safety: Guidance for Food Businesses: Interim Guidance. Available online: https://www.who.int/publications/i/item/covid-19-and-food-safety-guidance-for-food-businesses (accessed on 31 July 2021).
161. Butterick, M.; Charlwood, A. HRM and the COVID-19 pandemic: How can we stop making a bad situation worse? *Hum. Resour. Manag. J.* **2021**, 1–10. [CrossRef]
162. Dyal, J.W.; Grant, M.P.; Broadwater, K.; Bjork, A.; Waltenburg, M.A.; Gibbins, J.D.; Hale, C.; Silver, M.; Fischer, M.; Steinberg, J.; et al. COVID-19 Among Workers in Meat and Poultry Processing Facilities —19 States, April 2020. *MMWR. Morb. Mortal. Wkly. Rep.* **2020**, *69*. [CrossRef] [PubMed]
163. Leone, L.A.; Fleischhacker, S.; Anderson-Steeves, B.; Harper, K.; Winkler, M.; Racine, E.; Baquero, B.; Gittelsohn, J. Healthy Food Retail during the COVID-19 Pandemic: Challenges and Future Directions. *Int. J. Environ. Res. Public Health* **2020**, *17*, 7397. [CrossRef] [PubMed]
164. Zuber, S.; Brüssow, H. COVID 19: Challenges for virologists in the food industry. *Microb. Biotechnol.* **2020**, *13*, 1689–1701. [CrossRef]
165. Shahbaz, M.; Bilal, M.; Moiz, A.; Zubair, S.; Iqbal, H.M. Food Safety and COVID-19: Precautionary Measures to Limit the Spread of Coronavirus at Food Service and Retail Sector. *J. Pure Appl. Microbiol.* **2020**, *14*, 749–756. [CrossRef]
166. Al Amin, M.; Arefin, M.S.; Alam, M.R.; Ahammad, T.; Hoque, M.R. Using Mobile Food Delivery Applications during COVID-19 Pandemic: An Extended Model of Planned Behavior. *J. Food Prod. Mark.* **2021**, *27*, 105–126. [CrossRef]
167. Nguyen, T.H.D.; Vu, D.C. Food Delivery Service During Social Distancing: Proactively Preventing or Potentially Spreading Coronavirus Disease–2019? *Disaster Med. Public Health Prep.* **2020**, *14*, e9–e10. [CrossRef] [PubMed]
168. Ouabdesselam, L.; Sayad, A. Food Packaging and COVID-19. *Eur. J. Basic Med Sci.* **2021**, *11*, 23–26. [CrossRef]
169. Kitz, R.; Walker, T.; Charlebois, S.; Music, J. Food packaging during the COVID-19 pandemic: Consumer perceptions. *Int. J. Consum. Stud.* **2021**. [CrossRef] [PubMed]

170. Riddell, S.; Goldie, S.; Hill, A.; Eagles, D.; Drew, T.W. The effect of temperature on persistence of SARS-CoV-2 on common surfaces. *Virol. J.* **2020**, *17*, 145. [CrossRef]
171. Xue, X.; Ball, J.K.; Alexander, C.; Alexander, M.R. All Surfaces Are Not Equal in Contact Transmission of SARS-CoV-2. *Matter* **2020**, *3*, 1433–1441. [CrossRef]
172. Fiorillo, L.; Cervino, G.; Matarese, M.; D'Amico, C.; Surace, G.; Paduano, V.; Fiorillo, M.T.; Moschella, A.; La Bruna, A.; Romano, G.L.; et al. COVID-19 Surface Persistence: A Recent Data Summary and Its Importance for Medical and Dental Settings. *Int. J. Environ. Res. Public Health* **2020**, *17*, 3132. [CrossRef]
173. Kampf, G.; Todt, D.; Pfaender, S.; Steinmann, E. Persistence of coronaviruses on inanimate surfaces and their inactivation with biocidal agents. *J. Hosp. Infect.* **2020**, *104*, 246–251. [CrossRef]
174. Chi, Y.; Wang, Q.; Chen, G.; Zheng, S. The Long-Term Presence of SARS-CoV-2 on Cold-Chain Food Packaging Surfaces Indicates a New COVID-19 Winter Outbreak: A Mini Review. *Front. Public Health* **2021**, *9*, 650493. [CrossRef]
175. Choi, H.; Chatterjee, P.; Coppin, J.D.; Martel, J.A.; Hwang, M.; Jinadatha, C.; Sharma, V.K. Current understanding of the surface contamination and contact transmission of SARS-CoV-2 in healthcare settings. *Environ. Chem. Lett.* **2021**, *19*, 1935–1944. [CrossRef]
176. Saha, N.C. COVID 19—Its Impact on Packaging Research. *J. Packag. Technol. Res.* **2020**, *4*, 217–218. [CrossRef] [PubMed]

 nutrients

Article

Impact of School Closures, Precipitated by COVID-19, on Weight and Weight-Related Risk Factors among Schoolteachers: A Cross-Sectional Study

Jill R. Silverman [1,*] and Branden Z. Wang [2]

1 Department of Nutrition Science and Wellness, Farmingdale State College, 2350 Broadhollow Road, Farmingdale, NY 11735, USA
2 Institute of Human Nutrition, Columbia University Irving Medical Center, New York, NY 10027, USA; bw2642@cumc.columbia.edu
* Correspondence: silverj@farmingdale.edu

Abstract: The school closures, precipitated by the COVID-19 pandemic, required teachers to convert their entire classroom curricula to online formats, taught from home. This shift to a more sedentary teaching environment, coupled with the stresses related to the pandemic, may correlate with weight gain. In total, 52% of study participants reported weight gain, with a higher prevalence observed among kindergarten and elementary school teachers when compared to high school teachers ($p < 0.05$). Deviations in physical activity, emotional eating, and dietary patterns were assessed among 129 teachers (using the Leisure Time Exercise Questionnaire, the Dutch Eating Behavioral Questionnaire, and a short-form Food Frequency Questionnaire, respectively) to uncover possible associations with the observed weight gain. Increases in sedentariness ($p < 0.005$), emotional eating ($p < 0.001$), the consumption of potatoes, fries, breads, cheese, cake ($p < 0.05$), chips, candy, ice-cream, and soft drinks ($p < 0.005$) were all positively correlated with weight gain. Decreases in exercise frequency ($p < 0.001$), and the consumption of fruits ($p < 0.05$) and beans ($p < 0.005$), were also positively correlated with weight gain. Weight gain, observed among teachers during school closures, was associated with changes in diet, emotional eating and physical activity.

Keywords: coronavirus; exercise; emotional eating; pandemic; quarantine; questionnaire; sedentariness; New York

Citation: Silverman, J.R.; Wang, B.Z. Impact of School Closures, Precipitated by COVID-19, on Weight and Weight-Related Risk Factors among Schoolteachers: A Cross-Sectional Study. *Nutrients* **2021**, *13*, 2723. https://doi.org/10.3390/nu13082723

Academic Editor: William B. Grant and Ronan Lordan

Received: 15 July 2021
Accepted: 5 August 2021
Published: 7 August 2021

1. Introduction

The coronavirus pandemic has caused significant disruption in the lifestyles of Americans. At the beginning of March 2020, there was a huge increase in the number of confirmed cases, hospitalizations, and deaths resulting from the virus [1], precipitating severe restrictions nationwide [2]. In New York state, mandatory shelter-in-place guidelines were implemented, and people were only allowed to leave their homes for food or medical reasons. As with many other occupations, schoolteachers were required to make the rapid and unplanned move to working from home [3]. For teachers, who spend most of their days standing in classrooms interacting with their students, this meant adapting lesson plans to an online format, significantly increasing the amount of time spent at home in front of their computer screens [4].

The impact of the mandatory shelter-in-place restrictions on changes in weight, dietary habits, and physical activity have been demonstrated by several studies [5–8]. In addition, studies have shown that an increase in unstructured time, a result of the COVID-19 school closures, can result in an increase in weight gain-related risk factors among children [9,10]. However, the effect of these school closures on teachers has not been determined. It was hypothesized that the unstructured time and stress associated with the shift to a virtual classroom and working from home, might lead to deviations in food intake, activity levels, and emotional eating, and that these changes may correlate to weight gain among teachers.

Time spent sitting is greatly associated with increased rates of obesity [11]. During a traditional school year, teachers spend, on average, 12–16 h a day devoted to the direct instruction of students, curriculum development, and administrative duties [12]. With the closing of schools and switch to teaching virtually from home, teachers were no longer actively moving within the school building. Therefore, the change in the teaching environment may have resulted in an increase in sedentary activities and possible weight gain.

Furthermore, the increase in stress due to the pandemic and subsequent school closings may have triggered emotional eating; the use of food for comfort when experiencing negative emotions. Increased levels of stress have been reported during the pandemic [13–15]. Stressors associated with the pandemic include fear of infection, frustration, boredom, isolation, home schooling of children, loss of unemployment, and financial loss [16–23]. Previous studies have demonstrated that stress is correlated with deviations in emotional eating, eating patterns, and food choices [24–28]. Furthermore, a positive association between stress and an increased appetite for palatable, calorically dense foods has also been reported [29,30] with a preference for "comfort foods" (foods high in calories, sugar, and fat) [31–33]. It has been observed that emotional eaters tend to ingest greater amounts of sweet, fatty, and salty foods during difficult times [34,35]. The cause of these changes in food consumption may be due to alterations in cortisol or satiety hormones, which are negatively affected during periods of chronic stress [36–39]. Therefore, an observed increase in the consumption of these calorically dense "comfort" foods could also be associated with weight gain among teachers.

Not only has stress been associated with emotional eating, but several studies have demonstrated that stress is correlated with a decrease in physical activity and an increase in sedentariness. Further, it was indicated that both objective and subjective indicators of stress were associated with the observed reduction in physical activity [40–42].

Cumulatively, all of these factors have the potential to induce weight gain. Therefore, the purpose of this study was to determine if the mandatory school closures (16 March–26 June 2020), a result of the COVID-19 pandemic, were associated with weight gain among schoolteachers, and if so, did the weight gain correlate with changes in food consumption, emotional eating or physical activity.

2. Materials and Methods

A cross-sectional study was conducted on public-school teachers in Long Island, a suburban area in the east of New York City. Teachers were recruited in June and July 2020, via social media (Long Island teacher Facebook pages) and mass emails, to take part in this study. Informed consent was obtained from all 129 subjects prior to their access to the survey. Inclusion criteria were: (a) full-time teacher in a public school in Long Island, NY, during the 2019–2020 school year (b) not pregnant or lactating (c) have internet access. The study was completed in accordance with the IRB of Farmingdale State College, NY, USA.

2.1. Study Design

Once subjects voluntarily agreed to participate in the study, they received the link to a Qualtrics (Seattle, WA, USA) survey consisting of 52 questions. The descriptive information collected from the subjects included age, gender, marital status, employment status, duration of years teaching, income, ethnicity, and self-reported height and weight. Participants weight in pounds was used to determine changes in body weight. Subjects were also asked how many total people, and how many people under the age of 18 years, were residing in the home during the 3-month quarantine period, and which grade they were teaching during the school closures. In most school districts on Long Island, kindergarten classrooms are in the same building as elementary school grades—first through fifth—and will be referred to as K-5. Middle school (MS) encompasses grades 6 through 8, and high school (HS) grades 9 through 12. Both MS and HS are, customarily, housed within their own separate school buildings. A short-form food-frequency questionnaire (SFFFQ) was used to assess food intake. The participants were asked to report the frequency of consumption of alcoholic and non-alcoholic beverages, fruits, vegetables, starches, fiber-rich foods, high-fat and high-sugar foods, meats, and dairy products prior to (before 16 March 2020) and during (up until 26 June 2020) school closures. Frequency of consumption was measured using a scale ranging from less than once a month to 2 or more times per day. SFFFQs have been found to be reliable and validated for measuring food consumption [43]. The Leisure-Time Exercise Questionnaire (LTEQ) was used to assess changes in the frequency and intensity of physical activity over the 3-month period. Participants were asked, on average, how many hours per week they engaged in the following exercises: strenuous exercise (heart beats rapidly), moderate exercise (not exhausting), mild exercise (minimal effort). Examples of exercises from each of the 3 categories were provided. The LTEQ has been found to be reliable and valid [44]. The frequency of emotional eating was measured using the Dutch Eating Behavior Questionnaire (DEBQ). The DEBQ contains 13 questions that could determine whether certain emotional states trigger eating (emotional eating). The participants were asked to report on the frequency of their emotional eating prior to and during school closures using a 5-answer scale ranging from "never" to "very often." The DEBQ has been tested for reliability and validity [45].

2.2. Statistics

Text answers from the SFFFQ, LTEQ, and DEBQ were converted to a number, based on the indicated frequency. Nearly all data, converted and non-converted, were non-parametric and not normally distributed. Therefore, Wilcoxon signed-rank tests were used to measure any changes in dietary habits, physical activity, and emotional eating during school closures. Spearman's rank-order correlation test was used to find correlation. One-way ANOVA was used for parametric data. Dunn's Kruskal–Wallis test was used to compare the differences between subgroups nonparametrically. Median and interquartile ranges are reported for the comparisons with significant *p*-value. All statistical analyses were performed using R. Significance for all tests was set as $p < 0.05$.

3. Results

3.1. Baseline Characteristics

Table 1 depicts the baseline characteristics of the participants, separated by grade taught during the 2019–2020 school year and the demographic questions that were included in the survey. As can be seen in the table, approximately 51% of the respondents taught in K-5 classrooms during the shelter-in-place period. Totals of 26% and 23% taught in middle school and high school classrooms, respectively. The majority of respondents were white females, similar to the national demographics for teachers (80% white and 77% female) [46].

Table 1. Baseline characteristics of participants.

	K5 Education	Middle School	High School	All
Grade teaching	66 (51%)	34 (26%)	29 (23%)	129
Year teaching				
Below 10 years	19 (29%)	11 (32%)	8 (28%)	38 (30%)
Between 10 to 20 years	20 (30%)	16 (47%)	12 (41%)	48 (37%)
More than 20 years	27 (41%)	7 (21%)	9 (31%)	43 (33%)
Gender				
Female	62 (94%)	32 (94%)	22 (76%)	116 (90%)
Male	4 (6%)	2 (6%)	5 (17%)	11 (9%)
Prefer not to respond	0 (0%)	0 (0%)	2 (7%)	2 (2%)
Age group (year)				
20–29	8 (12%)	8 (24%)	2 (7%)	18 (14%)
30–39	11 (17%)	9 (27%)	9 (31%)	29 (23%)
40–49	28 (42%)	12 (35%)	9 (31%)	49 (38%)
50–59	15 (23%)	5 (15%)	6 (21%)	26 (20%)
60+	4 (6%)	0 (0%)	3 (10%)	7 (5%)

Table 1. *Cont.*

	K5 Education	Middle School	High School	All
Race				
White	59 (89%)	31 (91%)	25 (86%)	115 (89%)
Asian	2 (3%)	0 (0%)	2 (7%)	4 (3%)
Black or African American	1 (2%)	0 (0%)	0 (0%)	1 (1%)
Other or prefer not to respond	4 (6%)	3 (9%)	2 (7%)	9 (7%)
Change in body weight (lbs.)				
Mean	3.7	0.2	−1.2	1.7
SD	8	11.3	9	9.4
Household income (USD)				
below USD 100,000	13 (20%)	4 (12%)	7 (24%)	24 (19%)
between USD 100,000 and USD 150,000	15 (23%)	9 (27%)	4 (14%)	28 (22%)
above USD 150,000	31 (46%)	19 (56%)	16 (55%)	65 (50%)
Prefer not to respond	8 (12%)	2 (6%)	2 (7%)	12 (9%)

3.2. Weight Change

To determine if the school closures, and complete shift from the school classroom to the home classroom, were associated with weight change among the teachers, we compared the mean body weight prior to (M = 157.09, SD = 33.5) and during (M = 158.8, SD = 33.6) the school closures. As depicted in Figure 1A, there was no significant change in body weight observed among the 129 participants. However, when we separated the teachers by grade taught (Figure 1B), the subgroup analysis revealed that participants who taught K-5 (Mdn = 4, IQR = −2–8) had a significant gain in weight compared to participants who taught HS (Mdn = −2, IQR = −7–5) (Kruskal–Wallis statistic = 6.781, $p = 0.03$).

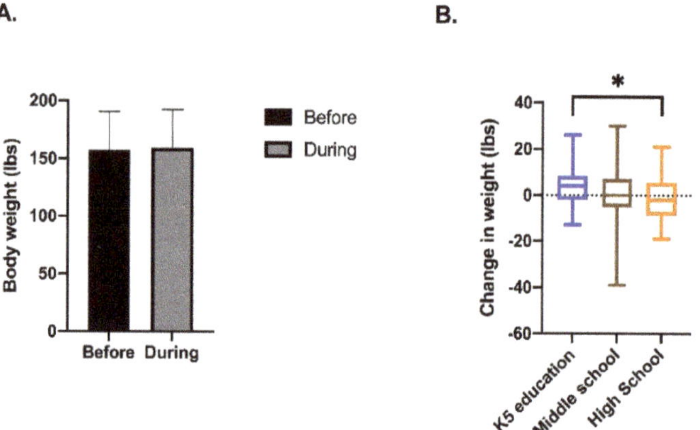

Figure 1. Body weight change (in pounds) for all participants (**A**) and separated by grade taught (**B**). (**A**) reports the mean and standard deviation of body weight for participants before and during the school closures. (**B**) reports Median and IQR. Kruskal-Wallis rank-sum tests were used to determine significance. * indicates p value < 0.05.

To further explore these changes in weight, we separated the teachers into five groups: lost more than 10 pounds; gained more than 10 pounds; lost less than 10 pounds; gained less than 10 pounds; no change in weight (12%, 18%, 27%, 34%, and 9%, respectively). When we analyzed the amount of weight gained and weight lost by teachers according to grade taught, 23% of K-5 teachers reported a weight gain of 10 or more pounds compared to 18% and 7% of MS and HS teachers, respectively. Further, 12% of MS and 24% of HS teachers lost 10 or more pounds, whereas only 6% of K-5 teachers reported this amount of weight loss.

3.3. Lifestyle Factors Associated with Weight Gain

As previously mentioned, increases in the consumption of "comfort" foods and emotional eating have been reported during times of stress. Figure 2A illustrates a significant increase in the consumption of alcohol, soft drinks, and calorically dense foods (cereal, chips, potatoes, fries, candy, ice cream, cakes, rice, pizza) among the participants during the school closures. In contrast, considerable reductions in the consumption of raw vegetables and whole-grain bread products, were observed. Regarding emotional eating there were significant increases in food consumption for all 13 categories of emotional eating measured by the Dutch Eating Behavior Questionnaire (DEBQ) (Figure 2B).

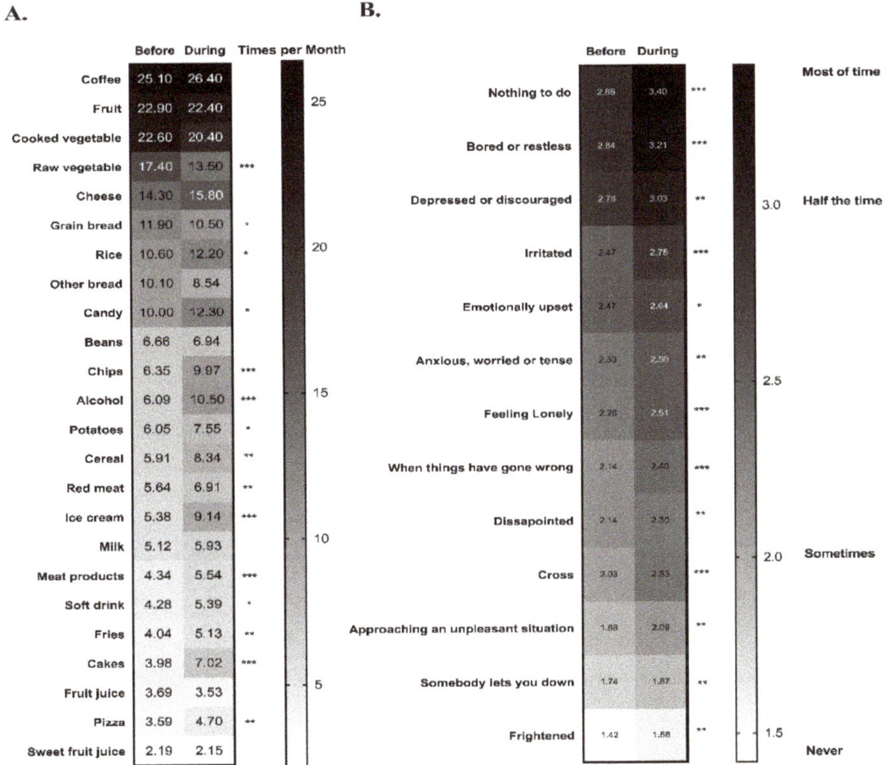

Figure 2. Heat map indicating changes in food consumption and emotional eating. Means of the frequencies of foods consumed (**A**) and emotional eating (**B**) were reported for before school closures (before 16 March) and during school closures (until 26 June). Darker colored cells represent higher frequency of the parameter measured. Wilcoxon signed-rank tests were used, and *, **, *** indicating p-value < 0.05, 0.005, 0.0005, respectively.

In order to determine if these variations in food intake and emotional eating were associated with changes in weight, we conducted a Spearman's rank-order correlation test. Table 2 illustrates that an increase in the consumption of calorically dense foods (chips, potatoes, fries, bread products, cheese, candy, ice cream, cake, and soft drinks) was positively correlated with weight gain. In contrast, increased consumption of fresh fruits and beans was negatively correlated with weight gain. Further, an increase in emotional eating, regardless of the cause, was positively associated with weight gain, with a particularly strong correlation seen between weight gain and eating when irritated. Lastly,

Table 2 also shows that an increase in the frequency of exercise at any intensity (strenuous, moderate, mild/light) was negatively correlated with weight gain, whereas sedentary activities were positively correlated with weight gain.

Table 2. Factors affecting weight gain. Spearman's rank-order correlation test were used to measure significance.

Change in Food Intake	Correlation Coefficient	p-Value
Fresh fruit	−0.17	0.049 *
Chips	0.25	0.001 **
Potatoes	0.2	0.03 *
French fries, home fries, or hash browns	0.22	0.01 *
Other bread products	0.21	0.02 *
Cheese (do not include cheese on pizza)	0.18	0.02 *
Beans or pulses (lentils, green peas)	−0.26	0.003 **
Candy or chocolate (not sugar-free)	0.25	0.004 **
Ice cream or other frozen desserts	0.29	0.005 *
Cakes, scones, pies, pastries, biscuits, brownies	0.24	0.04 *
Soft drinks (not diet)	0.19	0.0005 ***
Change in Physical Activity		
Strenuous exercise	−0.44	0.0005 ***
Moderate exercise	−0.31	0.0005 ***
Mild/Light exercise	−0.18	0.04 *
Sedentary activities	0.26	0.003 **
Change in Emotional Eating		
Irritated	0.33	0.0005 ***
Depressed or discouraged	0.18	0.047 *
Approaching an unpleasant situation	0.22	0.01 *
Things have gone wrong	0.22	0.01 *
Anxious, worried, or tense	0.28	0.01 *
Emotionally upset	0.25	0.005 *
Nothing to do	0.22	0.01 *
Feeling lonely	0.2	0.02 *
Bored or restless	0.22	0.01 *

*, **, *** indicating p-value < 0.05, 0.005, 0.0005, respectively.

3.4. Factors Affecting Weight by Pounds Lost or Gained

To further determine the contribution of the observed changes in food intake and physical activity to weight gained, we performed nonparametric Dunn tests. As presented in Table 3, among participants who gained more than 10 pounds, we observed a significant increase in the consumption of chips, potatoes, fries, other bread products, candies, ice cream, cakes, and soft drinks when compared to the participants who lost more than 10 pounds. A significant increase in the consumption of chips, potatoes, other bread products, candy, ice cream, and soft drinks among those who gained more than 10 pounds when compared to teachers who lost less than 10 pounds was observed, as well. Regarding the contribution of physical activity to weight change, among teachers who lost any amount of weight there was a significant increase in the frequency of strenuous exercise when compared to the no change in weight and weight gain groups. Participants who gained more than 10 pounds reported a significant decrease in the frequency of both strenuous and moderate exercise, but a significant increase in mild/light exercise and sedentary activities when compared to teachers that lost more than 10 pounds.

Table 3. Changes in food consumption and exercise intensity according to weight change (pounds). One-way ANOVA rank tests were used. Significant changes are represented by non-overlapping letters across the rows.

	Lost > 10 Pounds	Lost < 10 Pounds	No Change	Gain < 10 Pounds	Gain > 10 Pounds
Food intake changes					
Cereal	0 (0–8.7) a	0 (0–9.5) a	0 (−0.4–0) b	0 (0–1.5) ab	0 (0–8) ab
Chips	0 (−1.5–0.8) a	0 (0–1.5) a	0 (0–1.9) ab	1.5 (0–8.4) b	8 (0–9.5) b
Potatoes	0 (−0.7–5.5) a	0 (0–0) a	0 (0–1.5) a	0 (0–0.4) a	1.5 (0–10) b
Fries	0 (−0.8–0) a	0 (0–1.5) ab	0 (0–1.5) ab	0 (0–1.5) b	0 (0–8) b
Other Breads	0 (−10.5–0) a	0 (−8–0) a	0 (−1.9–1.9) ab	0 (0.4–1.9) b	0 (0–5.5) b
Candies	0 (−4.8–1.5) a	0 (−5.5–1.5) a	0 (0–8.5) ab	0 (−0.4–8) ab	0 (0–10.5) b
Ice cream	0 (−1.5–1.5) a	0 (0–8) ab	0 (0–3) abc	0.8 (0–8) bc	1.5 (0–9.8) c
Cakes	0 (−1.5–0.8) a	0 (0–5.5) ab	0 (0–1.5) ab	0 (0–8) ab	0 (0–10.5) b
Soft drinks	0 (0–0) a	0 (0–0) a	0 (0–0) a	0 (0–0) a	0 (0–5.5) b
Physical activity changes					
Strenuous exercise	3 (0.5–5) a	1 (0–3) ab	0 (−0.3–1) bc	0 (−0.3–0) c	0 (−1.5–0) c
Moderate exercise	3 (0–4) a	2 (0–3) a	1 (0.8–2) a	0 (−2–3) ab	−1 (−3–1.5) b
Mild/light exercise	1 (0–3) a	0 (0–0.5) bc	0 (0–3.3) ab	0 (0–1) abc	0 (0–1) c
Sedentary activities	2 (−0.5–5) a	3 (1.5–6) ab	2.5 (2–5) ab	3 (2–5) ab	5 (4–6) b

3.5. Food Consumption among Teachers of Different Grade Levels

As noted above, we revealed a significant weight gain among the K-5 teachers, with a likelihood of 1.51 (95%CI: 1.06–2.14, *p*-value = 0.02). Further, when a chi-square test of independence was performed to determine if there was a correlation between grade taught and weight gain, we observed a significant correlation between these variables, χ^2 (1, n = 129) = 4.81, p = 0.03. To further elucidate which factors contributed to the significant weight gain observed among K-5 teachers, we performed pairwise comparisons using Dunn's all-pairs tests on food intake. When evaluating the data from the SFFFQ separated by grade taught (Table 4), we noticed a significant increase in the consumption of chips and meat products among K-5 teachers when compared to HS teachers (p = 0.03 and 0.02, respectively). In addition, the reported consumption of fruits was significantly reduced among K-5 teachers when compared to both MS and HS teachers (p = 0.008 and =0.03, respectively). Further, a considerable reduction in the intake of raw vegetables was observed among the K-5 teachers when compared to MS teachers (p = 0.03).

Table 4. Changes in foods consumed by grade taught. One-way ANOVA rank tests were used. Significant changes are represented by non-overlapping letters across the rows.

	K5 Education	Middle School	High School
Fruits	0 (−9.5–0) a	0 (0–9.9) b	0 (0–8) b
Raw Vegetable	−8 (−9.9–0) a	0 (−6.8–1.1) b	0 (−8–0) ab
Chips	1.5 (0–9.5) a	0 (0–2.6) ab	0 (−1.5–1.5) b
Meat-products	0 (0–3) a	0 (0–1.1) ab	0 (0–0) b

3.6. Eating Patterns among Teachers of Different Grade Levels

Lastly, we examined whether there were any specific changes in eating patterns among the teachers according to grade taught during the school closures. We observed a considerable increase in the frequency of eating while working in K-5 teachers (Mdn = 9, IQR = 0–29) when compared to both the MS and HS teachers (Mdn = 0 and 0, IQR = −17–12.8 and −17–8, p = 0.001 and 0.003, respectively). Kruskal–Wallis χ^2 (2, n = 129) = 14.38, p = 0.0008. In addition, a significant increase in the consumption of foods consumed when things had gone wrong was observed among K-5 teachers (Mdn = 0, IQR = 0–1), in contrast to MS and HS teachers (Mdn = 0 and 0, IQR = 0–0 and 0–0, p = 0.03 and 0.01, respectively). Kruskal–Wallis χ^2 (2, n = 129) = 8.79, p = 0.01.

4. Discussion

The COVID-19 pandemic put enormous stress on the nation; economically, emotionally, and physiologically. Although the long-term health effects of the pandemic, and subsequent quarantine, are yet to be elucidated, we were able to identify some of the short-term effects on weight, overall eating habits and frequency of emotional eating and physical activity among kindergarten to 12th grade schoolteachers in Long Island, NY. Although no significant change in weight was observed when comparing all 129 teachers during the three-month school closures, subgroup analysis demonstrated a significant weight gain among K-5 teachers when compared to non-K-5 teachers. Specifically, we noted that 23% of K-5 teachers gained 10 or more pounds compared to 18% and 7% of MS and HS teachers, respectively. Further, 12% of MS and 24% of HS teachers lost 10 or more pounds, whereas only 6% of K-5 teachers reported losing this amount of weight.

To uncover potential risk factors associated with the observed weight gain, we conducted a Spearman's rank-order correlation test. The results revealed that changes in food intake, emotional eating, and physical activity were all independently and significantly correlated with weight change. Regarding food intake, an increase in the consumption of calorically dense foods, such as chips, potatoes, fries, bread products, cheese, candy, ice cream, cake, and soft drinks, was positively correlated with weight gain. Studies have shown that acute stress can lead to disinhibited eating behaviors, with food choices predominantly favoring sugary and fatty foods [30,33,35,47,48]. Therefore, the stress associated with the coronavirus pandemic, and subsequent school closures, may explain the reported increase in the consumption of these highly palatable, rewarding "comfort" foods.

Due to the school closures and shift to an online classroom, daily routines have changed dramatically. In addition to the adaptation of teaching from home, many teachers were engaged in the homeschooling of their own children. Due to these deviations from their typical daily routines, foods consumed during this three-month period may have largely been shaped by convenience [49–51]. According to Locher et al. [52], convenience is another characteristic of "comfort food". As reported in the New York Times [53], during the pandemic, many people did relax their usual food rules and reached for the comfort foods of their childhood, macaroni and cheese, chips, cookies. The need for quick and easy-to-prepare foods became essential as people tried to squeeze in a meal between Zoom meetings. In addition, the shift to working from home meant food was always readily available. Therefore, the shift to working from home combined with the convenience of easily accessible, palatable, and emotionally comforting foods may further explain some of the weight gain observed among the teachers.

In contrast to the positive association seen between weight gain and the intake of these comfort foods, a negative association was observed between weight gain and the consumption of fresh fruit and beans. Previous studies have illustrated that protein is the most satiating macronutrient, whereas fatty foods are the least satiating. Therefore, due its weaker role of promoting satiation and its high palatability, the increase in the ingestion of fatty foods, such as chips, fries, cheese, and ice cream, demonstrated in this study, may have led to an increase in overall caloric consumption and, consequently, weight gain. Legumes, on the other hand, are low in fat and high in protein and fiber. As with protein-rich foods, fiber, which is found in legumes, as well as fruits, is another contributor to satiety. Therefore, a decrease in the consumption of satiety-inducing beans and fruit may also further explain some of the weight gain observed [54–56].

Several studies revealed significantly higher levels of stress reported by teachers during quarantine, due to difficulties adjusting to distance education and the increased workload associated with working from home [57,58]. As previously mentioned, stress is associated with less physical activity and more sedentary behavior, which may contribute to weight gain [40–42]. Table 2 reveals a significant decrease in the frequency of both strenuous and moderate exercise, and a significant increase in the frequencies of mild/light exercise and sedentary activities among participants who gained more than 10 pounds when compared to participants who lost more than 10 pounds. Further, an increase in the occurrence of exercise at any intensity (strenuous, moderate, mild/light) was negatively correlated with weight gain, whereas an increase in sedentariness was positively correlated with weight gain. Specifically, we noted that teachers who lost weight (1 to 10+ pounds) had a significant increase in the frequency of strenuous exercise compared to teachers who gained weight (1 to 10+ pounds).

As exhibited in Figure 1B, we saw a significant increase in weight gain among K-5 teachers when compared to teachers of older grade levels. Although several studies showed heightened levels of stress among teachers during school closures [57,58], Ozamiz-Etxebberia et al. [59] reported that the highest levels of stress, and stress-precipitated anxiety, were observed among K-5 teachers. Interestingly, these findings contradict non-quarantine conditions in which high school teachers report being most affected by stress [60]. The increase in stress exhibited by the K-5 teachers may be attributed to the age of the students they teach; younger students (aged 4 to 10) generally require more care and guidance than older students. With schools closed, and face-to-face interaction no longer possible, these teachers may feel that they are no longer able to adequately carry out these duties of care, resulting in heightened stress. As previously mentioned, stress is correlated with an increase in emotional eating, the consumption of comfort foods, and sedentariness; all risk factors for weight gain.

Although teachers of every grade level, spend similar amounts of time dedicated to the education of their students, their daily responsibilities in the classroom greatly differ. Kindergarten and elementary teachers work with children during their first years of school. A significant amount of movement and exploration occur in the classrooms of younger children; teachers need to use a lot of play, games, and hands-on teaching activities to keep 4- to 10-year-old students engaged. In addition, K-5 teachers ambulate frequently throughout the day as they shift between the academic classroom stations, escort students to different activities located throughout the school building, and monitor lunch, recess, and bus dismissal [61,62]. With school closures, days of frequent activity and movement throughout the school building were no longer occurring, perhaps playing a role in the weight gain seen among the K-5 teachers.

Lastly, as previously stated, a significant increase in eating while working was observed among K-5 teachers when compared to MS and HS teachers. Eating while engaged in other activities is an example of mindless eating and, as Wansink [63] has demonstrated, mindless eating can result in a failure to respond to internal satiety cues and, subsequently, causes increased potential of overeating. Therefore, this increase in mindless eating during school closures may provide further explanations in regard to the weight gain detected in this group.

At this point, over 43% of Americans are fully vaccinated [64]. However, with the appearance of new and more contagious COVID variants around the world [65], it seems that future pandemics, and mandatory shelter-in-place regulations, may be an inevitable threat. If these results were extrapolated to the general population, it could result in the adaptation of proactive measures regarding food consumption, emotional eating, and frequency of physical activity during any future quarantines.

5. Conclusions

In conclusion, these findings provide the first look at the impact of the COVID-19 school closures on weight and weight-related behaviors among schoolteachers. Weight gain among study participants was independently and significantly associated with increases in the consumption of "comfort" foods, emotional eating and sedentariness. These observed changes in weight-related behaviors among the teachers may have been precipitated by increased stress due to changes in daily schedules, the rapid shift to an online classroom, and worries about the virus itself. Greater levels of stress were reported among kindergarten and elementary (K-5) schoolteachers than middle and high school teachers during the school closures, which may help explain the greater amount of weight gain observed among K-5 teachers. This study provides a starting point for future research looking at the impact of quarantining on behaviors affecting weight. Follow-up studies are needed to investigate whether weight gained during the pandemic was lost once the quarantine restrictions were lifted and to determine if the weight gain was associated with an increase in the incidence of metabolic disorders.

6. Limitations

The present study had several caveats. The retrospective design of the study allowed us to estimate associations only and the self-reported questions for weight may have been affected by bias. With the use of social media as the means of recruitment, it could be argued that the study sample was not an adequate representation of the population. In addition, future studies involving a more diverse population would be beneficial. Lastly, it should be mentioned that the existence of unknown confounding factors may have contributed to weight gain (e.g., certain medications. medical conditions). Although these are possible limitations, the current study still provides unique information about the potential impact of quarantine on weight and weight-related risk factors.

Author Contributions: Conceptualization, J.R.S.; methodology, J.R.S.; software, B.Z.W.; validation, J.R.S., B.Z.W.; formal analysis, B.Z.W.; investigation, J.R.S.; resources, J.R.S.; data curation, J.R.S., B.Z.W.; writing—original draft preparation, J.R.S., B.Z.W.; writing—review and editing, J.R.S.; visualization, J.R.S.; supervision, J.R.S.; project administration, J.R.S. Both authors have read and agreed to the published version of the manuscript.

Funding: This research received no external funding.

Institutional Review Board Statement: Ethical review and approval were waived for this study since it did not require direct human interaction. All data was collected using an online questionnaire.

Informed Consent Statement: Informed consent was obtained from all subjects involved in the study.

Conflicts of Interest: The authors declare no conflict of interest.

References

1. Price-Haywood, E.G.; Burton, J.; Fort, D.; Seoane, L. Hospitalization and Mortality among Black Patients and White Patients with Covid-19. *N. Engl. J. Med.* **2020**, *382*, 2534–2543. [CrossRef]
2. Gostin, L.O.; Hodge, J.G., Jr.; Wiley, L.F. Presidential Powers and Response to COVID-19. *JAMA* **2020**, *323*, 1547–1548. [CrossRef] [PubMed]
3. New York State. Available online: https://www.governor.ny.gov/news/governor-cuomo-signs-new-york-state-pause-executive-order (accessed on 14 June 2021).
4. Education Week. Available online: https://www.edweek.org/technology/assessing-the-impact-of-covid-19-on-ed-tech-use/2020/06 (accessed on 17 June 2021).
5. Ammar, A.; Brach, M.; Trabelsi, K.; Chtourou, H.; Boukhris, O.; Masmoudi, L.; Bouaziz, B.; Bentlage, E.; How, D.; Ahmed, M.; et al. Effects of COVID-19 Home Confinement on Eating Behaviour and Physical Activity: Results of the ECLB-COVID19 International Online Survey. *Nutrients* **2020**, *12*, 1583. [CrossRef]
6. Brooks, S.K.; Webster, R.; Smith, L.E.; Woodland, L.; Wessely, S.; Greenberg, N.; Rubin, G.J. The psychological impact of quarantine and how to reduce it: Rapid review of the evidence. *Lancet* **2020**, *395*, 912–920. [CrossRef]
7. Kriaucioniene, V.; Bagdonaviciene, L.; Rodríguez-Pérez, C.; Petkeviciene, J. Associations between Changes in Health Behaviours and Body Weight during the COVID-19 Quarantine in Lithuania: The Lithuanian COVIDiet Study. *Nutrients* **2020**, *12*, 3119. [CrossRef]
8. Thomson, B. The COVID-19 Pandemic: A Global Natural Experiment. *Circulation* **2020**, *142*, 14–16. [CrossRef] [PubMed]
9. Pietrobelli, A.; Pecoraro, L.; Ferruzzi, A.; Heo, M.; Faith, M.; Zoller, T.; Antoniazzi, F.; Piacentini, G.; Fearnbach, S.N.; Heymsfield, S.B. Effects of COVID-19 Lockdown on Lifestyle Behaviors in Children with Obesity Living in Verona, Italy: A Longitudinal Study. *Obesity* **2020**, *28*, 1382–1385. [CrossRef] [PubMed]
10. Rundle, A.G.; Park, Y.; Herbstman, J.B.; Kinsey, E.W.; Wang, Y.C. COVID-19–Related School Closings and Risk of Weight Gain Among Children. *Obesity* **2020**, *28*, 1008–1009. [CrossRef] [PubMed]
11. Hamilton, M.T.; Hamilton, D.G.; Zderic, T.W. Role of Low Energy Expenditure and Sitting in Obesity, Metabolic Syndrome, Type 2 Diabetes, and Cardiovascular Disease. *Diabetes* **2007**, *56*, 2655–2667. [CrossRef] [PubMed]

12. Ed Tech Magazine. Available online: https://edtechmagazine.com/k12/article/2013/08/how-many-hours-do-educators-actually-work (accessed on 18 June 2021).
13. Albott, C.S.; Wozniak, J.R.; McGlinch, B.P.; Wall, M.H.; Gold, B.S.; Vinogradov, S. Battle Buddies: Rapid Deployment of a Psychological Resilience Intervention for Health Care Workers During the COVID-19 Pandemic. *Anesth. Analg.* **2020**, *131*, 43–54. [CrossRef] [PubMed]
14. Rodgers, R.F.; Lombardo, C.; Cerolini, S.; Franko, D.L.; Omori, M.; Fuller-Tyszkiewicz, M.; Linardon, J.; Courtet, P.; Guillaume, S. The impact of the COVID -19 pandemic on eating disorder risk and symptoms. *Int. J. Eat. Disord.* **2020**, *53*, 1166–1170. [CrossRef]
15. Shah, M.; Sachdeva, M.; Johnston, H. Eating disorders in the age of COVID-19. *Psychiatry Res.* **2020**, *290*, 113122. [CrossRef]
16. Elbay, R.Y.; Kurtulmuş, A.; Arpacıoğlu, S.; Karadere, E. Depression, anxiety, stress levels of physicians and associated factors in Covid-19 pandemics. *Psychiatry Res.* **2020**, *290*, 113130. [CrossRef]
17. Lippi, G.; Henry, B.M.; Bovo, C.; Sanchis-Gomar, F. Health risks and potential remedies during prolonged lockdowns for coronavirus disease 2019 (COVID-19). *Diagnosis* **2020**, *7*, 85–90. [CrossRef]
18. Liu, C.H.; Zhang, E.; Wong, G.T.F.; Hyun, S.; Hahm, H. Factors associated with depression, anxiety, and PTSD symptomatology during the COVID-19 pandemic: Clinical implications for U.S. young adult mental health. *Psychiatry Res.* **2020**, *290*, 113172. [CrossRef] [PubMed]
19. Ozamiz-Etxebarria, N.; Dosil-Santamaria, M.; Picaza-Gorrochategui, M.; Idoiaga-Mondragon, N. Stress, anxiety, and depression levels in the initial stage of the COVID-19 outbreak in a population sample in the northern Spain. *Cad. Saude Publica* **2020**, *36*, e00054020. [CrossRef] [PubMed]
20. Pedrozo-Pupo, J.C.; Pedrozo-Cortés, M.J.; Campo-Arias, A. Perceived stress associated with COVID-19 epidemic in Colombia: An online survey. *Cad. Saude Publica* **2020**, *36*, e00090520. [CrossRef]
21. Qiu, J.; Shen, B.; Zhao, M.; Wang, Z.; Xie, B.; Xu, Y. A nationwide survey of psychological distress among Chinese people in the COVID-19 epidemic: Implications and policy recommendations. *Gen. Psychiatry* **2020**, *33*, e100213. [CrossRef] [PubMed]
22. Rajkumar, R.P. COVID-19 and mental health: A review of the existing literature. *Asian J. Psychiatry* **2020**, *52*, 102066. [CrossRef]
23. Rehman, U.; Shahnawaz, M.G.; Khan, N.H.; Kharshiing, K.D.; Khursheed, M.; Gupta, K.; Kashyap, D.; Uniyal, R. Depression, Anxiety and Stress Among Indians in Times of Covid-19 Lockdown. *Community Ment. Health J.* **2020**, *57*, 42–48. [CrossRef]
24. Devonport, T.J.; Nicholls, W.; Fullerton, C. A systematic review of the association between emotions and eating behaviour in normal and overweight adult populations. *J. Health Psychol.* **2017**, *24*, 3–24. [CrossRef]
25. Hou, F.; Xu, S.; Zhao, Y.; Lu, Q.; Zhang, S.; Zu, P.; Sun, Y.; Su, P.; Tao, F. Effects of emotional symptoms and life stress on eating behaviors among adolescents. *Appetite* **2013**, *68*, 63–68. [CrossRef]
26. Michels, N.; Sioen, I.; Braet, C.; Eiben, G.; Hebestreit, A.; Huybrechts, I.; Vanaelst, B.; Vyncke, K.; De Henauw, S. Stress, emotional eating behaviour and dietary patterns in children. *Appetite* **2012**, *59*, 762–769. [CrossRef]
27. Nguyen-Rodriguez, S.T.; Chou, C.-P.; Unger, J.B.; Spruijt-Metz, D. BMI as a moderator of perceived stress and emotional eating in adolescents. *Eat. Behav.* **2008**, *9*, 238–246. [CrossRef]
28. Torres, S.J.; Nowson, C. Relationship between stress, eating behavior, and obesity. *Nutrients* **2007**, *23*, 887–894. [CrossRef] [PubMed]
29. Chaput, J.-P.; Drapeau, V.; Poirier, P.; Teasdale, N.; Tremblay, A. Glycemic Instability and Spontaneous Energy Intake: Association With Knowledge-Based Work. *Psychosom. Med.* **2008**, *70*, 797–804. [CrossRef] [PubMed]
30. Lemmens, S.G.; Rutters, F.; Born, J.M.; Westerterp-Plantenga, M.S. Stress augments food 'wanting' and energy intake in visceral overweight subjects in the absence of hunger. *Physiol. Behav.* **2011**, *103*, 157–163. [CrossRef] [PubMed]
31. Ulrich-Lai, Y.M.; Fulton, S.; Wilson, M.; Petrovich, G.; Rinaman, L. Stress exposure, food intake and emotional state. *Stress* **2015**, *18*, 381–399. [CrossRef] [PubMed]
32. van Strien, T.; Gibson, L.; Baños, R.; Cebolla, A.; Winkens, L.H. Is comfort food actually comforting for emotional eaters? A (moderated) mediation analysis. *Physiol. Behav.* **2019**, *211*, 112671. [CrossRef]
33. Zellner, D.A.; Loaiza, S.; Gonzalez, Z.; Pita, J.; Morales, J.; Pecora, D.; Wolf, A. Food selection changes under stress. *Physiol. Behav.* **2006**, *87*, 789–793. [CrossRef] [PubMed]
34. Camilleri, G.M.; Méjean, C.; Kesse-Guyot, E.; Andreeva, V.; Bellisle, F.; Hercberg, S.; Péneau, S. The Associations between Emotional Eating and Consumption of Energy-Dense Snack Foods Are Modified by Sex and Depressive Symptomatology. *J. Nutr.* **2014**, *144*, 1264–1273. [CrossRef]
35. Oliver, G.; Wardle, J.; Gibson, E.L. Stress and Food Choice: A Laboratory Study. *Psychosom. Med.* **2000**, *62*, 853–865. [CrossRef] [PubMed]
36. Adam, T.C.; Epel, E.S. Stress, eating and the reward system. *Physiol. Behav.* **2007**, *91*, 449–458. [CrossRef] [PubMed]
37. Born, J.M.; Lemmens, S.G.T.; Rutters, F.; Nieuwenhuizen, A.; Formisano, E.; Goebel, R.; Westerterp-Plantenga, M.S. Acute stress and food-related reward activation in the brain during food choice during eating in the absence of hunger. *Int. J. Obes.* **2009**, *34*, 172–181. [CrossRef] [PubMed]
38. Chao, A.M.; Jastreboff, A.M.; White, M.; Grilo, C.M.; Sinha, R. Stress, cortisol, and other appetite-related hormones: Prospective prediction of 6-month changes in food cravings and weight. *Obesity* **2017**, *25*, 713–720. [CrossRef]
39. Yau, Y.H.; Potenza, M.N. Stress and eating behaviors. *Minerva Endocrinal.* **2013**, *38*, 255–267.
40. American Psychological Association. Available online: https://www.apa.org/news/press/releases/stress/index (accessed on 7 June 2021).

41. Lutz, R.S.; Stults-Kolehmainen, M.; Bartholomew, J. Exercise caution when stressed: Stages of change and the stress–exercise participation relationship. *Psychol. Sport Exerc.* **2010**, *11*, 560–567. [CrossRef]
42. Stults-Kolehmainen, M.A.; Sinha, R. The Effects of Stress on Physical Activity and Exercise. *Sports Med.* **2013**, *44*, 81–121. [CrossRef]
43. Cleghorn, C.L.; Harrison, R.A.; Ransley, J.K.; Wilkinson, S.; Thomas, J.; Cade, J.E. Can a dietary quality score derived from a short-form FFQ assess dietary quality in UK adult population surveys? *Public Health Nutr.* **2016**, *19*, 2915–2923. [CrossRef] [PubMed]
44. Godin, G.; Jobin, J.; Bouillon, J. Assessment of leisure time exercise behavior by self-report: A concurrent validity study. *Can. J. Public Health* **1986**, *77*, 359–362. [CrossRef]
45. Van Strien, T.; Frijters, J.E.R.; Bergers, G.P.A.; Defares, P.B. The Dutch Eating Behavior Questionnaire (DEBQ) for Assessment of Restrained, Emotional, and External Eating Behavior. *Int. J. Eat. Disord.* **1986**, *5*, 295–315. [CrossRef]
46. National Education Association. Available online: https://www.nea.org/advocating-for-change/new-from-nea/who-average-us-teacher (accessed on 28 July 2021).
47. Kuo, L.E.; Czarnecka, M.; Kitlinska, J.B.; Tilan, J.U.; Kvetnansky, R.; Zukowska, Z. Chronic Stress, Combined with a High-Fat/High-Sugar Diet, Shifts Sympathetic Signaling toward Neuropeptide Y and Leads to Obesity and the Metabolic Syndrome. *Ann. N. Y. Acad. Sci.* **2008**, *1148*, 232–237. [CrossRef]
48. Rutters, F.; Nieuwenhuizen, A.; Lemmens, S.G.T.; Born, J.M.; Westerterp-Plantenga, M.S. Acute Stress-related Changes in Eating in the Absence of Hunger. *Obesity* **2009**, *17*, 72–77. [CrossRef] [PubMed]
49. The Atlantic. Available online: https://www.theatlantic.com/health/archive/2020/05/why-theres-no-flour-during-coronavirus/611527/ (accessed on 8 June 2021).
50. The New York Times. Available online: https://www.nytimes.com/2020/03/17/dining/food-shortage-coronavirus.html (accessed on 17 May 2021).
51. Warde, A. Convenience food: Space and timing. *Br. Food J.* **1999**, *101*, 518–527. [CrossRef]
52. Locher, J.L.; Yoels, W.C.; Maurer, D.; Van Ells, J. Comfort Foods: An Exploratory Journey into the Social and Emotional Significance of Food. *Food Foodways* **2005**, *13*, 273–297. [CrossRef]
53. The New York Times. Available online: https://www.nytimes.com/2020/04/07/business/coronavirus-processed-foods.html. (accessed on 10 June 2021).
54. McCrory, M.A.; Hamaker, B.R.; Lovejoy, J.C.; Eichelsdoerfer, P.E. Pulse Consumption, Satiety, and Weight Management. *Adv. Nutr.* **2010**, *1*, 17–30. [CrossRef] [PubMed]
55. Ramdath, D.; Renwick, S.; Duncan, A.M. The Role of Pulses in the Dietary Management of Diabetes. *Can. J. Diabetes* **2016**, *40*, 355–363. [CrossRef] [PubMed]
56. Westerterp-Plantenga, M.; Nieuwenhuizen, A.; Tomé, D.; Soenen, S.; Westerterp, K. Dietary Protein, Weight Loss, and Weight Maintenance. *Annu. Rev. Nutr.* **2009**, *29*, 21–41. [CrossRef] [PubMed]
57. Besser, A.; Lotem, S.; Zeigler-Hill, V. Psychological Stress and Vocal Symptoms among University Professors in Israel: Implications of the Shift to Online Synchronous Teaching during the COVID-19 Pandemic. *J. Voice* **2020**. [CrossRef] [PubMed]
58. Ng, K.C. Replacing Face-to-Face Tutorials by Synchronous Online Technologies: Challenges and pedagogical implications. *Int. Rev. Res. Open Distrib. Learn.* **2007**, *8*. [CrossRef]
59. Ozamiz-Etxebarria, N.; Mondragon, N.I.; Santamaria, M.D.; Gorrotxategi, M.P. Psychological Symptoms During the Two Stages of Lockdown in Response to the COVID-19 Outbreak: An Investigation in a Sample of Citizens in Northern Spain. *Front. Psychol.* **2020**, *11*, 1491. [CrossRef] [PubMed]
60. Gallegos, W.L.A.; Cahua, J.C.H.; Canaza, K.D.C. Síndrome de Burnout en profesores de escuela y universidad: Un análisis psicométrico y comparativo en la ciudad de Arequipa. *Propósitos Represent.* **2019**, *7*, 72–110. [CrossRef]
61. Chron. Available online: https://work.chron.com/average-hours-firstyear-kindergarten-teacher-works-per-week-27723.html (accessed on 7 June 2021).
62. Rentzou, K. Prevalence of burnout syndrome of Greek child care workers and kindergarten teachers. *Education* **2013**, *43*, 249–262. [CrossRef]
63. Wansink, B. *Mindless Eating: Why We Eat More than We Think*; Random House: Manhattan, NY, USA, 2006.
64. Our World in Data. Available online: https://ourworldindata.org/covid-vaccinations?country=USA (accessed on 23 June 2021).
65. John Hopkins Medicine. Available online: https://www.hopkinsmedicine.org/health/conditions-and-diseases/coronavirus/a-new-strain-of-coronavirus-what-you-should-know (accessed on 23 June 2021).

Article

Food Enrichment with *Glycyrrhiza glabra* Extract Suppresses ACE2 mRNA and Protein Expression in Rats—Possible Implications for COVID-19

Daniela Jezova [1,†], Peter Karailiev [1,†], Lucia Karailievova [1,†], Agnesa Puhova [1] and Harald Murck [2,3,*]

1 Institute of Experimental Endocrinology, Biomedical Research Center, Slovak Academy of Sciences, 84505 Bratislava, Slovakia; Daniela.Jezova@savba.sk (D.J.); peter.karailiev@savba.sk (P.K.); lucia.karailievova@savba.sk (L.K.); Agnesa.Puhova@savba.sk (A.P.)
2 Department of Psychiatry and Psychotherapy, Philipps-University Marburg, 35039 Marburg, Germany
3 Murck-Neuroscience, Westfield, NJ 07090, USA
* Correspondence: murck@staff.uni-marburg.de; Tel.: +1-2012941195
† These authors contributed equally to this work.

Abstract: Angiotensin converting enzyme 2 (ACE2) is a key entry point of severe acute respiratory syndrome coronavirus 2 (SARS-CoV-2) virus known to induce Coronavirus disease 2019 (COVID-19). We have recently outlined a concept to reduce ACE2 expression by the administration of glycyrrhizin, a component of *Glycyrrhiza glabra* extract, via its inhibitory activity on 11beta hydroxysteroid dehydrogenase type 2 (11betaHSD2) and resulting activation of mineralocorticoid receptor (MR). We hypothesized that in organs such as the ileum, which co-express 11betaHSD2, MR and ACE2, the expression of ACE2 would be suppressed. We studied organ tissues from an experiment originally designed to address the effects of *Glycyrrhiza glabra* extract on stress response. Male Sprague Dawley rats were left undisturbed or exposed to chronic mild stress for five weeks. For the last two weeks, animals continued with a placebo diet or received a diet containing extract of *Glycyrrhiza glabra* root at a dose of 150 mg/kg of body weight/day. Quantitative PCR measurements showed a significant decrease in gene expression of ACE2 in the small intestine of rats fed with diet containing *Glycyrrhiza glabra* extract. This effect was independent of the stress condition and failed to be observed in non-target tissues, namely the heart and the brain cortex. In the small intestine we also confirmed the reduction of ACE2 at the protein level. Present findings provide evidence to support the hypothesis that *Glycyrrhiza glabra* extract may reduce an entry point of SARS-CoV-2. Whether this phenomenon, when confirmed in additional studies, is linked to the susceptibility of cells to the virus requires further studies.

Keywords: COVID-19; glycyrrhizin; mineralocorticoid receptor; toll like receptor 4; angiotensin converting enzyme; aldosterone

Citation: Jezova, D.; Karailiev, P.; Karailievova, L.; Puhova, A.; Murck, H. Food Enrichment with *Glycyrrhiza glabra* Extract Suppresses ACE2 mRNA and Protein Expression in Rats—Possible Implications for COVID-19. *Nutrients* **2021**, *13*, 2321. https://doi.org/10.3390/nu13072321

Academic Editor: William B. Grant

Received: 19 May 2021
Accepted: 1 July 2021
Published: 6 July 2021

1. Introduction

The coronavirus pandemic 2019 (COVID-19) has clearly revealed the need to search for new therapeutic options including natural products as food supplements [1,2]. The angiotensin converting enzyme 2 (ACE2) serves as an entry point for the severe acute respiratory syndrome coronavirus 2 (SARS-CoV-2), which leads to COVID-19. Therefore, reducing ACE2 expression would reduce the number of access points of the virus to the body during primary infection, and potentially the spread inside the body. Cells which are susceptible to infection with SARS-CoV-2 appear to be primarily type II pneumocytes, intestinal absorptive enterocytes, and nasal goblet secretory cells [3]. The identification of mechanisms to reduce membrane ACE2 expression at these cells may be valuable.

We have recently proposed that glycyrrhizin, a key component of an extract from *Glycyrrhiza glabra*, may have such an effect [4]. A beneficial effect and potential mechanisms of action of glycyrrhizin, or components from *Glycyrrhiza glabra*, have been reviewed by several groups independently of our original suggestion [5–7]. Glycyrrhizin is metabolized into the systemically active metabolite glycyrrhetinic acid. Via this metabolite, glycyrrhizin inhibits an enzyme called 11-beta-hydroxysteroid dehydrogenase type 2 (11betaHSD2) [8]. Its inhibition allows cortisol to access mineralocorticoid receptors (MR) in aldosterone specific peripheral tissues, including the kidney, lung, intestinal, nasal and endothelial cells, in which it would otherwise have been prevented from doing so. In other words, an inhibition of this enzyme leads to an aldosterone-like activation at the MR by cortisol and may resemble the effects of high aldosterone levels in these organs. This may be relevant, as compounds which reduce plasma aldosterone, including angiotensin-converting enzyme inhibitors and angiotensin receptor antagonists, increase the expression of ACE2 [9]. Conversely, MR activation leads to a downregulation of ACE2, as demonstrated in the kidney [10].

Such action could therefore be a mechanism which could be employed to reduce ACE2 expression and, therefore, access of the virus to specific cells. ACE2 is an enzyme [3,11,12], not a receptor, but serves as a receptor for viral particles. This is important to keep in mind in order to avoid confusion regarding nomenclature. The confusion between the term "ACE2" and "ACE2 receptor" present in the literature arose due to the fact that ACE2 serves as a receptor for SARS-CoV-2, therefore being correctly called the "SARS-CoV-2 receptor" and not the "ACE2 receptor" [3,11,12]. A relevant tissue expresses, besides ACE2, both 11betaHSD2 and the MR. This includes lung and nasal, as well as intestinal epithelial cells. From a mechanistic perspective it is important that the small intestine, in particular ileum cells, co-expresses MR, 11betaHSD2 and ACE2, implying that the ileum could serve as an entry point for Cov-SARS-2 and be a target for 11betaHSD2 inhibition; it is at least a model organ to test for the effect of glycyrrhizin on ACE2.

The downstream consequences of reduced ACE2 expression are somewhat controversial. ACE2 activity is generally protective, including for lung tissue [13]. It protects by converting angiotensin II to angiotensin1–7 [14] as well as by suppressing the consequences of the activation of the receptor for endotoxin (LPS), i.e., the toll-like receptor 4 (TLR4) and, as a consequence, related inflammation in the lung (endotoxin storm) [15]—ACE2 overexpression inhibited the LPS induced inflammation in the mentioned study. Therefore, the reduced expression of ACE2 could be regarded as concerning. The anti-inflammatory ACE2-system is, however, balanced against the pro-inflammatory classical ACE [16], which leads to an increase in the pro-inflammatory mediator angiotensin II. Inhibition of 11betaHSD2 by glycyrrhizin or glycyrrhetinic acid suppresses the classical renin–angiotensin–aldosterone system (RAAS), i.e., reduces the plasma concentrations of renin, angiotensin and aldosterone [8,17] and increases cortisol/corticosterone locally [18]. This inhibition of the classical RAAS and activation of glucocorticoid receptors may therefore add to a potential beneficial effect of glycyrrhizin via the reduction of the pro-inflammatory angiotensin II [19]. Furthermore, a direct anti-inflammatory effect of *Glycyrrhiza glabra* extract and glycyrrhizin, via inhibition of TLR4 and inhibition of the release of high mobility group box 1 (HMGB1), has been described [20,21]. Such actions would counteract the consequences of ACE2 suppression on inflammation. In accordance, glycyrrhizin has protective effects in acute respiratory distress syndrome induced by the TLR4 activator LPS in mice [20].

The objective of this retrospective analysis is to explore the capability of *Glycyrrhiza glabra* extract to reduce ACE2 expression in the small intestine (ileum), as a target tissue with active 11betaHSD2 and MR expression, in comparison to non-target tissues (brain and heart), to provide mechanistical evidence that *Glycyrrhiza glabra* extract may have clinical benefits via reduced expression of ACE2. The tissues were obtained in an already performed study in rats, which was designed to identify the effects of *Glycyrrhiza glabra* extract on the stress response.

2. Materials and Methods

2.1. Animals

Forty-eight male Sprague-Dawley rats (Velaz, Prague, Czech Republic) weighing 225–250 g at the beginning of the experiments were used. The rats were allowed to habituate to the housing facility for 5 days. The animals were housed under standard laboratory conditions with free access to food and water. A constant 12:12 h light–dark cycle was maintained with light on at 07.00 h and off at 19.00 h. Temperature was maintained at 22 ± 2 °C and humidity at 55 ± 10%. All experimental procedures were approved by the Animal Health and Animal Welfare Division of the State Veterinary and Food Administration of the Slovak Republic (permission No. Ro 2291/18-221/3) and conformed to the NIH Guidelines for Care and Use of Laboratory Animals.

2.2. Study Design

This was not originally designed to study the effects of *Glycyrrhiza glabra* root extract on ACE2 expression, but instead, was based on data from an already performed study in rats, which addressed the effects of *Glycyrrhiza glabra* extract on the stress response (report in preparation). However, given the urgent need to identify treatments for COVID-19, we were motivated to address a hypothesis, which was formulated earlier [4]. With respect to the nature of this study, obvious limitations had to be accepted, in particular the unavailability of lung tissue at the time of raising these questions.

Following the habituation to the animal facility, the rats were randomly assigned to the control groups (n = 24) and to groups of animals exposed to chronic mild stress (n = 24). The model of chronic mild stress was based on seven different stress stimuli [22]. These involved social isolation (animal alone in the cage), unknown cage mate (the animal shared the cage with a rat from another cage), stroboscopic light (light flashes with frequency of 5 flashes/s), cage tilt (cages were tilted to 45 degrees from the horizontal), wet cage (water surface reached 2 cm above the bottom of the cage), continuous lighting (lighting for 24 h) and water deprivation. These stimuli were applied for 12 h each, in a randomized order, i.e., two conditions per day for 5 weeks. Control animals were housed undisturbed in a different room under the same light and temperature conditions. They had free access to food and water.

2.3. Treatment

The control rats as well as rats exposed to chronic mild stress were randomly assigned to one of the two groups: animals fed a diet with extract of *Glycyrrhiza glabra* (n = 12) and animals fed a placebo diet (n = 12). The extract of *Glycyrrhiza glabra* roots (Gall-Pharma GmbH, Judenburg, Austria) (Batch. no. P17092209) contained 6.25% of glycyrrhizinic acid. Water was used as a solvent during the extraction. The extract was mixed into the placebo diet at a dose of 150 mg/kg/day (SSNIFF Specialdiäten GmbH, Soest, Germany). The dose was selected according to [23]. In this study behavioral effects of *Glycyrrhiza glabra* extract were observed, ensuring that a relevant plasma level had been reached. Unfortunately, we were not able to measure plasma levels directly. The placebo diet (SSNIFF Specialdiäten GmbH, Soest, Germany) consisted of carbohydrates (65%), protein (24%) and fat (11%).

As mentioned above, the experiments lasted for 5 weeks. All animals received normal control diet for the first 3 weeks. The rats assigned for *Glycyrrhiza glabra* were fed the diet containing extract of *Glycyrrhiza glabra* for the next two weeks.

2.4. Organ Collection

Following 5 weeks of experimental procedures, the animals were quickly decapitated with a guillotine between 08.00 and 10.30 h in the morning. The brain was quickly removed from the skull and the prefrontal cortex was dissected on an ice-cold plate. The heart was removed and rinsed in 0.9% NaCl solution. The left heart ventricle was cut from the whole heart. Subsequently, the small intestine was removed from the body and all samples were frozen and stored at −70 °C until analyzed.

2.5. ACE2 mRNA and Protein Quantification

The gene expression of ACE2 was measured in the small intestine, the prefrontal cortex and the left heart ventricle by quantitative PCR. In the case of small intestine, its content was removed before tissue homogenization. Total RNA extraction, transcription of mRNA into cDNA as well as gene expression quantification was performed as described previously [24]. Primer BLAST NCBI software was used to design primers specific for the studied genes as well as reference genes (Table 1).

Table 1. Oligonucleotide sequences used in quantitative PCR.

Gene	Sense	Sequence 5′→3′
ACE2	Forward Reverse	ACCCTTCTTACATCAGCCCTACTG TGTCCAAAACCTACCCCACATAT
UQCRFS1—reference gene	Forward Reverse	ACAGTGGGCCTGAATGTTCC CACGGCGATAGTCAGAGAAGTC
TfR1—reference gene	Forward Reverse	ATACGTTCCCCGTTGTTGAGG GGCGGAAACTGAGTATGGTTGA
HPRT1—reference gene	Forward Reverse	CGTCGTGATTAGTGATGATGAAC CAAGTCTTTCAGTCCTGTCCATAA

ACE2: Angiotensin converting enzyme 2; *UQCRFS1*: Ubiquinol-cytochrome c reductase, Rieske iron-sulfur polypeptide 1; *TfR1*: Transferrin receptor protein 1; *HPRT1*: Hypoxanthine Phosphoribosyltransferase 1.

The concentration of ACE2 protein in the small intestine was determined by Rat Ace2 ELISA Kit (cat. no. ER0609, FineTest, Wuhan Fine Biotech Co., Ltd., Wuhan, Hubei, China). The intestines were thawed and their contents were removed. The samples were then frozen in liquid nitrogen in a mortar and pulverized by a pestle. A pre-test with the mentioned ELISA kit was performed to determine the optimal amount of tissue to be used in the subsequent analysis. We found out that 12.5 mg of powdered tissue (1/8 of the recommended amount) was the most favourable amount that fitted well into the kit's standard curve. The powdered tissue was suspended in 900 μL of PBS (according to the manufacturer's protocol) and was left at room temperature for 30 min. The samples were centrifuged at 5000× *g* for 5 min at 4 °C. From this point, the analysis was performed according to the manufacturer's protocol with 100 μL of sample supernatant put into the plate wells. The results are expressed as ng/mg of tissue.

2.6. Hormone Measurements

The trunk blood was collected and the plasma used for the analyses. Plasma corticosterone was measured by double-antibody radioimmunoassay (MP Biomedicals, Solon, OH, USA). Both intra- and inter-assay coefficients of variation (CVs) were <5%. Plasma renin activity was measured using angiotensin I radioimmunoassay kit (Immunotech, Marseille, France). The intra- and inter-assay CVs were 11.3% and 20.9%, respectively. Serum aldosterone was analyzed by a coated-tube radioimmunoassay (RIAZENco Aldosterone kit, ZenTech, Liège, Belgium), according to the manufacturer's instructions. The intra- and inter-assay CVs were 3.8% and 6.2%, respectively.

2.7. Statistical Analyses

The software package used for the statistical analysis was Statistica 7 (Statsoft, Tulsa, OK, USA). The values were checked for normality of distribution using the Shapiro-Wilks test. Data not normally distributed were Winsorized to normalize the distributions before analyses. Data from the gene and protein expression of ACE2 were analyzed by two way analysis of variance (ANOVA) with main factors of treatment (*Glycyrrhiza glabra* extract vs. placebo) and stress (chronic mild stress vs. control). Only the data of food intake, which were recorded in time, were analysed by a repeated measures ANOVA for factor time, treatment and stress. For post hoc comparisons, the Tukey post hoc test was chosen as this test is appropriate for two-way ANOVA and is stricter in comparison with other tests, such as Fisher least significant difference (LSD). Results are expressed as means ± standard error of the mean (SEM). The overall level of statistical significance was set as $p < 0.05$.

3. Results

In the small intestine, a tissue with known high activity of 11β-HSD2, the gene expression of ACE2 was significantly lower in rats fed the diet with *Glycyrrhiza glabra* extract compared to rats fed the placebo diet (Figure 1A). Two-way ANOVA revealed a significant main effect of treatment ($F_{(1,44)} = 4.41$; $p = 0.0415$) on concentrations of mRNA coding for ACE2 in the small intestine. The effect of stress was not statistically significant.

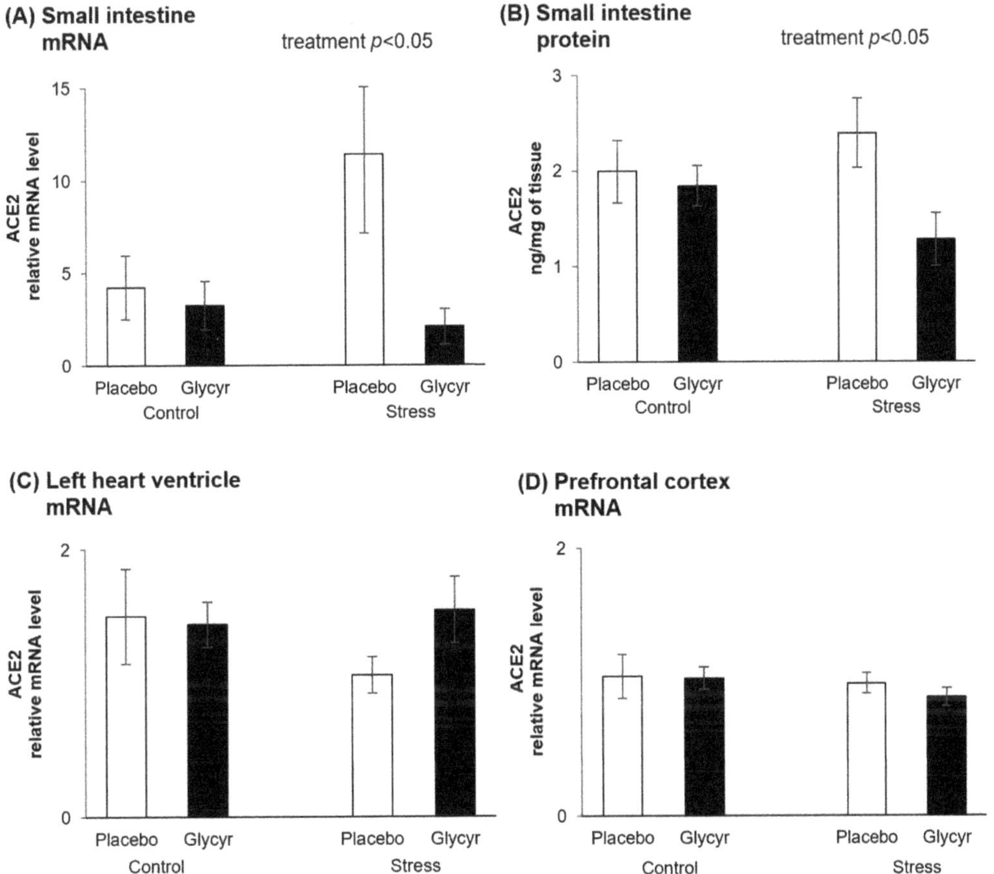

Figure 1. ACE2 gene expression in the small intestine (**A**), ACE2 protein expression in the small intestine (**B**), ACE2 gene expression in the left heart ventricle (**C**) and ACE2 gene expression in the prefrontal cortex (**D**) of rats treated with *Glycyrrhiza glabra* extract or placebo with or without exposure to chronic mild stress. Each value represents mean ± standard error of the mean (SEM) ($n = 12$ rats/group). Statistical significance as revealed by two-way ANOVA.

To verify if the observed changes in intestinal ACE2 mRNA production correlate with the protein levels, ACE2 protein concentrations were measured by ELISA. The protein concentrations of ACE2 in the small intestine were significantly lower in rats fed the diet with *Glycyrrhiza glabra* extract compared to rats fed the placebo diet (Figure 1B). Two way ANOVA revealed a significant main effect of treatment ($F_{(1,44)} = 4.46$, $p = 0.0403$)

on concentrations of ACE2 protein in the small intestine. The effect of stress was not statistically significant.

Concentrations of mRNA coding for ACE2 in the left heart ventricle were not affected by *Glycyrrhiza glabra* extract treatment (Figure 1C). The difference between the concentrations in the stressed groups was not statistically significant. Similarly, the gene expression of ACE2 in the prefrontal cortex was unchanged (Figure 1D).

The statistical analysis of corticosterone concentrations in plasma revealed a significant interaction between the main factors of treatment and stress ($F_{(1,44)} = 10.23$; $p = 0.0030$). The post hoc analysis showed that plasma corticosterone was significantly increased in the control group, which received *Glycyrrhiza glabra* extract vs. placebo ($p = 0.0186$; Figure 2a). This was not found in the stressed group, which showed a numerical reduction of corticosterone with the administration of *Glycyrrhiza glabra* extract which failed to be statistically significant. Aldosterone concentrations were numerically but not significantly reduced (Figure 2b). Plasma renin activity was unchanged (Figure 2c).

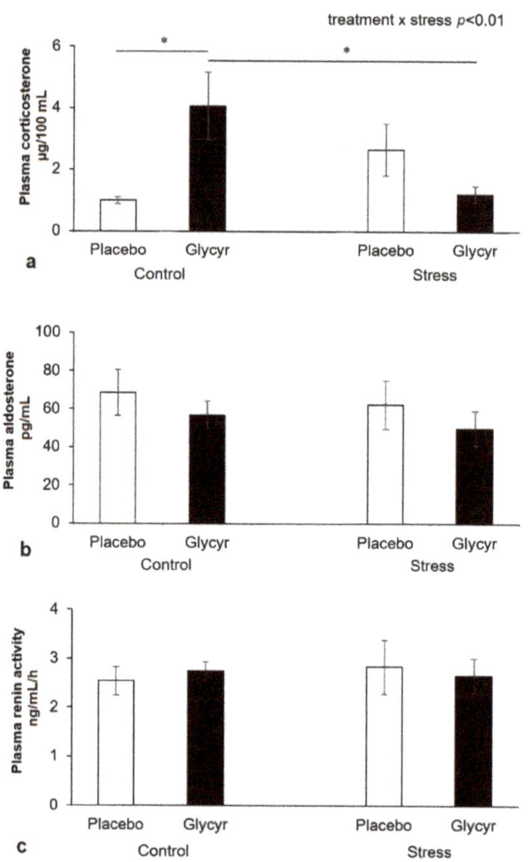

Figure 2. Concentrations of plasma corticosterone (**a**), concentrations of plasma aldosterone (**b**) and plasma renin activity (**c**) in rats treated with *Glycyrrhiza glabra* extract or placebo with or without exposure to chronic mild stress. Each value represents the mean ± standard error of the mean (SEM) (n = 12 rats/group). Statistical significance as revealed by two-way ANOVA followed by Tukey post-hoc test: * $p < 0.05$.

To check the potential influence of *Glycyrrhiza glabra* extract on food intake and thus on the dose of the drug ingested, we measured the food intake in two 2-day time intervals. The food intake was not significantly affected by the treatment. Chronic mild stress induced a significant reduction of food intake. Repeated measures ANOVA revealed a significant main effect of stress ($F_{(1,44)} = 43.43$; $p < 0.001$), as well as time ($F_{(1,44)} = 40.59$; $p < 0.001$) on food intake (Table 2).

Table 2. The effect of treatment with *Glycyrrhiza glabra* extract on average food intake of stressed and non-stressed animals in selected two-day time intervals. Each value represents the mean ± SEM (n = 12 rats/group). Statistical significance as revealed by repeated measures ANOVA.

Food Intake (g)	**Group**				**Statistical Significance**
	Control		**Stress**		
Treatment Day	**Placebo**	**Glycyr**	**Placebo**	**Glycyr**	
1–2	62.6 ± 1.2	65.2 ± 0.9	56.3 ± 0.6	59.8 ± 1.1	Treatment N.S. Stress $p < 0.001$ time $p < 0.001$
8–9	62.7 ± 0.9	61.7 ± 1.1	57.3 ± 0.8	53.8 ± 1.2	

N.S.: Not significant.

4. Discussion

The main finding of this study is the support for the hypothesis [4] that the treatment with *Glycyrrhiza glabra* reduces the expression of both gene and protein of ACE2 in tissue, which co-expresses 11betaHSD2 and MR, and may therefore reduce the cellular uptake and spread of SARS-CoV-2. Mechanistically, corticosterone acts as a mineralocorticoid to activate the MR in this situation and as a consequence reduces ACE2 expression. This effect was independent of the stress condition and failed to be observed in non-target tissues, such as the heart and the brain. Observed increase in plasma corticosterone in the control condition with the treatment of *Glycyrrhiza glabra* extract confirms target engagement.

The observed increase in glucocorticoid concentrations may also be partially responsible for an expected clinical benefit, as the synthetic glucocorticoid dexamethasone is clinically effective against COVID-19 symptoms [25], which is considered to be mediated via its anti-inflammatory effect. An alternative pathway of glycyrrhizin to affect inflammatory processes is via its activity to modify gut microbiota [26]. It has been reported that gut bacteria metabolize steroids into compounds, which modifies 11betaHSD2 [27] and may therefore have an impact on local ACE2 expression. Interestingly, ACE2 expression appears to affect the gut microbiome and, in turn, changes in gut microbiota may lead to changes in ACE2 expression [28].

The reduction in ACE2 mRNA levels as well as the ACE2 protein content in the small intestine revealed by feeding the rats with *Glycyrrhiza glabra* extract supplemented diet in the present study represents a novel original finding. Consistently with our hypothesis, the expression of ACE2 in tissues without evident 11betaHSD2 activity, such as the heart and brain cortex [29,30], remained unchanged. There are several review articles suggesting potential positive action of natural products on both prevention and treatment of the disease induced by SARS-CoV-2 [4,31–36]. With respect to supporting experimental evidence, effects of *Glycyrrhiza glabra* root extract on ACE2 expression have not been reported so far but the present data are consistent with the action of an extract of another plant, namely *Glycyrrhiza uralensis*, in the lung tissue of mice [37]. Other supporting data show that extract of *Glycyrrhiza glabra* or its main components may affect affinity interactions with ACE2 and/or viral proteases [38–40].

It may by suggested that reduced ACE2 expression induced by treatment with *Glycyrrhiza glabra* extract could have several beneficial implications for future clinical research. The most important future direction is to verify the protective effect of the extract and/or its main component glycyrrhizin against the entry of SARS-CoV-2 into the cells. Indeed, our preliminary results have shown a direct antiviral effect of glycyrrhizin on the replication of isolated human SARS-CoV-2 in a Vero E6 cell culture in a plaque-reduction inhibition test. These experiments revealed that depending on the concentration of glycyrrhizin added to the cell culture media, an inhibition of SARS-CoV-2 replication down to the detection limit of the assay was observed [41]. Supporting results were also reported by others [42]. Glycyrrhizin exerted a stronger effect when it was present in the cell culture media during the infection and subsequent incubation than when it was added after the virus infection step. Thus, the whole extract from *Glycyrrhiza glabra* or glycyrrhizin, which are generally regarded as safe, are promising dietary ingredients to help with prevention or early treatment of COVID-19 [41]. Preliminary clinical data on the positive effects of glycyrrhizin [43] or glycyrrhizin containing extracts [44] to treat patients with COVID-19 support the mechanistic data outlined here.

To discuss the present findings in the context of pulmonary diseases, the primarily targeted cells for glycyrrhizin treatment are lung epithelial cells (type II pneumocytes), which express 11betaHSD, both type 1 and 2 [45,46]. However, 11betaHSD2 appears to be upregulated in acute respiratory distress syndrome [47]. Aldosterone leads to an increase in alveolar clearance via an interaction with MR [48]. These observations support a role of 11betaHSD2 inhibition and resultant MR activation in lung protection from inflammatory stimuli. An independent confirmation of the role of MR in lung protection comes from clinical observations that the MR agonist fludrocortisone in combination with corticosteroids leads to a better clinical outcome in septic shock than corticosteroids alone [49]. It should, however, be noted that aldosterone via MR activation has, in many situations, pro-inflammatory effects, depending on tissue and other factors.

The present results are important also with respect to gastrointestinal problems related to COVID-19. It has been reported that a significant percentage of patients with COVID-19 experience gastrointestinal symptoms [50]. Approximately half of patients with confirmed COVID-19 have shown measurable SARS-CoV-2 RNA in their stool samples [51]. It is known that intestinal tissues contain the coronaviruses for weeks after the initial upper respiratory syndrome. Indeed, viral nucleic acid was found to be present in the feces after pharyngeal swabs became negative [52]. The present findings of reduced intestinal ACE2 expression by dietary supplementation with *Glycyrrhiza glabra* extract might attenuate virus accumulation in the gastrointestinal tract and thus contribute to the prevention of potential fecal–oral transmission of SARS-CoV-2.

An obvious limitation of this study is the lack of lung tissue as the original experimental design had a different goal. Another limitation of the presented concept is its focus on ACE2 only as an entry point for SARS-CoV-2. It has to be noted that other proteins, including CD209L and CD147 may serve this purpose. An interaction between these proteins and either the renin–angiotensin–aldosterone system, ACE2 or the effect of glycyrrhizin has not been reported.

5. Conclusions

In conclusion, the treatment with *Glycyrrhiza glabra* root extract leads to a significant reduction in the expression of ACE2 in the small intestine, which may serve as an entry point of SARS CoV-2. An important aspect of the current study is to motivate additional work, which needs to be performed to provide more conclusive evidence. Whether a similar effect exists in the lungs needs to be further explored, but it is plausible, given a similar receptor constellation of ACE2, 11betaHSD2 and MR. Whether this phenomenon, when confirmed in additional studies, is linked to the susceptibility of cells to the virus requires further studies.

Author Contributions: D.J. and H.M.: Conceptualisation, funding acquisition, writing–original draft, review and editing. D.J.: investigation, study design, methodology, formal analysis, supervision. P.K. and L.K.: tissue collection and analysis of quantitative PCR data and ELISA data, methodology, writing–review and editing. A.P.: concrete experimental design, animal handling, performance of experiments. All authors have read and agreed to the published version of the manuscript.

Funding: This research was funded by the Slovak Research and Development Agency, grant number APVV-18-0283.

Institutional Review Board Statement: The study was conducted according to the guidelines of the Declaration of Helsinki, and approved by the Animal Health and Animal Welfare Division of the State Veterinary and Food Administration of the Slovak Republic (protocol code Ro 2291/18-221/3, 8 August 2018).

Informed Consent Statement: Not applicable.

Data Availability Statement: The data that support the findings of this study are available from the corresponding author upon reasonable request.

Conflicts of Interest: HM is the owner of the consulting company Murck-Neuroscience LLC and holds a patent for the use of glycyrrhizin in the area of depression treatment. HM is also a full time employee of Aptinyx Inc, Evanston, IL, USA. The authors declare no conflict of interest.

References

1. Cawood, A.L.; Walters, E.R.; Smith, T.R.; Sipaul, R.H.; Stratton, R.J. A Review of Nutrition Support Guidelines for Individuals with or Recovering from COVID-19 in the Community. *Nutrients* **2020**, *12*, 3230. [CrossRef]
2. Hensel, A.; Bauer, R.; Heinrich, M.; Spiegler, V.; Kayser, O.; Hempel, G.; Kraft, K. Challenges at the Time of COVID-19: Opportunities and Innovations in Antivirals from Nature. *Planta Medica* **2020**, *86*, 659–664. [CrossRef] [PubMed]
3. Ziegler, C.G.K.; Allon, S.J.; Nyquist, S.K.; Mbano, I.M.; Miao, V.N.; Tzouanas, C.N.; Cao, Y.; Yousif, A.S.; Bals, J.; Hauser, B.M.; et al. SARS-CoV-2 Receptor ACE2 Is an Interferon-Stimulated Gene in Human Airway Epithelial Cells and Is Detected in Specific Cell Subsets across Tissues. *Cell* **2020**, *181*, 1016–1035.e19. [CrossRef]
4. Murck, H. Symptomatic Protective Action of Glycyrrhizin (Licorice) in COVID-19 Infection? *Front. Immunol.* **2020**, *11*, 1239. [CrossRef]
5. Huan, C.; Chen, C.; Xu, W.; Guo, T.; Pan, H.; Gao, S. Study on Antiviral Activities of Glycyrrhizin. *Int. J. Biomed. Eng. Clin. Sci.* **2020**, *6*, 68–70. [CrossRef]
6. Gomaa, A.; Abdel-Wadood, Y. The potential of glycyrrhizin and licorice extract in combating COVID-19 and associated conditions. *Phytomedicine Plus* **2021**, *1*, 100043. [CrossRef]
7. Li, F.; Liu, B.; Li, T.; Wu, Q.; Xu, Z.; Gu, Y.; Li, W.; Wang, P.; Ma, T.; Lei, H. Review of Constituents and Biological Activities of Triterpene Saponins from Glycyrrhizae Radix et Rhizoma and Its Solubilization Characteristics. *Molecules* **2020**, *25*, 3904. [CrossRef]
8. Cosmetic Ingredient Review Expert Panel. Final report on the safety assessment of Glycyrrhetinic Acid, Potassium Glycyrrhetinate, Disodium Succinoyl Glycyrrhetinate, Glyceryl Glycyrrhetinate, Glycyrrhetinyl Stearate, Stearyl Glycyrrhetinate, Glycyrrhizic Acid, Ammonium Glycyrrhizate, Dipotassium Glycyrrhizate, Disodium Glycyrrhizate, Trisodium Glycyrrhizate, Methyl Glycyrrhizate, and Potassium Glycyrrhizinate. *Int. J. Toxicol.* **2007**, *26*, 79–112. [CrossRef]
9. Esler, M.; Esler, D. Can angiotensin receptor-blocking drugs perhaps be harmful in the COVID-19 pandemic? *J. Hypertens.* **2020**. [CrossRef]
10. Fukuda, S.; Horimai, C.; Harada, K.; Wakamatsu, T.; Fukasawa, H.; Muto, S.; Itai, A.; Hayashi, M. Aldosterone-induced kidney injury is mediated by NFkappaB activation. *Clin. Exp. Nephrol.* **2011**, *15*, 41–49. [CrossRef]
11. Shang, J.; Wan, Y.; Luo, C.; Ye, G.; Geng, Q.; Auerbach, A.; Li, F. Cell entry mechanisms of SARS-CoV-2. *Proc. Natl. Acad. Sci. USA* **2020**, *117*, 11727–11734. [CrossRef] [PubMed]
12. Gheblawi, M.; Wang, K.; Viveiros, A.; Nguyen, Q.; Zhong, J.C.; Turner, A.J.; Raizada, M.K.; Grant, M.B.; Oudit, G.Y. Angiotensin-Converting Enzyme 2: SARS-CoV-2 Receptor and Regulator of the Renin-Angiotensin System: Celebrating the 20th Anniversary of the Discovery of ACE2. *Circ. Res.* **2020**, *126*, 1456–1474. [CrossRef]
13. Kaparianos, A.; Argyropoulou, E. Local renin-angiotensin II systems, angiotensin-converting enzyme and its homologue ACE2: Their potential role in the pathogenesis of chronic obstructive pulmonary diseases, pulmonary hypertension and acute respiratory distress syndrome. *Curr. Med. Chem.* **2011**, *18*, 3506–3515. [CrossRef]
14. Tikellis, C.; Thomas, M.C. Angiotensin-Converting Enzyme 2 (ACE2) Is a Key Modulator of the Renin Angiotensin System in Health and Disease. *Int. J. Pept.* **2012**, *2012*, 256294. [CrossRef] [PubMed]
15. Ye, R.; Liu, Z. ACE2 exhibits protective effects against LPS-induced acute lung injury in mice by inhibiting the LPS-TLR4 pathway. *Exp. Mol. Pathol.* **2020**, *113*, 104350. [CrossRef]
16. Ingraham, N.E.; Lotfi-Emran, S.; Thielen, B.K.; Techar, K.; Morris, R.S.; Holtan, S.G.; Dudley, R.A.; Tignanelli, C.J. Immunomodulation in COVID-19. *Lancet Respir. Med.* **2020**. [CrossRef]

17. Armanini, D.; Lewicka, S.; Pratesi, C.; Scali, M.; Zennaro, M.C.; Zovato, S.; Gottardo, C.; Simoncini, M.; Spigariol, A.; Zampollo, V. Further studies on the mechanism of the mineralocorticoid action of licorice in humans. *J. Endocrinol. Investig.* **1996**, *19*, 624–629. [CrossRef] [PubMed]
18. MacKenzie, M.A.; Hoefnagels, W.H.; Jansen, R.W.; Benraad, T.J.; Kloppenborg, P.W. The influence of glycyrrhetinic acid on plasma cortisol and cortisone in healthy young volunteers. *J. Clin. Endocrinol. Metab.* **1990**, *70*, 1637–1643. [CrossRef]
19. Benicky, J.; Sanchez-Lemus, E.; Honda, M.; Pang, T.; Orecna, M.; Wang, J.; Leng, Y.; Chuang, D.M.; Saavedra, J.M. Angiotensin II AT1 receptor blockade ameliorates brain inflammation. *Neuropsychopharmacology* **2011**, *36*, 857–870. [CrossRef]
20. Yu, Z.; Ohtaki, Y.; Kai, K.; Sasano, T.; Shimauchi, H.; Yokochi, T.; Takada, H.; Sugawara, S.; Kumagai, K.; Endo, Y. Critical roles of platelets in lipopolysaccharide-induced lethality: Effects of glycyrrhizin and possible strategy for acute respiratory distress syndrome. *Int. Immunopharmacol.* **2005**, *5*, 571–580. [CrossRef]
21. Seo, E.H.; Song, G.Y.; Kwak, B.O.; Oh, C.S.; Lee, S.H.; Kim, S.H. Effects of Glycyrrhizin on the Differentiation of Myeloid Cells of the Heart and Lungs in Lipopolysaccharide-Induced Septic Mice. *Shock* **2017**, *48*, 371–376. [CrossRef]
22. Duncko, R.; Kiss, A.; Skultetyova, I.; Rusnak, M.; Jezova, D. Corticotropin-releasing hormone mRNA levels in response to chronic mild stress rise in male but not in female rats while tyrosine hydroxylase mRNA levels decrease in both sexes. *Psychoneuroendocrinology* **2001**, *26*, 77–89. [CrossRef]
23. Chakravarthi, K.K.; Avadhani, R. Beneficial effect of aqueous root extract of Glycyrrhiza glabra on learning and memory using different behavioral models: An experimental study. *J. Nat. Sci. Biol. Med.* **2013**, *4*, 420–425. [CrossRef] [PubMed]
24. Buzgoova, K.; Graban, J.; Balagova, L.; Hlavacova, N.; Jezova, D. Brain derived neurotrophic factor expression and DNA methylation in response to subchronic valproic acid and/or aldosterone treatment. *Croat. Med. J.* **2019**, *60*, 71–77. [CrossRef] [PubMed]
25. Johnson, R.M.; Vinetz, J.M. Dexamethasone in the management of covid -19. *BMJ* **2020**, *370*, m2648. [CrossRef] [PubMed]
26. Xu, X.; Gong, L.; Wang, B.; Wu, Y.; Wang, Y.; Mei, X.; Xu, H.; Tang, L.; Liu, R.; Zeng, Z.; et al. Glycyrrhizin Attenuates Salmonella enterica Serovar Typhimurium Infection: New Insights into Its Protective Mechanism. *Front. Immunol.* **2018**, *9*, 2321. [CrossRef] [PubMed]
27. Morris, D.J.; Ridlon, J.M. Glucocorticoids and gut bacteria: "The GALF Hypothesis" in the metagenomic era. *Steroids* **2017**, *125*, 1–13. [CrossRef] [PubMed]
28. Richards, E.M.; Pepine, C.J.; Raizada, M.K.; Kim, S. The Gut, Its Microbiome, and Hypertension. *Curr. Hypertens. Rep.* **2017**, *19*, 36. [CrossRef] [PubMed]
29. Gray, G.A.; White, C.I.; Castellan, R.F.; McSweeney, S.J.; Chapman, K.E. Getting to the heart of intracellular glucocorticoid regeneration: 11beta-HSD1 in the myocardium. *J. Mol. Endocrinol.* **2017**, *58*, R1–R13. [CrossRef]
30. Wyrwoll, C.S.; Holmes, M.C.; Seckl, J.R. 11beta-hydroxysteroid dehydrogenases and the brain: From zero to hero, a decade of progress. *Front. Neuroendocrinol.* **2011**, *32*, 265–286. [CrossRef]
31. Shahzad, F.; Anderson, D.; Najafzadeh, M. The Antiviral, Anti-Inflammatory Effects of Natural Medicinal Herbs and Mushrooms and SARS-CoV-2 Infection. *Nutrients* **2020**, *12*, 2573. [CrossRef]
32. Silveira, D.; Prieto-Garcia, J.M.; Boylan, F.; Estrada, O.; Fonseca-Bazzo, Y.M.; Jamal, C.M.; Magalhaes, P.O.; Pereira, E.O.; Tomczyk, M.; Heinrich, M. COVID-19: Is There Evidence for the Use of Herbal Medicines as Adjuvant Symptomatic Therapy? *Front. Pharmacol.* **2020**, *11*, 581840. [CrossRef]
33. Zhong, L.L.D.; Lam, W.C.; Yang, W.; Chan, K.W.; Sze, S.C.W.; Miao, J.; Yung, K.K.L.; Bian, Z.; Wong, V.T. Potential Targets for Treatment of Coronavirus Disease 2019 (COVID-19): A Review of Qing-Fei-Pai-Du-Tang and Its Major Herbs. *Am. J. Chin. Med.* **2020**, *48*, 1051–1071. [CrossRef]
34. Jalali, A.; Dabaghian, F.; Akbrialiabad, H.; Foroughinia, F.; Zarshenas, M.M. A pharmacology-based comprehensive review on medicinal plants and phytoactive constituents possibly effective in the management of COVID-19. *Phytother. Res.* **2021**, *35*, 1925–1938. [CrossRef] [PubMed]
35. Pastor, N.; Collado, M.C.; Manzoni, P. Phytonutrient and Nutraceutical Action against COVID-19: Current Review of Characteristics and Benefits. *Nutrients* **2021**, *13*, 464. [CrossRef] [PubMed]
36. Merarchi, M.; Dudha, N.; Das, B.; Garg, M. Natural products and phytochemicals as potential anti-SARS-CoV-2 drugs. *Phytother. Res.* **2021**. [CrossRef]
37. Wu, C.Y.; Lin, Y.S.; Yang, Y.H.; Shu, L.H.; Cheng, Y.C.; Liu, H.T. Potential Simultaneous Inhibitors of Angiotensin-Converting Enzyme 2 and Transmembrane Protease, Serine 2. *Front. Pharmacol.* **2020**, *11*, 584158. [CrossRef]
38. Li, Y.; Chu, F.; Li, P.; Johnson, N.; Li, T.; Wang, Y.; An, R.; Wu, D.; Chen, J.; Su, Z.; et al. Potential effect of Maxing Shigan decoction against coronavirus disease 2019 (COVID-19) revealed by network pharmacology and experimental verification. *J. Ethnopharmacol.* **2021**, *271*, 113854. [CrossRef] [PubMed]
39. Li, X.; Qiu, Q.; Li, M.; Lin, H.; Cao, S.; Wang, Q.; Chen, Z.; Jiang, W.; Zhang, W.; Huang, Y.; et al. Chemical composition and pharmacological mechanism of ephedra-glycyrrhiza drug pair against coronavirus disease 2019 (COVID-19). *Aging* **2021**, *13*, 4811–4830. [CrossRef]
40. Van de Sand, L.; Bormann, M.; Alt, M.; Schipper, L.; Heilingloh, C.S.; Steinmann, E.; Todt, D.; Dittmer, U.; Elsner, C.; Witzke, O.; et al. Glycyrrhizin Effectively Inhibits SARS-CoV-2 Replication by Inhibiting the Viral Main Protease. *Viruses* **2021**, *13*, 609. [CrossRef]

41. Klempa, B.; Slavikova, M.; Murck, H.; Jezova, D. Direct inhibitory effects of glycyrrhizin on isolated human SARS-CoV-2. *Psychiatrie* **2021**, *25*, 6.
42. Gowda, P.; Patrick, S.; Joshi, S.D.; Kumawat, R.K.; Sen, E. Glycyrrhizin prevents SARS-CoV-2 S1 and Orf3a induced high mobility group box 1 (HMGB1) release and inhibits viral replication. *Cytokine* **2021**, *142*, 155496. [CrossRef]
43. Ding, H.; Deng, W.; Ding, L.; Ye, X.; Yin, S.; Huang, W. Glycyrrhetinic acid and its derivatives as potential alternative medicine to relieve symptoms in nonhospitalized COVID-19 patients. *J. Med. Virol.* **2020**, *92*, 2200–2204. [CrossRef]
44. Luo, H.; Tang, Q.L.; Shang, Y.X.; Liang, S.B.; Yang, M.; Robinson, N.; Liu, J.P. Can Chinese Medicine Be Used for Prevention of Corona Virus Disease 2019 (COVID-19)? A Review of Historical Classics, Research Evidence and Current Prevention Programs. *Chin. J. Integr. Med.* **2020**, *26*, 243–250. [CrossRef]
45. Suzuki, S.; Tsubochi, H.; Darnel, A.; Suzuki, T.; Sasano, H.; Krozowski, Z.S.; Kondo, T. Expression of 11 beta-hydroxysteroid dehydrogenase type 1 in alveolar epithelial cells in rats. *Endocr. J.* **2003**, *50*, 445–451. [CrossRef] [PubMed]
46. Suzuki, T.; Sasano, H.; Suzuki, S.; Hirasawa, G.; Takeyama, J.; Muramatsu, Y.; Date, F.; Nagura, H.; Krozowski, Z.S. 11Beta-hydroxysteroid dehydrogenase type 2 in human lung: Possible regulator of mineralocorticoid action. *J. Clin. Endocrinol. Metab.* **1998**, *83*, 4022–4025. [CrossRef]
47. Suzuki, S.; Tsubochi, H.; Ishibashi, H.; Suzuki, T.; Kondo, T.; Sasano, H. Increased expression of 11 beta-hydroxysteroid dehydrogenase type 2 in the lungs of patients with acute respiratory distress syndrome. *Pathol. Int.* **2003**, *53*, 751–756. [CrossRef] [PubMed]
48. Suzuki, S.; Tsubochi, H.; Suzuki, T.; Darnel, A.D.; Krozowski, Z.S.; Sasano, H.; Kondo, T. Modulation of transalveolar fluid absorption by endogenous aldosterone in adult rats. *Exp. Lung. Res.* **2001**, *27*, 143–155. [CrossRef] [PubMed]
49. Archontakis Barakakis, P.; Palaiodimos, L.; Fleitas Sosa, D.; Benes, L.; Gulani, P.; Fein, D. Combination of low-dose glucocorticosteroids and mineralocorticoids as adjunct therapy for adult patients with septic shock: A systematic review and meta-analysis of randomized trials and observational studies. *Avicenna. J. Med.* **2019**, *9*, 134–142. [CrossRef]
50. Cheung, K.S.; Hung, I.F.N.; Chan, P.P.Y.; Lung, K.C.; Tso, E.; Liu, R.; Ng, Y.Y.; Chu, M.Y.; Chung, T.W.H.; Tam, A.R.; et al. Gastrointestinal Manifestations of SARS-CoV-2 Infection and Virus Load in Fecal Samples From a Hong Kong Cohort: Systematic Review and Meta-analysis. *Gastroenterology* **2020**, *159*, 81–95. [CrossRef]
51. Guo, M.; Tao, W.; Flavell, R.A.; Zhu, S. Potential intestinal infection and faecal-oral transmission of SARS-CoV-2. *Nat. Rev. Gastroenterol. Hepatol.* **2021**, *18*, 269–283. [CrossRef] [PubMed]
52. Cuicchi, D.; Lazzarotto, T.; Poggioli, G. Fecal-oral transmission of SARS-CoV-2: Review of laboratory-confirmed virus in gastrointestinal system. *Int. J. Colorectal. Dis.* **2021**, *36*, 437–444. [CrossRef] [PubMed]

MDPI
St. Alban-Anlage 66
4052 Basel
Switzerland
Tel. +41 61 683 77 34
Fax +41 61 302 89 18
www.mdpi.com

Nutrients Editorial Office
E-mail: nutrients@mdpi.com
www.mdpi.com/journal/nutrients

www.ingramcontent.com/pod-product-compliance
Lightning Source LLC
LaVergne TN
LVHW070924170726
843515LV00005B/861